# International Conference and Early Career Scientists School on Environmental Observations, Modelling and Information Systems (ENVIROMIS-2016)



IOP Conference Series: Earth and Environmental Science
Volume 48

Tomsk, Russia
11 – 16 July 2016

**Editors:**

**Evgeny Gordov**

ISBN: 978-1-5108-3379-1
ISSN: 1755-1307

**Printed from e-media with permission by:**

Curran Associates, Inc.
57 Morehouse Lane
Red Hook, NY 12571

proceedings.com

**Some format issues inherent in the e-media version may also appear in this print version.**



Printed by Curran Associates, Inc. (2017)

For permission requests, please contact the Institute of Physics at the address below.

Institute of Physics
Dirac House, Temple Back
Bristol BS1 6BE UK

Phone: 44 1 17 929 7481
Fax: 44 1 17 920 0979

techtracking@iop.org

**Additional copies of this publication are available from:**

Curran Associates, Inc.
57 Morehouse Lane
Red Hook, NY 12571 USA
Phone: 845-758-0400
Fax: 845-758-2633
Email: curran@proceedings.com
Web: www.proceedings.com

# Table of contents

## Volume 48

**International Conference and Early Career Scientists School on Environmental Observations, Modelling and Information Systems (ENVIROMIS-2016)**

**11–16 July 2016, Tomsk, Russian Federation**

**Accepted papers received: 14 November 2016**
**Published online: 26 November 2016**

### Preface



### Papers

# Preface

These Conference Proceedings contain the selected papers of the 9th multidisciplinary Conference ENVIROMIS-2016 comprising elements of Early Career Scientists School held at Institute of Monitoring of Climatic and Ecological Systems SB RAS, Tomsk, Russia, on July 11-16, 2016.

The conference was devoted to the state-of-the-art and usage of modern environmental observation techniques, computational and information technologies for monitoring and assessment of Northern Eurasia environment variations including those caused by the Global Climate Change. During the conference, a total of 111 manuscripts were presented and discussed in a series of successive presentations. All the contributions in this volume have been peer reviewed according to rigorous international standards. However the authors take full responsibility for the content of their papers.

The conference was organized by Institute of Monitoring of Climatic and Ecological Systems SB RAS, Institute of Numerical Mathematics RAS, Research Computing Center of Moscow State University and Tomsk State University. It was supported by Russian Foundation for Basic Research and by P.P. Shirshov Institute of Oceanology RAS under Megagrant of the Ministry of Education and Science of the Russian Federation «Extra-tropical hydrological cycle in the current and future climate: uncertainties and predictability».

We are grateful to all of the reviewers for maintaining the standards and quality of the manuscripts throughout the reviewing process.

Editors

Professor Evgeny Gordov

RAS Corresponding Member Vasily Lykosov

# Peer review statement

All papers published in this volume of *IOP Conference Series: Earth and Environmental Science* have been peer reviewed through processes administered by the proceedings Editors. Reviews were conducted by expert referees to the professional and scientific standards expected of a proceedings journal published by IOP Publishing.

# Multidisciplinary ENVIROMIS conferences as a tool connecting climate information producers and users

**E P Gordov[1,2], V N Lykosov[3,4], Yu E Gordova[1] and E Yu Genina[1]**

[1] Institute of Monitoring of Climatic and Ecological Systems SB RAS, 10/3 Akademichesky pr., Tomsk, 634055, Russia
[2] National Research Tomsk State University, 36, Lenin ave., 634050, Tomsk, Russia
[3] Institute of Numerical Mathematics RAS, 8, Gubkina str., 119333, Moscow, Russia
[4] Lomonosov Moscow State University, Research Computing Center, 1, bld.4, Leninskie Gory, 119234, GSP-1, Moscow, Russia

e-mail: gordov@scert.ru

**Abstract.** The paper describes a regular multidisciplinary conference and early carrier scientists' school ENVIROMIS as a tool for facilitation of cooperation of specialists in different fields of environmental sciences and users of research results. The conference addresses critically important issues occurring in environmental sciences, namely, crossing disciplinary borders of this multidisciplinary research area, young scientists' training and diminishing a gap between science and decision-making. The viability of the approach to the Schools organization is proved by the fact that single event turned into a series, quite a few young participants successfully defended their PhD thesis and a number of researchers became Doctors of Science, several international multidisciplinary projects, initiated at these conferences, have been successfully carried out. Researchers from Russia and foreign countries show undiminishing interest to these conferences.

## 1. Introduction

Environmental sciences, that study the Earth system, cover all spheres that provide human habitat, namely, atmosphere, hydrosphere, and lithosphere. Nowadays the most urgent tasks for environmental sciences are ongoing and projected climate changes and their consequences [1]. Global impact of warming is presented in the Intergovernmental Panel on Climate Changes (IPCC) Reports [1, 2]. As for the Russian territory, due to its size and location in high and polar latitudes within the largest continental mass of Eurasia, the country belongs to the Earth's regions with the highest observed and projected climate changes [3]. This warrants a need to study these changes in order to elaborate mitigation and adaptation measures in response to them as well as to quantify their feedbacks to the other parts of the Earth System.

Hydrometeorological Center of Russian Federation issued a report "Strategic outlook of climate change in Russian Federation for the period up to 2010-2015, and its impact on sectors of the economy of Russia" [4] which comprises the research result of 14 leading research institutions in Russia. According to this report there is an increase in the magnitude and frequency of extreme weather events (heat waves, cold periods, heavy rains or snow, storms, floods or drought), which leads to the increasing damage to ecosystems and infrastructure. All this together causes enormous economic damage and needs to be thoroughly studied.

Since environmental sciences have very large research scope, they face many problems. Among them the main ones are a need of multidisciplinarity in research and gaps between science and decision-making.

First problem arises from the fact that Earth system includes a great variety of components, processes and interactions, which, if investigating separately, should be studied by methods from physics, chemistry, hydro- and gas dynamics, biology, ecology, etc. Specialists in different environmental areas use different approaches and data storage formats. At the same time, new instruments and data acquisition technologies are introduced into practice, new climate models and climate projection scenarios are created, and various information-computational systems for environmental studies are developed. Such diversity poses in front of the scientific community a challenge: specialists working in different environmental fields have to unite and cooperate in order to get insight into the environmental processes, to understand their mechanisms, and to apply into practice their findings. Multidisciplinarity of problems under consideration demands a development of common terminology, which is clearly understood by scientific community. Thoughtful study of environmental processes requires a synergy of approaches that would create the entire pattern of the Earth system functioning (including its dynamics) in all its diversity. Moreover, there is a need to train young scientists to apply new technologies and to work within multidisciplinary teams. So, it is necessary to find a way to overcome these problems. Addressing the multidisciplinarity, was among the major motivations behind the ENVIROMIS Events organization

The second problem is in the shortage of professionals, who can implement the research results into practice. Very often decision-makers and stakeholders, who have to execute adequate measures to adapt to or to mitigate the adverse effects of climate change, do not have skills and knowledge to intelligently assess the accumulated data from climate analysis and to use them for strategy development. Thus, there is an acute need to make a bridge between scientific results and recommendations and their implementation into practical measures.

There are several ways to solve these problems on both global and national levels. Those are organization of specialized conferences and seminars, issue of dedicated journals and reports, educational activities, training of the users. At the global scale activity of IPCC was aimed at informing governments on the ongoing and future climate changes in order to let them make proper decisions [2]. For user training purposes, many courses are developed including short-term private participation courses and e-learning [5, 6].

Along with comprehensive multidisciplinary environmental studies, it is clear now that urgent actions should be undertaken to adapt to climate change and mitigate its negative effects. Climate change is a process that occurs gradually. This makes it possible to adapt and minimize its negative impact, to organize activities to adapt to changing conditions. Assessments of environmental, economic, political and social consequences of global climate change to each specific region should be a priority.

According to the accepted terminology of the IPCC, the "adaptation" process is "The process of adjustment to actual or expected climate and its effects. In human systems, adaptation seeks to moderate or avoid harm or exploit beneficial opportunities. In some natural systems, human intervention may facilitate adjustment to expected climate and its effects" (http://www.ipcc.ch/pdf/assessment-report/ar5/wg2/ar5_wgII_spm_en.pdf p. 5). Adaptation actions imply a wide range of measures, such as economical use of scarce water resources, modification of existing building codes by new calculations of the buildings' stability that account for future climate conditions and extreme weather events, carrying out flood control works to protect against rising sea levels, design drought-resistant crops, etc. Other methods of adaptation include improving seasonal weather forecasts, early warning about weather-related emergencies, improved risk management, accumulation of sufficient insurance funds, and extra-efforts to secure fresh water supply.

According to the IPCC, future vulnerability depends not only on climate change but also on the selected ways of societal development. Sustainable development can reduce vulnerability. To be successful, adaptation should be a part of sustainable development plans at the international and national levels. At the national level development of effective adaptation strategies should include: improving the scientific basis for decision-making; development of educational programs; improvement of practical training and raising public awareness of the environmental change problems, especially among young people. In addition, possible adaptation measures could include development of appropriate legislation and regulatory frameworks to sustain adaptation activities.

The present paper highlights an approach, our contribution to address the above problems. It had been developed and tested during past 15 years.. The approach consists in organization of a multidisciplinary conferences and early carrier scientists' schools, which could be a midpoint for get-togethers and cooperation of specialists in different fields of environmental sciences and users of their research results, (e.g., decision makers, local authorities, and stakeholders.

## 2. Approach

At present in Russia the Climate Doctrine of the Russian Federation (http://www.lse.ac.uk/GranthamInstitute/law/climate-doctrine-of-the-russian-federation/) is adopted and implemented, which is the basis for the formation and implementation of climate policy.

A necessary condition for the climate policy is a government support to ensure compliance with the world standards in: systematic climate observations; fundamental and applied research in the field of climate science and related fields; application of research results to assess the risks and benefits associated with the effects of climate change and to adapt to these effects.

According to the Doctrine a threefold approach should be realized. First of all, young specialists in climate sciences should be expansively trained. Global changes have caused the development of techniques of measurement and simulation characteristics of the environment, accompanied by an extension of the conceptual and mathematical apparatus. Available training programs in environmental science, as a rule, do not have time to adapt to such rapid changes in the content of the subject area [7]. Specialists need to create large databases of observations and / or modeling results, to develop and use computational models, and to be able to quickly exchange the information within the scientific team. And scientific teams become interdisciplinary for a broader outlook in climatic issues.

The lag in training programs leads to the situation when graduates of the environmental faculties superficially know mathematical modeling of environmental processes, do not have required skills in modeling, processing and analysis of observational data, are not able to work with the data fields of meteorological variables and be useful teammates within interdisciplinary research groups.

The second line of adaptation and mitigation measures noted in the Doctrine is a preparation of skilled decision-makers, who will develop and implement adaptation measures. The goal is not only to provide these groups with tools, skills, thematic information but to give them knowledge for understanding climate processes occurring in the region.

The third proposed line in the Doctrine requires rising of public awareness of processes that occurs and could occur in their habitat. Therefore, special internet-based resources and e-learning courses should be developed for the general public. Evidence-based information about current and expected climate changes in the region of residence will help the entire society to prepare to the expected consequences and stimulate the search for ways to adapt to them.

ENVIROMIS multidisciplinary conferences had been designed and were considered as sites for meetings and cooperation of specialists in different areas of environmental sciences and end-users: decision makers, local authorities, and stakeholders. At these conferences the state-of-the-art usage of modern environmental observation techniques, computational and information technologies for assessment, modeling and mitigation of environment variations under natural and anthropogenic pressures have been presented. They have been aimed at filling the gap between basic science achievements and their practical applications in the study domain (mostly Siberia)

Usually four types of report are delivered on ENVIROMIS conferences, namely, invited lectures that present state-of-the-art in the corresponding environmental science topics, invited reports that present recent achievements, as well as oral and poster reports, where scientists present results of particular studies. As a rule, lectures are given in the manner that is clearly understandable for all conference participants. Successive sessions allow participants to listen all reports they are interested in and take part in the poster sessions.

Besides, representatives of international funds and organizations have taken part in ENVIROMIS conferences. They present current international calls and terms of participation for Russian scientists. We have also invited leaders and officers of local ecological services who are interested in implementing new technologies in their practical work.

In the context of the Doctrine, ENVIROMIS conferences address a critically important issue – a generation gap occurring in Russia in Environmental Sciences. To mitigate this gap and its consequences and to facilitate professional skill growth of young scientists, a number of invited lectures by leading specialists are usually presented at the ENVIROMIS Conferences. The lectures are devoted to the novelty in the environmental research and allow young participants to get first-hand information on hot topics in the Earth System Sciences attracting them professionally to this very important research area. Supplementary educational component, which is routinely implemented at the Conferences, is to teach young scientists to clearly lay out their own results during short oral presentations of their posters and during the follow-up course of poster sessions.

In the framework of adaptation strategy, each ENVIROMIS Conference program includes a Workshop dedicated to one or another international program on environmental studies. Among them have been Northern Eurasia Earth Science Partner Initiative / Northern Eurasia Future Initiative (NEESPI/NEFI)), large regional programs (Siberian Integrated Regional Study), large international projects (Project Enviro–RISKS of $6^{th}$ EC Framework Program, project NELDA by NASA, and several projects of Asia-Pacific Network for Global Change (APN)). Project partners and all participants of the Conference get acquainted with projects' state and its results, discuss perspectives and initiate new ideas. Regional decision-makers and stakeholders are especially invited to participate in these Workshops.

## 3. Results

ENVIROMIS conferences have a quite long history. The first conference was organized in 2000. That was the first attempt to gather specialists from different environmental fields and establish interdisciplinary relations, as well as to enhance professional skills of early carrier scientists within interdisciplinary teamwork. Although information-computational technologies in environmental sciences were the core of the first conference program, its sections were devoted to different researches starting from field measurements in atmospheric sciences to modeling and monitoring processes in soil and vegetation. When organizing this event, the following approaches were tested:

- the conference as a set of successive sessions;
- a logical order through the conference program (not simply added sessions' programs);
- embedding in the conference program for each session hourly invited lectures presented by prominent authors on one of the actual problems of the Session topic.

Next conference, ENVIROMIS-2002, refined earlier approaches and allowed to develop the best way for organization of a multidisciplinary conference together with an early carrier scientists school in environmental sciences. When preparing this conference, a minimum set of thematic sections has been defined in the program. This allowed us to cover multidisciplinary areas and at the same time to eliminate the contradiction between the thematic breadth and depth of subjects' coverages. This approach was tested, refined, and put into practice of organization of all consequent ENVIROMIS conferences.

Follow the event trends, we can identify the major research areas and leading organizations in these areas, which have been always present at the ENVIROMIS conferences. These research areas are:

- hydrology (Lomonosov Moscow State University, Research Computing Center; P.P Shirshov Institute of Oceanology RAS, Moscow; Institute of Water and Ecological Problems SB RAS, Barnaul);
- climate changes, meteorology (Institute of Monitoring of Climatic and Ecological Systems SB RAS; V.E. Zuev Institute of Atmospheric Optics, Tomsk; V.B. Sochava Institute of Geography SB RAS, Irkutsk);

- climatic and meteorological modeling (Institute of Numerical Mathematics RAS, Moscow; Lomonosov Moscow State University, Research Computing Center; Siberian Regional Hydrometeorological Research Institute, Institute of Computational Mathematics and Mathematical Geophysics, Novosibirsk; Hydrometeorological Center of Russian Federation, Moscow);
- pollution transport (Institute of Computational Mathematics and Mathematical Geophysics, Novosibirsk; Novosibirsk State University; Tomsk State University);
- vegetation cover, biochemical cycles (Siberian Research Institute of Agriculture and Peat, Tomsk; V.N. Sukachev Institute of Forest SB RAS, Krasnoyarsk; A.M. Obukhov Institute of Atmospheric Physics RAS, Moscow; Department of Mathematics Komi SC UrD RAS, Syktyvkar);
- information-computational and instrumental infrastructure for environmental studies (Institute of Monitoring of Climatic and Ecological Systems SB RAS, Tomsk; Institute of Numerical Mathematics RAS, Moscow; Institute of Computational Technologies SB RAS, Novosibirsk; Institute of Computational Modeling SB RAS, Krasnoyarsk).

This biannual event organization, comprising elements of Early Career Scientists School and scientific conference , became well-known not only in Russia but also abroad. Scientists from leading foreign centers (for instance, NOAA, MIT Center for the Global Change Science, etc. (USA), National Institute of Polar Research (Japan), Potsdam Institute for Climate Impact Research (Germany), Danish Meteorological Institute (Denmark)) regularly participate in ENVIROMIS conferences, where they present studies carried out at these organizations. About 1400 researchers took part in ENVIROMIS conferences, among them there were 120 foreigners. Half of the ENVIROMIS conferences participants have been young scientists.

As already has been mentioned above, ENVIROMIS events contribute to professional skill growth. Within ENVIROMIS conferences' history, quite a few young participants, who were students and post-graduates, successfully defended thereafter their PhD theses and a number of researchers became Doctors of Science.

Besides, during ENVIROMIS events, several international multidisciplinary projects have been initiated. These are:

- Project EC INCO "Man-induced Environmental Risks: Monitoring, Management and Remediation of Man-made Changes in Siberia" (Enviro-RISKS);
- SB RAS supported "Siberia Integrated Regional Study" Project (SIRS);
- APN Project "Capacity Building to Study Interrelations between Atmospheric Composition, Anthropogenic Load and Climate Change in Northern Asia";
- APN Project: «Human Impact on Land-Cover Changes in the Heart of Asia»; and
- APN Project "Capacity building to study and address climate change induced extremes in Northern Asia".

Project EC INCO "Man-induced Environmental Risks: Monitoring, Management and Remediation of Man-made Changes in Siberia" (Enviro-RISKS) (http://project.risks.scert.ru/) was one of the largest projects discussed and initiated at one of the ENVIROMIS conferences. The project has been implemented by a consortium of 10 organizations:Danish Meteorological Institute (coordinator), Siberian Center for Environmental Research and Training, International Institute of Applied Systems Analysis, MPI for biogeochemistry, Institute of Numerical Mathematics RAS, V.N. Sukachev Institute of Forest SB RAS, "KAZGEOCOSMOS" Joint Stock Company, Ugra Research Institute of Information Technologies, Institute of Monitoring of Climatic and Ecological Systems SB RAS, and Institute of Computational Mathematics and Mathematical Geophysics SB RAS. The project target were revealing of anthropogenic risks in Siberian environment, understanding their influence on all ecosystem components, and development of recommendations for these risks mitigation by means of coordinating relevant projects and commensing integrated research focused on interrelations between global climate changes and Siberian environment dynamics. The project implementation resulted in development of a computational infrastructure, creation of data storage and thematic information-computational web-systems. Danish Meteorological Institute published five project report volumes (http://project.risks.scert.ru/project_results/project_reports/). Three academic events (ENVIROMIS-2006, CITES-2007, ENVIROMIS-2008) had been financially supported by this project.

SIRS (http://sirs.scert.ru/) Project had four lines of investigation foci:

- quantification of the terrestrial biota full greenhouse gas budget; in particular, documenting exchange of major biophilic elements between biota and atmosphere;
- monitoring and modeling of regional climate change impact;
- development of SIRS information-computational infrastructure; and
- development of a regional strategy of adaptation to and mitigation of the negative consequences of global change.

APN Project “Capacity Building to Study Interrelations between Atmospheric Composition, Anthropogenic Load and Climate Change in Northern Asia” (http://project.enviromis.scert.ru/apn/) encompassed regional activity oriented to the Northern Asia territory and had been fulfilled by institutions of four countries, Russia, Japan, Kazakhstan, and Uzbekistan. The project was relevant to the APN Science Agenda. It enhanced involvement of regional young scientists into this important study and assisted to their further career development. The Project contributed to the regional capacity building activities and to sustainable development in the region by its assertive data and scientific results dissemination activity. Within the project, a number of young scientists from APN target countries had been financially supported to participate in scientific-educational event CITES-2007.

APN Project «Human Impact on Land-Cover Changes in the Heart of Asia» (http://project.enviromis.scert.ru/asia/) was fulfilled by organizations from Russia, USA, and Mongolia. This project was focused only on four (though quite typical) regions in West Siberia and Mongolia in the light of investigation of human impact on land-cover change. Within the Project, land-cover change and disturbance indices maps were developed for the study regions. Investigations were carried out using satellite image analysis and field observations. Project results were disseminated at several international conferences, including CITES-2009, ENVIROMIS-2010, European Geoscience Union General Assembly 2010 and American Geophysical Union Fall Meeting 2010. Participation of the project team members in these events was financially supported.

The activity within the APN Project "Capacity building to study and address climate change induced extremes in Northern Asia" (http://project.enviromis.scert.ru/Capacity/) was aimed at involvement of regional research community, especially young scientists, in the studies of climatic and environmental changes focused on extremes with a special emphasis on their manifestations in Northern Asia. The project was carried out by organizations from Russia, USA, and China. Two international actions, namely, MAIRS/NEESPI/SIRS APN Workshop «Climate Change induced extremes in Northern Asia» and NEESPI/SIRS Workshop "Different aspects of the Northern Eurasia regional climate dynamics and its interrelations with global processes" were organized. Participation of project team members in these events was financially supported.

In order to disseminate research results among researchers and public, we publish ENVIROMIS conferences proceedings. Electronic versions of these proceedings are posted on the conference web-site (http://www.scert.ru/en/conferences/enviro/) and selected papers are published in special issues of journals “Computational Technologies”, 2006 [8]; “Atmospheric and Oceanic Optics”, 2006 [9]. Presentations of lectures and reports made at conferences are also posted on the relevant conference web-site (http://www.scert.ru/ru/conference/). Besides, all completed Projects have websites, where ENVIROMIS educational materials are posted.

## 4. Conclusion

The described above approach has shown its viability and proved that multidisciplinary ENVIROMIS conferences can contribute to a process of establishing relations between Earth Sciences Informatics (observations, models and infrastructure development), data producers, and users (analysis and climate change forecast, extreme phenomena prediction, decision making). This year, ENVIROMIS-2016 conference was the 9th in the series and some delivered papers are presented in this volume Proceedings.

It should be added that at different periods of time, these conferences were financially supported by INTAS, European Commission, WMO, and the Asia-Pacific Network for Global Change. They are also supported by the Russian Foundation for Basic Research. Despite the fact that this financial support is modest, researchers show undiminishing interest to these conferences. This allows us to hope that the conferences' organizers are on the right path..

The ENVIROMIS conference series is instrumental to strengthen interdisciplinary connections. At the same time, the efficiency of the decision-makers involvement leaves much to be desired. We plan a two-step approach to solve this problem. The first step will be in organizing dedicated workshops and trainings for decision-makers. The second step will be a development of supporting topical manuals for their e-learning. Key specialists – participants of ENVIROMIS conferences will be involved in preparation of these manuals.

**References**

[1] 2013 *IPCC Fifth Assessment Report on Climate Change* (London: Cambridge University Press)
[2] 2012 *Managing the Risks of Extreme Events and Disasters to Advance Climate Change Adaptation* A Special Report of Working Groups I and II of the Intergovernmental Panel on Climate Change ed. C B Field et al (Cambridge and New York: Cambridge University Press)
[3] 2014 *Second Assessment Report ROSHYDROMET on Climate Change and its Impacts on the Territory of the Russian Federation* (Moscow: Rosgidromet)
[4] 2014 *Strategic outlook of climate change in Russian Federation for the period up to 2010-2015, and its impact on sectors of the economy of Russia* (Moscow: Rosgidromet)
[5] von Winterfeldt D 2013 Bridging the gap between science and decision making *Proc. Natl. Acad. Sci. USA* **110** 14055–061 doi:10.1073/pnas.1213532110
[6] Walton P J, Yarker M B, Mesquita M D S and Otto F E L 2016 Helping to make sense of regional climate modeling *BAMS* 1173-1186
[7] Gordov E and Lykossov V 2008 Informational and computation technologies for environmental sciences: synthesis of science and education *Computational technologies* **13** 3–11
[8] Gordov E P, Kabanov M V, Lykossov V N 2006 Information-computational technologies for environmental science: young scientists training *Computational technologies* **11** Special issue 3–15
[9] Kabanov M V and Lykosov V N 2006 Monitoring and modeling of climatic changes in Siberia *Atmos. Ocean. Opt.* **19** 675–685

# Application of a principle of synchronicity to an analysis of climatic processes

**V A Tartakovsky[1], V A Krutikov[1], Y V Volkov[1,2] and N N Cheredko[1]**

[1]Institute of Monitoring of Climatic and Ecological Systems SB RAS
Akademichesky av. 10/3, Tomsk, 634055, Russia
[2]Tomsk Polytechnic University, Lenin av. 30, 634050, Tomsk, Russia
trtk@list.ru, atnik3@rambler.ru

**Abstract.** Geospheric processes are largely initiated and controlled by external influences. The Sun is the main source of energy, and it contributes to those processes directly and as a regulator of some cosmic influences. The forcing effects synchronize the oscillatory processes in the solar system. Specific processes differ in energy. However, they show similar essential signs. Components, which differ in coincidence and non-coincidence of the essential signs, are introduced. Based on a synchronous analysis of the monthly averaged Wolf numbers series and temperatures registered at 818 weather stations in the Northern Hemisphere (1955-2010), we show that the above-introduced components of the series under consideration have extreme properties. The histograms of the primary temperature series coincide with those of their components, except for a range of ± 3°C. The second initial moments of the Wolf numbers correspond to the climate geography and can be divided into two zones, the width and the distance between them are about one third of all possible changes. A synchronicity principle allows performing a decomposition of the original set into subsets containing strongly related elements. The relationship between the synchronization features of geospheric processes initiated by external forcing and the physical and geographical hierarchy allows solving some problems of classification. Some experimental results for the temperature field are presented. The approach has revealed some new properties of the solar-terrestrial relations. No inconsistencies with the known notions of climatic processes have been found.

Various aspects of the existence of life on Earth, ranging from biological to technological, are determined by the external space, in which the movement of the planet occurs. An urgent task is to identify the degree of the influence on the climate change processes, which is shown by natural factors, as well as to take this influence into account in climate models. This problem has not had a unique solution yet. This has been the subject of years-long debates among scientists of different specialities. The fifth report of the Intergovernmental Panel on Climate Change recognizes the contribution of natural external factors as a minor effect, which possesses a medium degree of reliability [1]. Numerous studies on this issue, on the other hand, talk about the significance of the natural factors impact [2, 3] and propose a number of possible mechanisms [4-9].

In this paper, we are interested in solar-terrestrial relations, including all cosmic effects of external influences. The Earth's climate system constantly exchanges matter and energy with the environment. Many scientific works present a list of supposed exchange mechanisms, which mostly remain in the field of hypotheses. Only some of them are included in climate models. External influences can affect the climate system directly, due to radiative forcing (i.e. due to changes in the solar constant) [4], or indirectly, through a variety of processes in the atmosphere and on the boundary of the medium [5, 9-13]. A large series of works are devoted to the influence of the solar system planets and their relative positions on periodic manifestations of solar activity [14, 15].

It is possible that, compared with the energy of the processes in the climate system itself, the direct impact due, for example, to changes in solar irradiance fluctuations in the situation of solar activity, is small [1, 16]. However, this effect is able to generate forced vibrations in the system that may have a resonant character.

The main forcing factor for processes on the Earth is solar energy. The sun makes a direct contribution as a regulator of cosmic influences. All natural systems, as well as the global climate system, are open. This or that degree of stability of intra-processes is provided by a number of controlling and energy redistributing mechanisms. The most difficult to model are feedback: positive ones strengthening the deviation from the initial state, and negative ones reducing it.

Oscillatory processes in the solar system are elements of interconnected subsystems and they are present in all components of the geospheres. They have been investigated extensively enough. However, the quality of forecasting the state of the environment with the help of the revealed patterns has not been improved yet. The diagnosis and prognosis of such complex multilevel systems dynamics is complicated due to the inadequate description of the processes and mechanisms of their interactions. This is related to the fact that they are characterized by a wide range of spatial and time scales, as well as a large number of differently oriented relations. In addition, all natural objects are characterized by elements of accumulation and response delay, as well as different degrees of resilience. This ultimately leads to a divergence of solving predictive equations. The error grows faster than a certain distance from the initial state. Modern forecasting models use huge computing resources. Despite this, there are still only short-acceptable forecasts. The problems of medium- and long-term prognosis are not solved. Climate changes are characterized only by possible scenarios whose connection with reality is problematic.

With reference to the above mentioned, it is appropriate to update the phenomenological approach and to use observational data for the study of the internal structure and dynamics of the climatic parameters and their relations with the forcing. A better understanding of these processes will create new conditions for simulations of climate systems and identification of their functioning patterns on the global and regional scales. New functional patterns, when identified empirically, largely determine further progress of physical and mathematical climate models.

Forcing external impacts generated in the solar system promote synchronization of oscillatory processes in the geospheres. An agreement of forcing impact components with various natural and climatic processes is formed. Previously, we determined synchronicity as an essential factor of solar-terrestrial relations. [17] Generalization of experimental facts allowed formulating a new principle: "an external forcing inherently initiates and synchronizes elementary processes in the geospheres. In our previous studies, we suggested various applications of this principle.

We believe that, at the predetermined level of the hierarchy, particular processes differ in energy but have similar essential features. This is reflected in the consistency of certain components of the processes compared. The latter are regarded as a presence of a useful signal, as a manifestation of a forcing influence. As the essential features, we chose the signs of the Fourier transform coefficients of the analyzed processes. For the processes under study, we introduce components, which differ in the coincidence and non-coincidence of the essential features: the CS- and NS-components, respectively. The components redistribute information from the original data without distortion, and new properties of the processes are identified. The procedure of paired separation of series into such components is called "decomposition by selected series." This decomposition selects components with an extreme correlation from a pair of series: CS-components with a positive correlation, NS-components with a negative one. These components are orthogonal between themselves. These properties, in particular, are the basis for selecting the signs of the Fourier transform coefficients as essential features of synchronicity.

A positive correlation of the CS-components and a negative correlation of the NS-components characterize their opposite directional action in a geosystem. It can be assumed that the components with the coincident essential features (CS) are a manifestation of a positive relation between exposure and reaction, whereas negative connections are characterized by the components with non-coincident essential features (NS). In the nature, these two situations are realized together. The Sun provides the global flow of energy to the Earth. However, in any given point there is both an inflow and an outflow of energy. It is known that the surface temperature conditions are formed as a balancing result of the energy flow from the Sun to the Earth, the processes of scattering of energy and its reflection from the Earth back into space and inner energy processes in the geospheres. Incoming solar energy is distributed unevenly over the surface. This triggers the redistribution mechanisms of heat (energy) and water through processes of large-scale atmospheric circulation, the processes of heat and moisture exchange in the atmosphere itself and between the atmosphere and the Earth's surface in various ways (conduction, evaporation and condensation, turbulence, adiabatic processes, convection and advection). Furthermore, along with the hierarchy of the regional level, there is an increase in the influence of local characteristics of the underlying surface that trigger different climate-regulating factors with the same energy redistribution mechanisms, but of a smaller scale. On the one hand, the same effect can be generated by various influencing factors. On the other hand, the effect of the same influence factor can lead to the opposite result. In this regard, a simple correlation of climate variability characteristics and, for example, solar activity parameters, generally gives no satisfactory consistent results. Thus, the existing Sun-Earth relations are hardly ever used in climate models.

The developed theory was applied to analyze the synchronicity of the Wolf numbers series and the temperature recorded by 818 weather stations of the Northern Hemisphere within the period between 1955 and 2010. [17] The selected series in this case was the series of Wolf numbers. They give a concept of fluctuation scales in solar activity and have a high correlation with other solar and cosmic manifestations [18]. This unique series of instrumental data on space should be regarded as a comprehensive indicator of solar activity changes.

Figure 1 shows the correlation coefficients of the initial series of solar activity and temperature $r$ and their respective components $\hat{r}$ and $\tilde{r}$. As can be seen, in 85% of cases the modulus of $r$ is less than 0.2 and it never reaches the values of $\hat{r}$ and $\tilde{r}$ for a particular month and a weather station. The correlation coefficients $\hat{r}$ and $\tilde{r}$ show the presence of a significant and unique relationship between the homonymous components of the temperature series and the Wolf numbers. In most cases, these coefficients are significant with a probability of at least 0.95 for the used small sample.

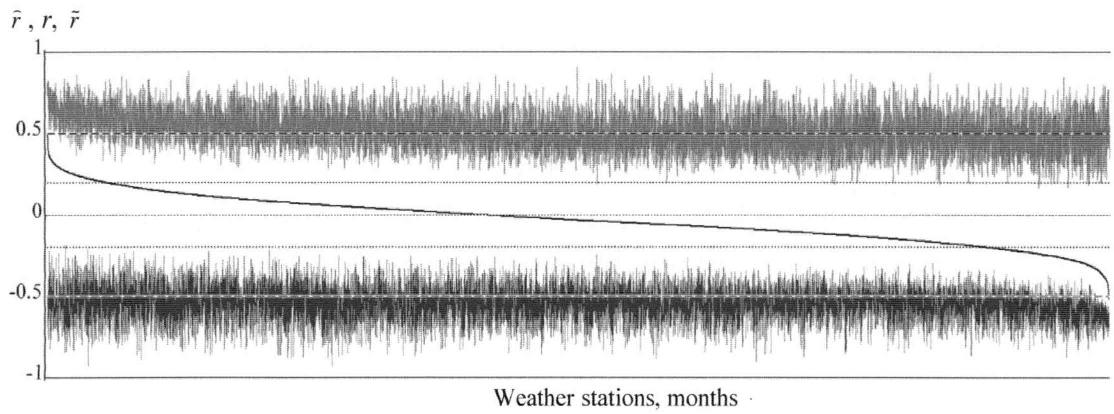


**Figure 1.** Correlation coefficients of Wolf numbers and temperature series and their CS- and NS-components: Red - $\hat{r}$ ; blue - $\tilde{r}$ ; black curve - $r$. All correlation coefficients are ordered with $r$ and range in descending order along the abscissa. Each value corresponds to one of 818 temperature series for each of the 12 months over 56 years. In total, there are 9816 points along the abscissa. Dotted lines designate the corridor of ± 0.2.

The histograms of the initial temperature series coincide with those of their components over long continuous temperature intervals, with the exception of the range of about ± 3°C (Figure 2). This pattern appears to be interesting for deeper exploration.

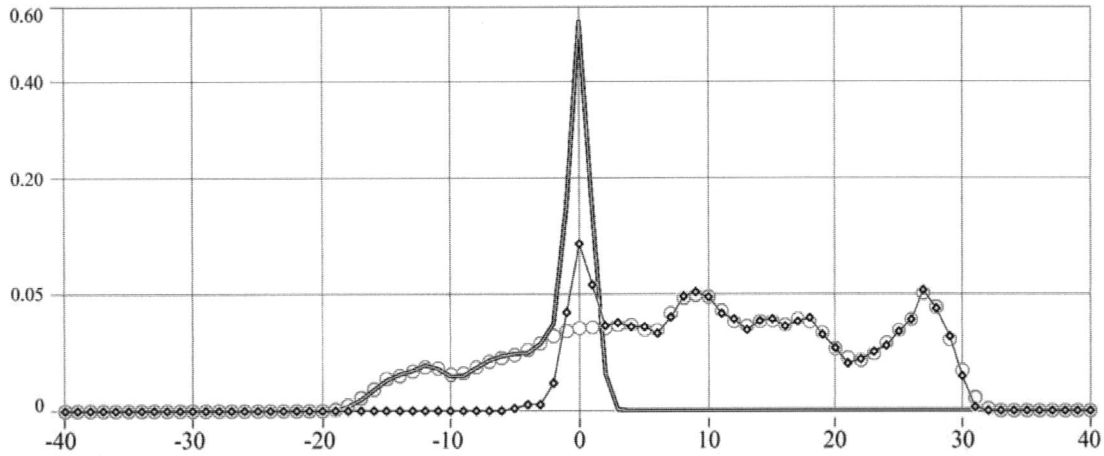


**Figure 2.** Histograms of average annual temperature series and their components within range from 1956 to 2010 at 818 weather stations: Temperatures during analyzed period - OO; CS-components - -◆-; NS-components - ═. The partitioning interval is 1 °C. The normalized mean-square difference over the convergence areas between the curves does not exceed 4%. Along the axes of ordinates, the power scale has the index of ½.

Over the monthly variation of the second initial moments of the Wolf numbers CS-component (Figure 3), there are two clearly defined zones of changes, the width, and the distance between them being about a third of the possible values. There is a regional asymmetry of receiving the solar signal by the temperature field. The upper zone, for the whole year, is taken by the initial moments of 400 stations (orange, green, light blue, dark blue). The moments of the remaining 418 stations are moved into the upper zone only in July and August. The rest of the year, they are observed within the lower zone.

Here, the climatic geography is well displayed. The stations of the upper zone are located in the most energy-active areas of the climate system (equatorial and tropical latitudes, areas of the greatest influence of the North Atlantic thermohaline conveyor and the warm Alaska current). Only the use of decomposition revealed the hidden components, which characterize the features of the solar energy assimilation.

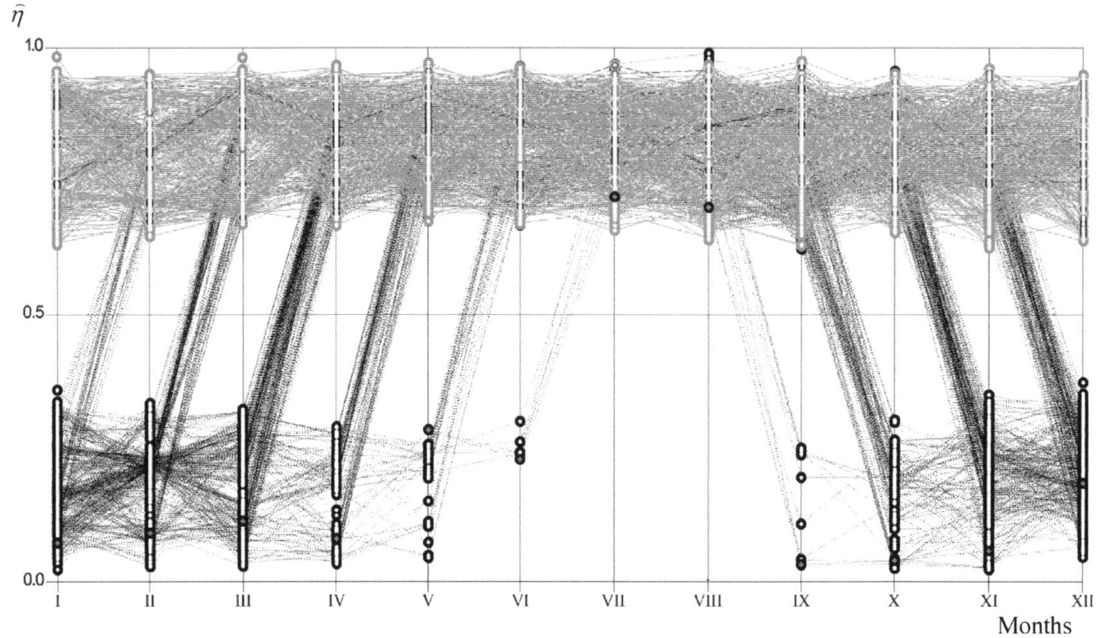


**Figure 3.** Normalized second initial moments of CS-components of Wolf numbers $\hat{\eta}$ from 1955 to 2010 at weather stations: Yellow and orange - 358 stations from 0.5 °N to 50 °N; green - 31 stations in the North Atlantic from 50 °N to 64.3 °N; blue - 11 stations of North-West Coast of America from 55 °N to 40.8 °N; black - 418 stations that do not coincide with the previous ones, from 80 °N to 50 °N; black circles with red centers - the most northerly weather station at 80.6 °N, the synoptic index 20046. The months are indicated by Roman numerals along the abscissa.

Based on the accepted synchronicity principle, it is possible to carry out decomposition of the original set into subsets containing strongly related elements. The connection between the elements of different subsets has to be weaker. Thus, if to compare the synchronization features of the geospheric processes initiated by external forcing actions with physical and geographical hierarchy, we can meet the challenges of climate classification. This can be done based on the hypothesis: a response to an external forcing action common to any physical and geographical unit is manifested in the similarity of essential features of all processes within this area.

For the temperature field we introduced into consideration the phase of temperature series as a factor of their common variation. The use of the phase rather than the amplitude is mainly due to the nonlinear distortion and noisiness of the time series. Specifically, the temperature changes, which characterize local climate peculiarities, are small if compared to the annual variation in temperature. However, these particular changes should be selected for classification. In the case of analyzing the fluctuation phase, the process of eliminating the annual temperature variation from the original series is reduced to the removal of only one linear function over the whole analyzed interval. Thus, applying the phase has advantages, and we propose to consider synchronization as phasing of the processes under study. For this purpose, we used algorithm based on the analytic signal formalism. This procedure was described by us in [19]. The presence of phase modulation in the temperature series is confirmed by the view of the input data power spectrum, where four modes at frequencies that are multiples of the carrier frequency are observed.

The basis of the hypothesis on the geographic conditionality for the features of the temperature signal phase modulation is the following. The amount of the incoming energy determined by the latitude is modulated by the rotation of the Earth, and then the phase of the original signal is transformed by large-scale circulation processes and local conditions. From time to time, under the influence of factors of different nature, there occur disturbing effects on the geospheres. This results in the changing nature of variations of different components parameters. A complicated form of phase modulation is formed. However, there is a consistency of these changes. The degree of the consistency can serve as a criterion of classification. Due to the cyclic rotation of the Earth around the Sun, the temperature values averaged over short periods have recurrences of a period corresponding to one calendar year. Therefore, they fit into the concept of vibration and it is possible to apply general laws of the vibration theory to them.

Previously, we tested the algorithm according to the surface temperature data obtained at 332 weather stations in Eurasia [19]. Now it is implemented to identify climate structures of the Northern Hemisphere (Figure 4) according to the data of 818 stations over the period of modern climate change (1956-2010).

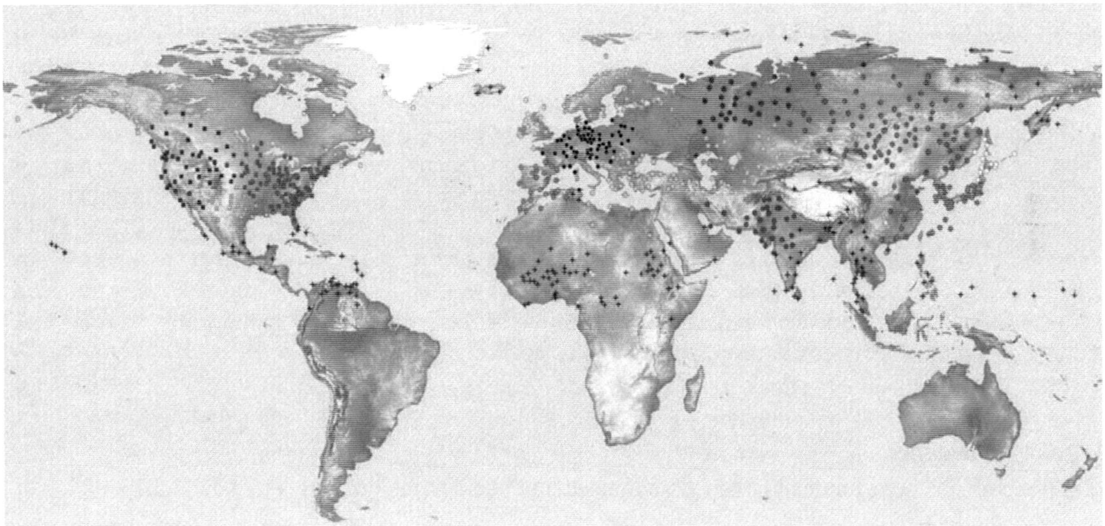

**Figure 4.** Classification of Northern Hemisphere temperature field (1955-2010).

The classification was performed by means of calculating the phase estimates for each temperature series of the total sample and identifying groups with consistent changes of these estimates there. As a measure of consistency, we chose the correlation coefficient. The groups are formed from weather stations, which have the temperature series phases correlated with the phases of the other weather stations stronger than the defined threshold value *r*. The condition for the completion of the iterative process is the lack of changes in the composition of the groups. The groups of phases that meet the predetermined level of correlation, on completion of the iterative process, form a "typical phase" as a limit within a group. The experiment conducted for a sample of 818 series of monthly temperatures showed the convergence of the iterative process up to the level of computational precision. With the increase in the fractionality of class division, we observe a greater influence of regional factors, but each class has information about global signals. The original 818-dimensional space of temperature series was reduced to a 17-dimensional space of typical phases corresponding to the climatic features of territories where temperature changes occur synchronously. Next, using the obtained typical phases, for the respective territory (a class of stations) we can define the basic patterns of response to the external signal and the geographical conditionality of its transformation.

The phases are not defined for some stations marked by the sign "+" (Figure 4). That is why these stations were not included in any of the classes. Isolation of these stations is consequence of that their temperature variations are belong to the class of broadband oscillations. This distinctive feature has a physical basis.

The advantage of the proposed method of classification, in the first place, is its objectivity, the use of precisely measured data and no need to use bioclimatic data. Secondly, it is the possibility of selecting the degree of differentiation of the tested field, as much as it is possible for the objective characteristics, depending on the task. It is obvious that the degree of differentiation is due to natural causes, and the threshold correlation coefficient is a tool for their identification. Despite the fact that the described algorithm was implemented only for the surface temperature data, climatic classes, in general, correspond to the laws of the classical Köppen climate classification. The comparison with this classification revealed some extension to the north of the warm climate borders. Thus, there is a reorganization of the spatial structure of the climatic parameters fields with the change of synchronous vibrations of climatic processes under the conditions of the changing climate. This is a consequence of the transition of the system to a new qualitative state. The phase, as a characteristic of the temperature fluctuations, allows objective determination of climate classes and provides an opportunity to continue to receive dynamic climate classifications. This is important against the background of changing climatic conditions and is of a great importance for strategic planning. The proposed approach can be used as an analytical framework for the study of climate change on any spatial scale using only the data on the surface temperature up to a level of quality predetermined by a researcher.

It is impossible to consider all factors when simulating variability of natural and climatic processes. However, knowing which particular factors should be taken into account and in what way, is necessary to optimization of models. So, taking into consideration the characteristics and mechanisms of consistency of the processes over different time periods and geographic territories could allow obtaining new prognostic features. There is a need for further detailed development of complex numerical algorithms and a software tool for solving the specified tasks that would be mostly aimed at the use of supercomputer technologies built into the overall system of monitoring the complex ecological and climate assessment of the state of the environment.

## Acknowledgements

This research has been funded by the Russian Academy of Sciences (project VIII.80.2.3).

## References

[1] IPCC 2013 *Climate Change 2013: The Physical Science Basis. Contribution of Working Group I to the Fifth Assessment Report of the Intergovernmental Panel on Climate Change* ed T F Stocker *et al* [Cambridge: Cambridge University Press]

[2] Soon W 2005 Variable solar irradiance as a plausible agent for multidecadal variations in the Arctic-wide surface air temperature record of the past 130 years *Geophys. Res. Lett.* **32** (16) L16712

[3] Skafetta N, West B 2006 Phenomenological solar contribution to the 1900-2000 global surface warming *Geophys. Res. Lett.* **33** (5) L05708

[4] Abdussamatov H I 2005 On long-term variations of the total irradiance and on probable changes of temperature in the Sun's core *Kinematics Phys. Celestial Bodies* **21** 471–7

[5] Barriopedro D, Garcia-Herrera R and Huth R 2008 Solar modulation of Northern Hemisphere winter blocking *J. Geophys. Res.* **113** D14118

[6] Raspopov O M, Veretenenko S V 2009 Solar activity and cosmic rays: influence on cloudiness and processes in the lower atmosphere (in memory and on the 75th anniversary of M I Pudovkin) *J. Geomagn. Aeron.* **49** (2) 147–55

[7] Gray L J et al. 2010 Solar influences on climate *Rev. Geophys.* **48** 1–53

[8] Zherebtsov G A, Kovalenko V A 2012 Solar activity effect on weather-climatic characteristics of the troposhere *Journal Solar-Terrestrial Physics* **21** (134) 98–106

[9] Veretenenko S V, Ogurtsov M G 2015 Nature of long-term correlations between cloud state and variations in galactic cosmic ray flux *J. Geomagn. Aeron.* **55** (4) 457–65

[10] Soukharev B E, Hood L L 2006 Solar cycle variation of stratospheric ozone: Multiple regression analysis of long-term satellite data sets and comparisons with models *J. Geophys. Res.* **111** D20314

[11] Ermakov V I, Okhlopkov V P and Stozhkov Yu I 2007 Influence of Cosmic Dust on the Earth's Cloudness, Albedo, and Climate *Vest. Mosk. Univers.* Fizika **5** 41–55

[12] Zherebtsov G A, Kovalenko V A, Molodykh S I, Rubtsova O A and Vasileva L A 2008 Impact of Helio- and Geophysical Disturbances on Thermobaric and Climatic Characteristics of the Earth's Troposphere *Cosmic Res.* **46** (4) 368–77

[13] Calisto M, Usoskin I, Rozanov E and Peter T 2011 Influence of Galactic Cosmic Rays on atmospheric composition and dynamics *Atmos. Chem. Phys.* **11** (9) 4547–56

[14] Okhlopkov V P 2014 The 11-Year Cycle of Solar Activity and Configurations of the Planets *Vest. Mosk. Univers. Fizika* **3** 50–5

[15] Scafetta N 2014 Discussion on the spectral coherence between planetary, solar and climate oscillations: a reply to some critiques *Astrophys Space Sci.* **354** 275–99

[16] Lean J L, Rind D H 2006 How natural and anthropogenic influences alter global and regional surface temperatures: 1889 to 2006 *Geophys. Res. Lett.* **35** L18701

[17] Tartakovsky V A 2015 Synchronous analysis of the Wolf numbers and temperature series from weather station in the Northern Hemisphere of the Earth *Atmos. Ocean. Opt.* **28** (2) 182–8

[18] Hathaway D H 2010 The Solar Cycle *Living Rev* Solar Phys **7** 1–65

[19] Tartakovsky V A, Krutikov V A, Volkov Yu V and Cheredko N N 2015 Climate classification by analysis of the phases of temperature series *Atmos. Ocean. Opt.* **28** (8) 711–7

# Hydrothermal conditions of South Eastern Siberia under the ongoing warming

**N N Voropay[1,2], E V Maksyutova[1] and A A Riazanova[2]**

[1] V.B. Sochava Institute of Geography SB RAS, Irkutsk
[2] Institute of monitoring of climatic and ecological systems SB RAS, Tomsk

**Abstract.** A great increase in air temperature has been observed since 1976. Siberia is a region with most severe ongoing climate change. To monitor the extreme weather events is important. To evaluate moisture conditions we used the D.A. Ped index ($S_i$). Monthly air temperature and precipitation data from 19 weather stations of South Eastern Siberia (50-60° N 90-120° E) were used for the index calculation during the vegetation period. During 1976-2010 the number of droughts in the study region was more than the number of excessive moisture periods. The maximal statistically significant trend (0.4-0.6 per 10 years) in Eastern Siberia was observed in May. The characteristics of the winter-spring period preceding the vegetation season were analyzed. Significant positive trends exist in the study area for the May temperature (0.5-0.9 °C per 10 years) and the May sum of positive temperatures (14-28 °C per 10 years). There are tendencies to increase the number of days with temperatures above zero in March (1-3 days per 10 years) and the sum of positive temperatures in April (5-16 °C per 10 years). The stable transition of air temperature over 0 °C shifts into early dates by 1-7 days every 10 years.

## 1. Introduction

The climate change issue attracts attention of researchers and is among the most important natural science issues. The research of climate trends of meteorological variables is presented in a number of publications. The review of recent publication shows that many scientists hold by an opinion that climate warming has been occurred in recent decades. However, some researchers believe that the temporary warming is occurring given the durable and slow cooling [1].

The air temperature is one of the most important indicators of climate change; therefore, particular qualities of the intra-annual fluctuations, the secular trend and interannual variability are well researched. The recent data shows that the surface air temperature increased by 0.6 ± 0.20 °C during the 20th century [2], [3]. However, the global warming was temporally heterogeneous. Three periods are singled out: warming up within the intervals of 1910-1945, the, low cooling within the 1946-1975 and intense warming since 1976.

These periods are characterized by spatial inhomogeneity, i.e. in some areas the warming is more intensive and in some regions even cooling than in others [4]. The largest positive trends in annual air temperature are observed mainly in the continental areas (Prebaikalie, Central Yakutia, the Far East) with values reaching 2-3.5 °C per 100 years [5], [6].

## 2. Methods

It should be mentioned that, the study of variations of single climatic elements furthered the study of the integral characteristics of the territory, which were different humidity or aridity indices, being in most cases combinations of air temperature and precipitation values. In Russia well-known Selyaninov hydrothermal coefficient is often used. But to study droughts in different landscapes index $S_i$, formula (1), suggested by D.A. Ped', is the most appropriate. One drought or excessive moistening of a particular region in a warm season is characterized not by the air temperature and precipitation values, but by their anomalies compared to climatic norms:

$$S = \frac{\Delta T}{\sigma_T} - \frac{\Delta R}{\sigma_R} \quad (1)$$

where $\Delta T, \Delta R$ are anomalies (deviations from the norm) of monthly air temperatures and precipitation, $\sigma_T, \sigma_R$ - mean-square deviations for the same parameters. The following conditions are used for determinations of the extreme events:

For droughts $\Delta T>0$ and $\frac{\Delta T}{\sigma_T}>0$, $\Delta R<0$ and $\frac{\Delta R}{\sigma_R}<0$, then $S_i>0$.

For excessive moistening $\Delta T<0$ and $\frac{\Delta T}{\sigma_T}<0$, $\Delta R>0$ and $\frac{\Delta R}{\sigma_R}>0$, then $S_i<0$.

We estimate the intensity of drought according the next rules: weak $1.0 < S_i < 2.0$; medium $2.0 \leq S_i < 3.0$; severe $S_i \geq 3.0$. The intensity of excessive moistening was divided into low $-2.0 < S_i < -1.0$; medium $-3.0 < S_i \leq -2.0$; high $S_i \leq -3.0$.

Daily temperature and precipitation obtained at weather stations during the vegetation period (May-September) were used for index $S_i$ calculation. Only weather stations with minimum (less than 5%) amount of missing data were used. Index $S_i$ was calculated for 19 weather stations located at the south of East Siberia (within rectangle with coordinates 50-60° N, 90-120° E) for a period of intense global warming (1976-2010) (Figure 1).

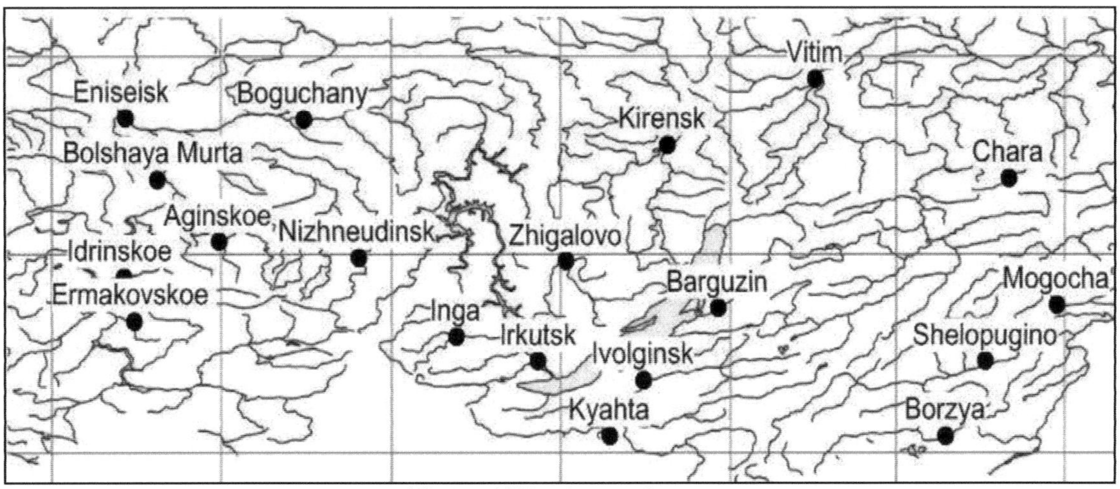


**Figure 1.** Weather station map.

Administratively, the study area includes the southern regions of the Krasnoyarsky Krai, the Irkutsk Region, the Republic of Buryatia, the Chita Region. The study area can be divided to three physical geographic regions (Southwest of Central Siberian Plateau, Predbaikalie, and Zabaikalie). Usually weather stations are located in the river valleys. Mountain areas in the present paper were not investigated. Only Inga station is located in the foothills of the Eastern Sayan Ridge.

We studied long-term variations of index $S_i$ for vegetation period (from May to September) for 1976-2010. May is a critical month for drought. During cyclonical activity periods in July and August a sufficient or extreme moistening is usually registered. Averaged values for the vegetation period do not reflect actual extreme events revealed for single months. Then, the territory covered by droughts in different years was evaluated. The maximum drought area was observed in May.

## 3. Results

Analysis hydrothermal conditions for the study area showed that the repeatability of the weak drought is more than a repeatability of the low excessive moistening during the vegetation period. But the repeatability of periods of medium and high excessive moistening is higher than the repeatability of periods with medium and severe drought.

Forecasting, of a single meteorological characteristic and complex indices such as $S_i$, is one of the main tasks in climatology. Its reliability is determined by the sufficient information base, which includes long-term data of instrumental observations from weather stations. However, data from the majority of Siberia stations were registered only from 1930`s. The Irkutsk station had more information on air temperature and precipitation since the late 19$^{th}$ century. We calculate the index for the long-term weather station Irkutsk since 1882. Significant trends were found in May. However, the amplitude of interannual fluctuations increases in recent decades. In different months this increase occurs due to the prevalence of extreme deviations of either positive or negative sign. For example, the positive extremes prevail in June, negative - in July, August and September. Thus, we can talk about a redistribution of dry and wet periods within the vegetation season. The number of weak atmospheric droughts has increased in all months of the vegetation season. The number of periods with weak excessive moisture is decreases. The extreme events with medium intensity have complicated the structure of changes.

Comparison of the $S_i$ indices for the long period (1891-2010) with the $S_i$ indices for the period of the most intense warming (1976-2010) have shown that more than half of the cases of extreme events (droughts or excessive moistening) are observed in the last 35 years (Table 1). For example, in September from 1892 to 2010, twenty-one weak droughts were observed, and most of the weak droughts (10-11 cases) were recorded after 1976. The severe droughts and the high excessive moistening have the highest influence on the terrestrial ecosystems. In particular, in May and June during the period 1976-2010 in Irkutsk three or four severe droughts observed no one severe drought was registered in the previous period. Repeatability of periods of excessive moistening increased only in July (medium intensity) and in August (high intensity).

**Table 1.** Repeatability of $S_i$ index (number of cases) in Irkutsk for different periods.

| | | | | | | | | | | | | | | | |
|---|---|---|---|---|---|---|---|---|---|---|---|---|---|---|---|
| | **Drought** | | | | | | | | | | | | | | |
| | $1<S_i<2$ - Weak | | | | | $2\leq S_i<3$ - Medium | | | | | $S_i \geq 3,0$ - Severe | | | | |
| ***Period*** | V | VI | VII | VIII | IX | V | VI | VII | VIII | IX | V | VI | VII | VIII | IX |
| **1891-2010** | 14 | 17 | 17 | 21 | 21 | 7 | 6 | 8 | 6 | 6 | 4 | 3 | 3 | 3 | 0 |
| **1976-2010** | 5 | 5 | 5 | 10 | 11 | 3 | 2 | 5 | 3 | 4 | 4 | 3 | 1 | 1 | 0 |
| | **Excessive moistening** | | | | | | | | | | | | | | |
| | $1<S_i<2$ - Weak | | | | | $2\leq S_i<3$ - Medium | | | | | $S_i \geq 3,0$ - High | | | | |
| ***Period*** | V | VI | VII | VIII | IX | V | VI | VII | VIII | IX | V | VI | VII | VIII | IX |
| **1891-2010** | 16 | 19 | 18 | 20 | 13 | 8 | 8 | 7 | 9 | 10 | 2 | 0 | 3 | 1 | 1 |
| **1976-2010** | 4 | 5 | 1 | 5 | 2 | 0 | 3 | 4 | 0 | 1 | 0 | 0 | 1 | 0 | 1 |

Tendencies of changes in $S_i$ index were calculated for the 35-year period (1976-2010). Trend values in most cases are statistically insignificant (Figure 2).

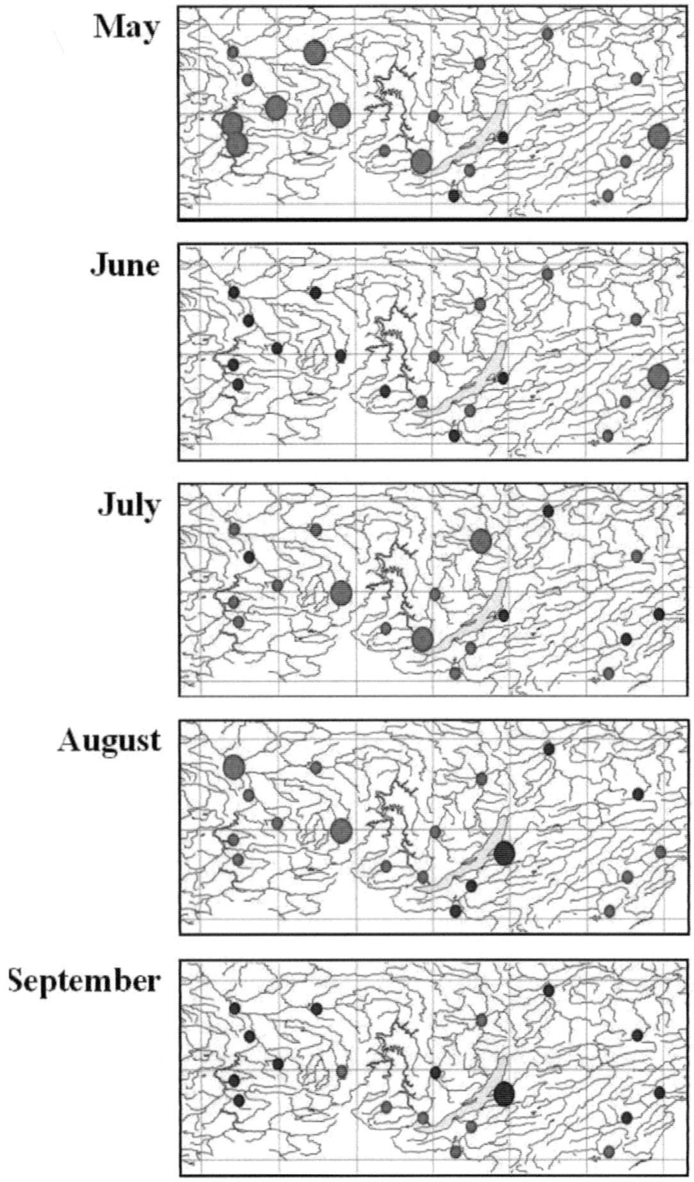


**Figure 2.** Trends of $S_i$ index for the period 1976-2010
statistically significant trend: ● - positive, ● - negative ($p<0.05$)
statistically insignificant trend: • - positive, • - negative ($p>0.05$).

The maximum of statistically significant trends of the $S_i$ were registered in May. At six of the nineteen weather stations the increase of the $S_i$ was observed with rates 0.4-0.6 per 10 years. In July, the number of droughts increases only at three weather stations of Predbaikalie (Nizhneudinsk, Irkutsk, and Kirensk) and at weather station Barguzin in Zabaikalie. In June, aridity increase at the weather station Mogocha, in August - at the weather stations Eniseisk and Nizhneudinsk. In

September, a statistically significant increase in the number of droughts was not observed. At Barguzin station (Zabaikalie) increase of the moistening is found in August and September. We have studied changes in area averaged indices $S_i$ for regions. Increasing of aridity in the Southwest of Central Siberian Plateau is observed only in May, in Predbaikalie - in May, July and August. In Zabaikalie statistically significant trends were not observed.

To reveal influences of winter periods on the vegetation period the following climatic characteristics of the winter-spring period, which was preceded by the vegetation period, were analyzed:

Precipitation in November-March, mm ($R_{XI-III}$)
April Precipitation, mm ($R_{IV}$)
May Precipitation, mm ($R_V$)
March temperature, °C ($T_{III}$)
April temperature, °C ($T_{IV}$)
May temperature, °C ($T_V$)
The sum of air temperatures above 0 °C in March, ° C ($\Sigma T>0_{III}$)
The sum of air temperatures above 0 °C in April, °C ($\Sigma T>0_{IV}$)
The sum of air temperatures above 0 °C in May, °C ($\Sigma T>0_V$)
The number of days with temperatures above 0 °C in March ($N_{T>0,\ III}$)
The number of days with temperatures above 0 °C in April ($N_{T>0,\ IV}$)
The number of days with temperatures above 0 °C in May ($N_{T>0,\ V}$)
Date of transition of air temperature through 0 °C ($D_{T>0}$)
Maximal snow depth, cm (H)
Date of final loss of snow ($D_{H<0}$)

We have analyzed the tendencies for the above listed characteristics. Statistically significant positive trends of the May temperature (0.5-0.9 °C per 10 years) and of the sum of positive temperatures in May (14-28 °C per 10 years) exists at the whole study area. The number of days with temperatures above zero increases in March (1-3 days per 10 years) at the Southwest of Central Siberian Plateau, and the sum of positive temperatures increases in April (5-16 °C per 10 years) in Zabaikalie. These changes can considered as a consequence of changes in the dates of stable transition of air temperature through 0°C. All weather stations show a shift to earlier dates (1-7 days per 10 years). Statistically significant changes (3-7 days per 10 years) were observed at a half of the studied stations.

Graphs of long-term changes of some climate characteristics are presented for some weather stations (Figure 3). These characteristics have statistically significant trends on most weather stations. Weather stations are located in different parts of the study area (Bolshaya Murta - Southwest of Central Siberian Plateau, Zhigalovo - Predbaikalie, Ivolginsk – Zabaikalie). The linear trend equations are shown on the figures. Positive trends in most cases (except the date of the spring transition temperature through 0°C) were registered. The trend of precipitation for the cold period is positive at many weather stations, but it is statistically significant only at three stations (Bolshaya Murta, Aginskoe, Nizhneudinsk).

Conditions of the preceding cold period have a negligent effect on $S_i$ index, which was calculated for the five months of the vegetation season. This is confirmed by the correlation analysis. In most cases, the linear correlation coefficients are not statistically significant. As expected, the maximum correlation was observed between the characteristics of the winter-spring period and indices $S_i$ in May. The statistically significant correlation coefficients (0.47-0.85) were obtained between $S_i$ index for May and hydrothermal conditions of the same month (the precipitation - 17 stations, the monthly air temperature - 18 stations, the sum of air temperatures above 0°C - 17 stations). Also, $S_i$ index in May depends on the precipitation from November to March (5 cases) and on the sum of temperatures above 0°C in March (5 cases).

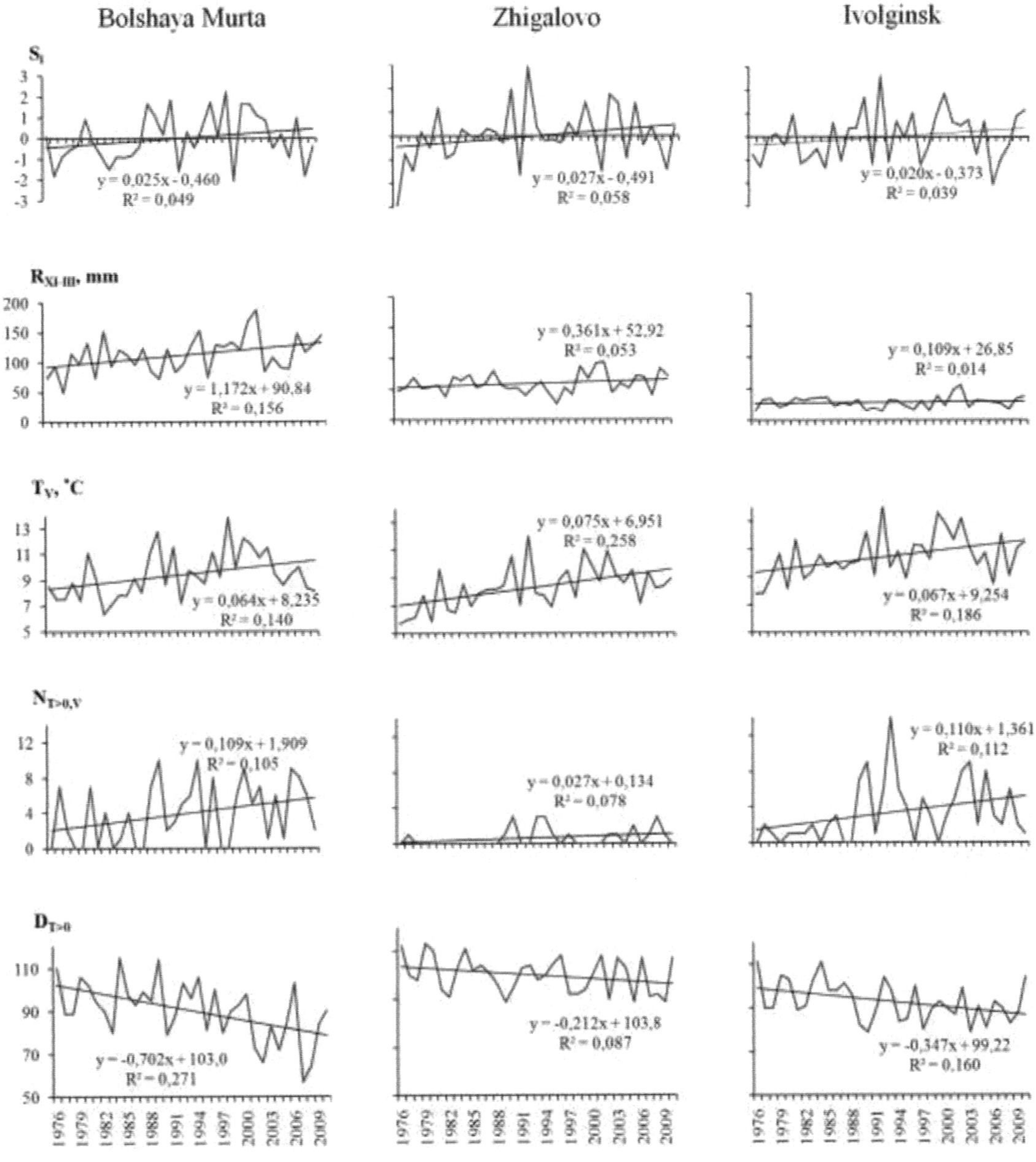


**Figure 3.** Long-term changes of the $S_i$ index in May and some climate characteristics for some weather stations.

Characteristics of the cold period have almost no effect on the indices for June and July. Some increase in correlation coefficients observed in August. The higher correlation coefficients will be observed in the study of correlation of parameters for the current and the previous months. It is claimed by our previous study.

Long-term variations of the $S_i$ index and the characteristics of the preceding cold period at the weather station Chara (Zabaikalie) are shown at Figure 4. Medium droughts were registered in 1993 and 2002, and medium excessive moistening events were observed in 1983 and 2006. The average value of each characteristic for long-term period (1976-2010) is shown in the first row of Table 2.

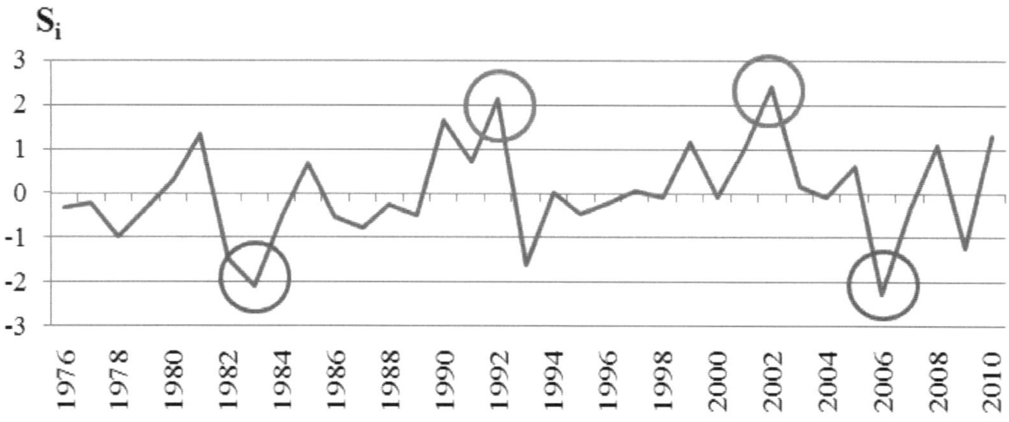


**Figure 4.** Long-term variations of the $S_i$ index in May at the weather station Chara (Zabaikalie).

The mirroring change of deviations for five climatic characteristics occurs at weather station in years with the excessive moistening and droughts. As shown before, the precipitation of the cold period, May precipitation May temperature , as well as the sum of the temperature and the number of days with temperatures above 0°C in May contribute to the moistening conditions in May. Red and blue cells are marked positive and negative deviations from the mean.

**Table 2.** Deviations *(italics)* of characteristics of the preceding cold period at the weather station Chara (Zabaikalie) from the mean.

| Year | $S_{i,V}$ | $R_{XI-III}$ | $R_{IV}$ | $R_V$ | $T_{III}$ | $T_{IV}$ | $T_V$ | $\Sigma T>0_{IV}$ | $\Sigma T>0_V$ | $N_{T>0,IV}$ | $N_{T>0,V}$ | $D_{T>0}$ | H | $D_{H<0}$ |
|---|---|---|---|---|---|---|---|---|---|---|---|---|---|---|
| **Mean (1976-2010)** | | 19.9 | 13.5 | 44.1 | -16.8 | -3.6 | 5.6 | 16.5 | 178.4 | 7.3 | 28.3 | 113.2 | 16.1 | 107.7 |
| **1983** | -2.9 | 17.1 | 4.9 | 95.2 | -14.9 | -5.9 | 4.5 | 0.0 | 142,2 | 0.0 | 27.0 | 122.0 | 11.0 | 102.0 |
| | | ***2.8*** | ***8.6*** | ***-51.1*** | ***-1.9*** | ***2.3*** | ***1.1*** | ***16.5*** | ***36.2*** | ***7.3*** | ***1.3*** | ***-8.8*** | ***5.1*** | ***5.7*** |
| **1992** | 2.12 | 25.7 | 10.7 | 14.7 | -15.9 | -4.1 | 8.3 | 2.6 | 256.0 | 3.0 | 31.0 | 120.0 | 14.0 | 90.0 |
| | | ***-5.8*** | ***2.8*** | ***29.4*** | ***-0.9*** | ***0.5*** | ***-2.6*** | ***13.9*** | ***-77.7*** | ***4.3*** | ***-2.7*** | ***-6.8*** | ***2.1*** | ***17.7*** |
| **2002** | 2.4 | 20.8 | 11.8 | 21,5 | -9.7 | -1.5 | 9.3 | 24.1 | 289.8 | 9.0 | 30.0 | 113.0 | 2.0 | 96.0 |
| | | ***-0.9*** | ***1.8*** | ***22.6*** | ***-7.1*** | ***-2.1*** | ***-3.7*** | ***-7.6*** | ***-111.4*** | ***-1.7*** | ***-1.7*** | ***0.2*** | ***14.1*** | ***11.7*** |
| **2006** | -2.27 | 12.8 | 12.0 | 68.8 | -18.1 | 0.2 | 4.6 | 54.8 | 145.2 | 18.0 | 28.0 | 105.0 | 18.0 | 100.0 |
| | | ***7.1*** | ***1.5*** | ***-24.7*** | ***1.3*** | ***-3.9*** | ***1.0*** | ***-38.3*** | ***33.2*** | ***-10.7*** | ***0.3*** | ***8.2*** | ***-1.9*** | ***7.7*** |

The other deviations are distributed more spontaneously. The date of final loss of snow cover and the maximal snow depth do not change the $S_i$ index even in May. Perhaps this is due to the fact that Eastern Siberia is characterized by small depth of snow. Snow cover is completely disappears in March or early April. Snow cover characteristics have small effect on the characteristics of atmospheric moisture in May. The role of the snow cover will be more significant if we use information about soil moisture and soil-atmospheric phenomena. If we analyze the changes in the index in the neighboring territories, the snow height of impact could be significant. For example it may be in Western Siberia, where the average snow depth is much greater. The area of this study is situated at the center of the Asian continent, is characterized by sharply continental climate. Here, the temperature is a major contributor to the change of aridity.

## 4. Conclusions

The following conclusions were obtained in the analysis of changes of hydrothermal conditions in the south of Eastern Siberia:

- The frequency of weak drought and weak excessive moistening periods is almost constant. The number of drought and excessive moistening with medium and severe intensity in the 1976-2010 has increased compared with 1891-1975.

- Maximum change in hydrothermal conditions in the south of Eastern Siberia occurred in May. Seven out of nineteen weather stations have positive trends of Si over the last 35 years.

- From the considered climatic characteristics of the preceding cold period, only the precipitation during the cold period and the sum of positive temperatures for March affect the atmospheric drought / excessive moistening at the beginning of the vegetation season.

- For a comprehensive analysis of drought / excessive moistening during the vegetation season in addition to the meteorological station data, in particular the conditions of the preceding cold period, in the future we should use the characteristics of the soil and vegetation. We will evaluate not only atmospheric but also atmospheric and soil drought / excessive moistening periods. In addition to the expanded Ped` index $S_i$ enhanced by soil component we can use other complex indices (eg the Palmer index).

**Acknowledgements**
The authors thank the Russian Foundation for Basic Research for support of this work under the grant № 14-05-00502.

**References**

[1] Kotlyakov V M 2005 Earth's climate over the past 400 000 years, and place of the modern era *Proc. XII Congr. of the Russian Geographical Society (St Petersburg)* vol 5 pp 297–301 (in Russian)

[2] Stocker T F, Qin D, Plattner G-K, Tignor M, Allen S K, Boschung J, Nauels A, Xia Y, Bex V and Midgley P M 2013 *Climate Change 2013: The Physical Science Basis. Contribution of Working Group I to the Fifth Assessment Report of the Intergovernmental Panel on Climate Change* (Cambridge: Cambridge University Press and NY, USA) p 1535

[3] Izrael Y A, Gruza G V, Kattsov V M and Meleshko V P 2001 Global climate change. The role of anthropogenic influence *Meteorol.Hydrol.* **5** 5–21

[4] Anisimov O A, Belolutskaya M A and Lobanov V A 2003 Recent changes in climate at high latitudes of the northern hemisphere *Meteorol. Hydrol.* **1** 18–30

[5] Novorotsky P V 2004 Long-term changes in air temperature and precipitation in the basin of the Lower Amur *Meteorol. Hydrol.* **5** 55–62

[6] Gruza G V and Rankova E Ya 2004 Detection of climate change: condition, variability and extremity of climate *Meteorol. Hydrol.* **4** 50–66

# Influence of the North Atlantic dipole on climate changes over Eurasia

**I V Serykh**

Shirshov Institute of Oceanology of the Russian Academy of Sciences,
117997, Russia, Moscow, Nakhimovsky prospect, 36

E-mail: iserykh@ocean.ru

**Abstract.** In this paper, some hydrophysical and meteorological characteristics of negative (1948-1976 and 1999-2015) and positive (1977-1998) phases of the Pacific Decadal Oscillation (PDO) and Interdecadal Pacific Oscillation (IPO) in the North Atlantic and Eurasia are constructed and investigated. Specifically, the near-surface temperature, sea-level atmospheric pressure, wind speed, heat content of the upper 700 m ocean layer, water temperature and salinity at various depths, the latent and sensible heat fluxes from the ocean to the atmosphere are analyzed. The fields obtained are in good agreement and complement each other. This gives important information about the hydrometeorological conditions in the region under study. Analysis of these data has shown that in the upper 1000 m North Atlantic layer there is a thermal dipole which can be interpreted as an oceanic analog of the atmospheric North Atlantic Oscillation (NAO). An index of the North Atlantic Dipole (NAD) as the difference between the mean heat contents in the upper 700 m oceanic layer between the regions (50°-70° N; 60°-10° W) and (20°-40° N; 80°-30° W) is proposed. A possible physical mechanism of the internal oscillations with a quasi-60-year period in the North Atlantics-Eurasia system of ocean-atmosphere interactions is discussed.

## 1. Introduction

There is some evidence that the current climate changes are mostly caused by anthropogenic factors [1, 2]. Natural factors, such as heat fluxes from the ocean to the atmosphere [3], also participate in this process. The existence of this effect is indirectly suggested by the fact that the heat content of the upper layer of North Atlantic waters changed considerably in the mid-1970s. This probably caused a change in the climatic scenario of this region [4]. An event of this type is the increase in the heat content of water in the Gulf Stream region in 1980-1995 [5]. These are hydrometeorological events of high energy. Therefore, the oscillations of the North Atlantic hydrophysical regime directly affect the climate of the North-American and Euro-Asian continents [6, 7].

The large-scale changes in the intensity and direction of heat fluxes of the ocean-atmosphere interaction affect the thermobaric characteristics of atmospheric centers of action, such as the Iceland cyclone and the Azores anticyclone, through the feedforward and feedback interrelations in the dynamics of the global climatic system [8]. The 1976/77 and 1998/99 PDO and IPO phase transitions caused global climatic shifts [9, 10, 11, 12, 13, 14]. These effects apparently caused a phase change in the NAO in the mid-1970s that affected the number of cyclones and the near-surface temperature in North Atlantic, as well as the transfer of heat from this region to the Euro-Asian continent [15]. In this paper, an attempt is made to determine the physical mechanism of these events, with most attention given to the thermohaline circulation of the North Atlantic Ocean.

## 2. Observations and method of data processing

The analysis is based on the most reliable independent sources among the most complete global average monthly databases [16]. Mean monthly fields of sea-level atmospheric pressure (HadSLP2) and near-surface temperature (CRUTEM4) processed by the Met Office Hadley Centre (United Kingdom) with a spatial resolution of 5°x5° for 1850-2015 are analyzed [17, 18]. To obtain climatic wind patterns at various levels, mean monthly fields with a resolution of 2.5°x2.5° are used from the NCEP/NCAR 1948-2015 reanalysis datasets [19]. The results obtained for the atmospheric pressure, wind, and near-surface temperature are verified with the data of the 1851-2011 NOAA CIRES 20th Century Reanalysis v2c (USA) [20], the 1900-2010 European ERA-20C Reanalysis [21], and the 1958-2013 Japanese JRA-55 Reanalysis [22]. Sea surface temperature (SST) was taken from the COBE SST2 of Japan Meteorological Agency with a resolution of 1°x1° for 1850-2014 [23].

A combined analysis is performed with 3-months means of heat content anomalies (with a resolution of 1°x1°) of the upper 700 m oceanic layer and mean water temperature anomalies at 16 depths (0, 10, 20, 30, 50, 75, 100, 125, 150, 200, 250, 300, 400, 500, 600, and 700 meters) for 1955-2015 produced by the National Oceanographic Data Center (NODC) [24]. To obtain a more accurate quantitative estimate, the oceanic temperature and salinity values at 24 different depths up to 1500 m are taken from the mean monthly Ishii Ocean Analyses Project data with a resolution of 1°x1° in 1945-2012 [25]. The depth of an upper active layer (for the regions of deep convection) is determined by "sigma-t" criterion [26] for 1900-2000 using the results of an experiment called Historical of the CMIP5 project with a combined ocean-atmosphere general circulation model called MPI-ESM-MR [27]. The upper active layer of the ocean is the near-surface ocean water layer subject to a direct action of atmospheric processes and affecting the atmosphere through turbulent energy and mass transfer. To analyze the latent and sensible heat fluxes from the ocean to the atmosphere in 1958-2015, OAFlux (NOAA) data with a resolution of 1°x1° are used [28].

The mean fields of the above characteristics for negative (1948-1976 and 1999-2015) and positive (1977-1998) PDO phases are calculated separately. The mean field for the previous period is subtracted from the mean field for the period chosen. The difference shows the changes that took place for the periods. A six-year smoothing moving average is used to construct the curves shown below. The seasonal variations are removed by subtracting the climatic values of the characteristics under investigation over the entire time interval.

## 3. Discussion of the analysis results

The field of difference for November-March mean sea level pressure (SLP) between the 1977-1998 and 1948-1976 periods (figure 1a) in the North Atlantic shows a negative anomaly northward and a positive anomaly southward of the 50th parallel. This corresponds to a positive phase of NOA. The field of difference for the November-March mean wind speed at 850 hPa between the 1977-1998 and 1948-1976 periods (figure 1c) has cyclonic circulation anomalies corresponding to the barometric structure (figure 1a), with center to the east of Greenland and in the Greenland Sea, as well as anticyclonic circulation anomalies, with center to the west of the Azores Islands. The pattern of wind field anomalies (figure 1c) indicates that in 1977-1998 there is (in comparison to the previous period) an increase in the north-eastern trade wind, southern wind in the Sargasso Sea zone, and western transport from the North Atlantic to the Euro-Asian continent along the 50° parallel north.

The field of difference of the mean SLP for November-March between the 1999-2015 and 1977-1998 periods (figure 1b) in the North Atlantic has anomalies mostly of the opposite sign (in comparison the field of figure 1a) corresponding to a negative phase of NAO: positive SLP anomalies in the Iceland cyclone and negative ones in the Azores anticyclone. The field of difference of the mean November-March wind speed at 850 hPa between the 1999-2015 and 1977-1998 periods (figure 1d) corresponds to the SLP field in figure 1b, and has anomalies that are typical for weakening of the western transport from the North Atlantic to the Euro-Asian continent along the 50° parallel north.

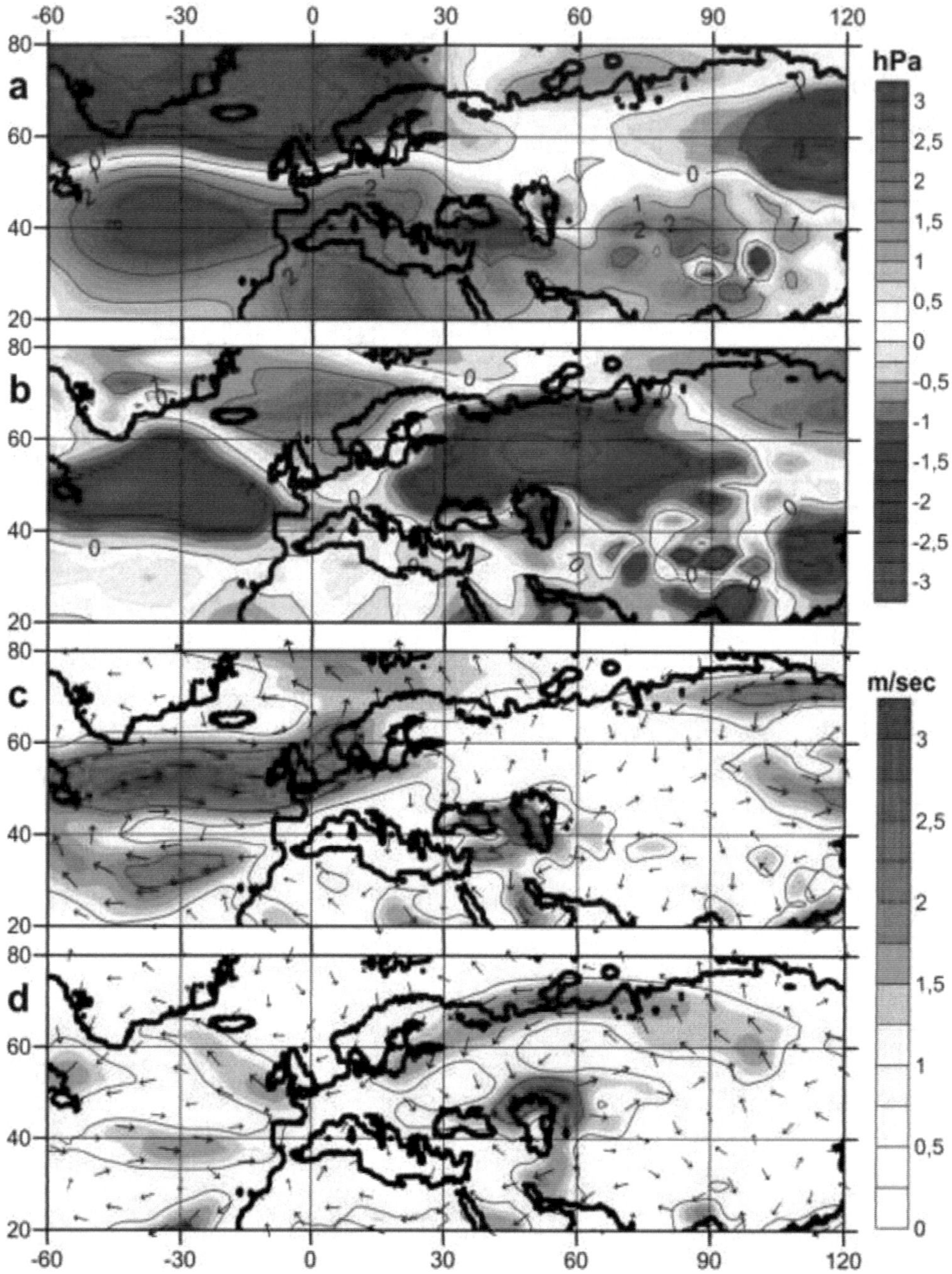


**Figure 1.** Fields of difference of mean sea-level atmospheric pressure for November-March between 1977-1998 and 1948-1976 (a), and between 1999-2015 and 1977-1998 (b); fields of difference of mean wind velocity at 850 hPa for November-March between 1977-1998 and 1948-1976 (c), and between 1999-2015 and 1977-1998 (d).

The field of mean November-March depths of the ocean mixed layer thickness defined by Sigma T (MLOTST) (figure 2a) shows that in the cold period of the year the major deep convection zones (depths more than 1 km) are located in the Labrador and Greenland Seas zone. Also, large MLOTST depths (200-300 m) are observed in the North-Atlantic current zone. The Labrador Sea and North-Atlantic current zones are separated by a small-depth (about 50 m) MLOTST band. This indicates that there is a considerable difference in the thermohaline conditions of these two zones. The field of mean MLOTST depths for the warm (May-September) time of the year (figure 2b) has much smaller MLOTST values in comparison to those of the cold period (figure 2a). This corresponds to weakening of the influence of the ocean on the atmosphere in the warm period of the year in comparison to that in the cold one.

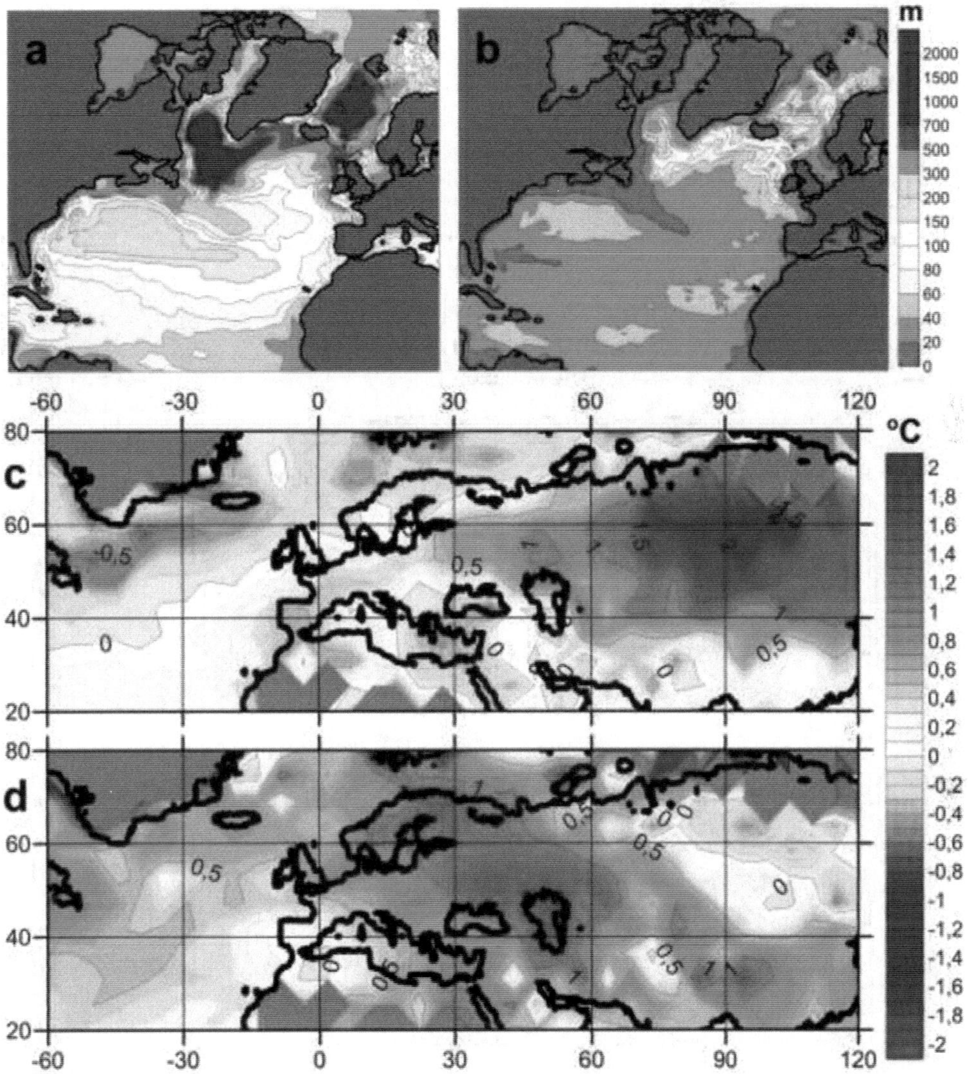


**Figure 2.** Fields of mean depths of the ocean mixed layer thickness for November-March (a) and May-September (b). Fields of difference of mean near surface temperatures for November-March between 1977-1998 and 1948-1976 (c), and between 1999-2015 and 1977-1998 (d).

The field of difference of the mean November-March near surface temperatures (NST) between 1977-1998 and 1948-1976 (figure 2c) is in good agreement with the corresponding fields of SLP and wind speed (figures 1a and 1c). Figure 2c clearly shows a zone of negative NST anomalies in the Labrador Sea, to the south of Greenland and to the north-east of the Newfoundland Island. This anomaly is in a region of western wind enhancement along the 50th parallel, from where this heat is transported to the Euro-Asian continent experiencing an abrupt NST growth since the mid-1970s. The field of difference of the November-March NST means between 1999-2015 and 1977-1998 (figure 2d) shows positive anomalies in North Atlantic and negative ones over a part of Siberia. In the author's opinion, this is due to a weakening of the western transport along the 50th parallel and decrease in the heat transfer from the northern part of the Atlantic Ocean to the Euro-Asian continent, which is typical for 1999-2015 (in comparison to 1977-1998) (figure 1d).

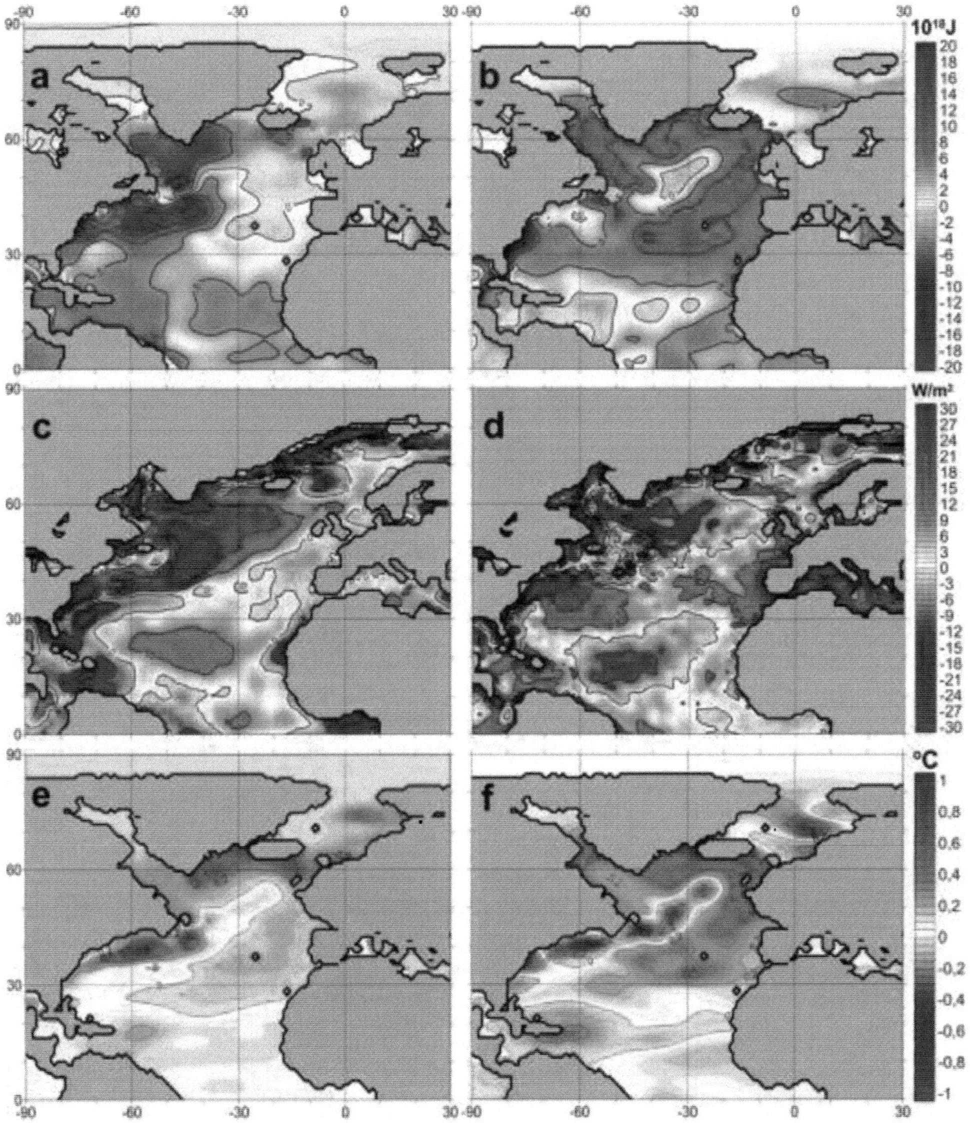


**Figure 3.** Fields of difference of the mean heat contents in the upper 700 m ocean layer between 1977-1998 and 1955-1976 (a) and between 1999-2015 and 1977-1998 (b). Fields of difference of the mean November-March latent and sensible heat ocean-atmosphere fluxes between 1977-1998 and 1958-1976 (c), and between 1999-2015 and 1977-1998 (d). Fields of difference of the mean ocean temperatures at a depth of 500 m between 1977-1998 and 1948-1976 (e) and between 1999-2012 and 1977-1998 (f).

Consider the upper 700 m ocean layer. The field of difference of the mean heat contents between 1977-1998 and 1955-1976 (figure 3a) shows the North Atlantic Ocean heat losses mostly to the north of the 50th parallel and gains to the south. The situation is different for the Labrador Sea zone between

1999-2015 and 1977-1998 (figure 3b): its heat content increased in 1999-2015 in comparison to that in 1977-1998 [29]. Thus, when one NAO phase changes for the opposite one, the heat is redistributed between the northern and southern parts of the North Atlantic. Hence, inter-decadal changes in the ocean heat content in the North Atlantic current and Labrador Sea zones after the removal of a linear trend probably caused by global climate warming are opposite in sign.

The field of difference of the mean November-March latent and sensible heat fluxes from the ocean to the atmosphere between 1977-1998 and 1958-1976 (figure 3c) demonstrates positive anomalies of heat transfer from the ocean to the atmosphere in the Labrador Sea zone. This corresponds to a decrease in the ocean heat content in this region (figure 3a). The field of difference of the mean November-March latent and sensible heat fluxes from the ocean to the atmosphere between 1999-2015 and 1977-1998 (figure 3d) shows negative anomalies of heat transfer from the ocean to the atmosphere in the Labrador Sea zone. This corresponds to an increase in the ocean heat content in this region (figure 3b).

The field of difference of the mean ocean temperatures at a depth of 500 m between 1977-1998 and 1948-1976 (figure 3e) shows negative anomalies in the Labrador Sea zone. This phenomenon may be explained by the enhancement of the cold Labrador current caused by the cyclonic wind anomaly (figure 1a), as well as by the regional cooling of the upper ocean layer due to the deep winter convection and the corresponding increase in the ocean heat transfer to the atmosphere (figure 3c). In figure 3e special attention must be given to the positive ocean temperature anomalies in the North Atlantic current zone. According to our calculations, these phenomena may be caused by the enhancement of the trade wind, as well as by the positive anomaly of the southern wind in the Sargasso Sea zone (figure 1c). Both factors have a "one-way" effect creating an additional flow of warm tropical waters to the Mexican Gulf. This affects the thermodynamic characteristics of the Gulf Stream current, which, in turn, affects the North Atlantic current. The anomalies of the field of difference of the mean ocean temperatures at a depth of 500 m between 1999-2012 and 1977-1998 (figure 3f) are almost opposite to those in figure 3e. In Figs. 3e and 3f between the regions of subtropical anti-cyclonic and subpolar cyclonic gyres, there is a dipole structure, which was detected by using sea level data [30, 31]. This structure could be called as the North Atlantic Dipole (NAD).

The fields of difference between the time periods of the mean temperature and salinity of the North Atlantic Ocean at various depths up to 1500 m show that NAD is most pronounced at the depths from 500 m to 700 m. In the present paper, a NAD index as the difference of the mean heat contents of the upper 700 m ocean layer between the regions (50°-70° N; 60°-10° W) and (20°-40° N; 80°-30° W) is proposed. It differs from the index of the Atlantic Multidecadal Oscillation (AMO), which is calculated as the mean surface temperature anomaly of the North Atlantic Ocean (20°-70° N; 80°-0° W) with lineal trend removing. It does not take into account the deep ocean temperature and the NAD dipole structure. Also, in calculating the AMO index there are problems associated with the removal of linear trend and the correction for the measurement device arrays affecting the resulting series. The NAD index proposed above is calculated as the difference between two regions and, hence, these problems have less effect on the series than when the index is calculated for only one of the regions [32, 33].

Figure 4a presents the NAD index, AMO index and the NAO index calculated as the difference of the mean SLP anomalies between the Azores maximum (20°-40° N; 40°-20° W) and the Iceland minimum (50°-70° N; 45°-25° W) regions. The NAD and NAO indices complement each other, since they are calculated, respectively, as the difference of the upper ocean layer heat content and the atmospheric pressure between almost the same regions. They characterize the largest part of the North Atlantic ocean-atmosphere system, and not just its atmospheric or oceanic component. The more than 50-years variation of the NAD and NAO indices (figure 4a) demonstrates the following interesting peculiarities: after the removal of inter-annual variations (influence of the El Niño – Southern Oscillation) the NAO index is ahead of the NAD index, although they have high negative correlation with each other. Thus, it can be concluded that on multidecadal timescales the atmosphere has a dominant role in the atmosphere-ocean system, which contradicts some earlier results that North

Atlantic Ocean control on atmosphere-ocean interaction on multidecadal timescales [8]. However, there is no large time shift between the NAD and NAO indices, whereas the phases of the AMO index variations are shifted considerably (by about 6 years) with respect to those of the NAO index (figure 4a) [34].

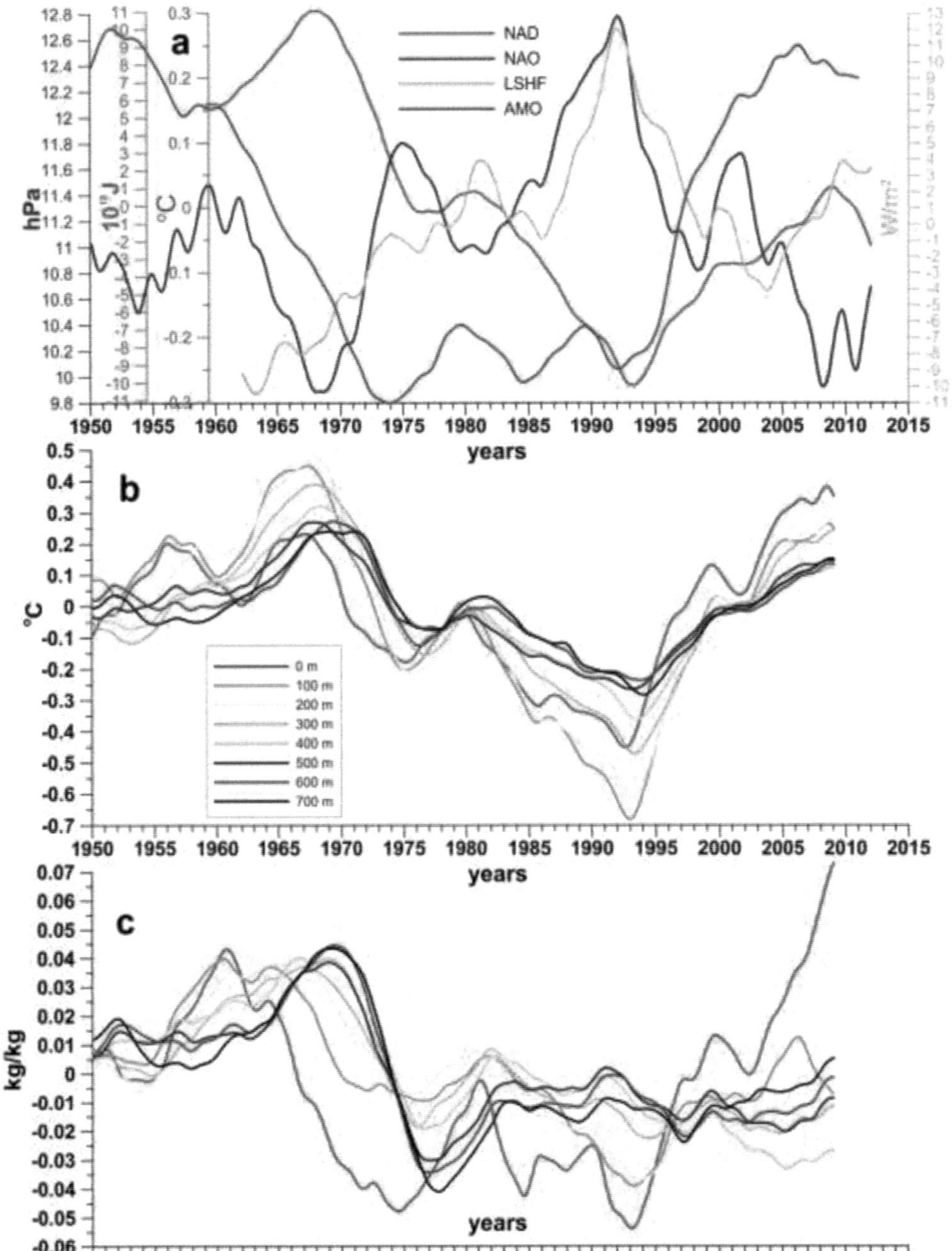


**Figure 4.** NAD (red), NAO (blue) and AMO (deep purple) indices (a). Time series of differences of mean values between (50°-70° N; 60°-10° W) and (20°-40° N; 80°-30° W), anomalies of latent and sensible heat fluxes from the ocean to the atmosphere (green) (a), anomalies of the ocean temperature (b) and salinity (c) at various depths.

The time series of the differences of the mean anomalies of latent and sensible heat fluxes from the ocean to the atmosphere between the regions (50°-70° N; 60°-10° W) and (20°-40° N; 80°-30° W) are presented in figure 4a. They demonstrate a high positive correlation with the NAO index and a negative one with the NAD index. Such, indeed, is the case, since with increasing latent and sensible heat fluxes from the ocean to the atmosphere the heat content of the ocean decreases, and with decreasing heat fluxes it increases.

Figures 4b and 4c show the time series of the differences of the mean ocean temperature and salinity anomalies at various depths between the regions (50°-70° N; 60°-10° W) and (20°-40° N; 80°-30° W). Notice that the temperature and salinity variations in the upper ocean layers are ahead of those in the lower layers. This again shows that the atmosphere on multidecadal timescales has a dominant role in the atmosphere-ocean system.

It follows from figure 4 that in the early 1990s there was a climatic shift in the North Atlantic: there is a strong bend in the series in 1992/93. This may be associated with the 1991 Pinatubo volcano eruption. After this event, starting from 1992/93, decreased latent and sensible heat fluxes from the ocean to the atmosphere and increased heat content in the North Atlantic have been observed [35]. These could lead to the "hiatus" in the Siberian climate warming in the cold time of the year observed over the 1999-2015 period [36].

## 4. Conclusions

A dipole structure of inter-decadal variations in the heat content of the ocean and heat fluxes from the ocean to the atmosphere has been detected in the North Atlantic. The following fact deserves special attention: the cyclonic and anti-cyclonic atmospheric circulation anomalies, as well as the decrease and increase in the ocean heat content, take place concurrently and quasi-synchronously in the Iceland minimum and Azores maximum regions. Owing to this, the western heat transport anomalies along the 50th parallel increase or decrease the transport of heat from the Atlantic Ocean to the Euro-Asian continent, and the climate in Europe and Siberia becomes more marine or more continental [37]. The very fast climate warming of the Euro-Asian continent that began in the 1970s may be associated with the enhanced heat transport from the North Atlantic in this period. This is evident from the fields and time series obtained in the present paper. The hiatus of this warming after 1999 may be due to the decreased heat transfer from the North Atlantic Ocean to the Eurasian territory.

**Acknowledgments**
This work was supported by the Russian Foundation for Basic Research, project no. 16-35-00111 mol_a.

## References

[1] IPCC 2007 *Climate Change 2007: The Physical Science Basis. Contribution of WG1 to the IV Assessment Report of the IPCC* (Cambridge UK/New York NY USA) p 996

[2] IPCC 2013 *Climate Change 2013: The Physical Science Basis. Contribution of WG1 to the V Assessment Report of the IPCC* (Cambridge UK/New York NY USA) p 1535

[3] Chu P C 2011 Global upper ocean heat content and climate variability Ocean Dynamics *Ocean Dynamics* **61** issue 8 pp 1189–1204

[4] Byshev V I Neiman V G Romanov Yu A and Serykh I V 2011 Phase variability of some characteristics of the present-day climate in the Northern Atlantic region *Doklady Earth Sciences* **438** 2 pp 887–892

[5] Zhai X and Sheldon L 2012 On the North Atlantic Ocean Heat Content Change between 1955–70 and 1980–95 *J. Climate* **25** pp 3619–3628

[6] Arguez A O'Brien J J and Smith S R 2009 Air temperature impacts over Eastern North America and Europe associated with low-frequency North Atlantic SST variability *Int. J. Climatol.* **29** pp 1–10

[7] Hurrell J W and Deser C 2009 North Atlantic climate variability: The role of the North Atlantic

Oscillation *J. Mar. Syst.* **78** 1 pp 28–41

[8] Gulev S K Latif M Keenlyside N Park W and Koltermann K P 2013 North Atlantic Ocean control on surface heat flux on multidecadal timescales *Nature* **499** 7459 pp 464–467

[9] Hare S R and Mantua N J 2000 Empirical evidence for North Pacific regime shifts in 1977 and 1989 *Progress in Oceanography* **47** 2–4 pp 103–145

[10] Folland C K Renwick J A Salinger M J and Mullan A B 2002 Relative influences of the Interdecadal Pacific Oscillation and ENSO in the South Pacific Convergence Zone *Geophysical Research Letters* **29** 13 pp 21–1–21–4

[11] Mantua N J and Hare S R 2002 The Pacific Decadal Oscillation *Journal of Oceanography* **58** 1 pp 35–44

[12] Chavez F P Ryan J Lluch-Cota S E and Ñiquen M C 2003 From Anchovies to Sardines and Back: Multidecadal Change in the Pacific Ocean *Science* **299** 5604 pp 217–221

[13] Bond N A Overland J E Spillane M and Stabeno P 2003 Recent shifts in the state of the North Pacific *Geophys. Res. Lett.* **30** 23

[14] Ding H Greatbatch R J Latif M Park W and Gerdes R 2013 Hindcast of the 1976/77 and 1998/99 Climate Shifts in the Pacific *J. Climate* **26** pp 7650–7661

[15] Jung T Hilmer M Ruprecht E Kleppek S Gulev S K and Zolina O 2003 Characteristics of the recent eastward shift of interannual NAO variability *J. Climate* **16** pp 3371–3382

[16] Blunden J and Arndt D S 2012 State of the Climate in 2011 *Bull. Amer. Meteor. Soc.* **93** 7 S1–S264

[17] Allan R J and Ansell T J 2006 A new globally-complete monthly historical gridded mean sea level pressure data set (HadSLP2): 1850-2004 *J. Climate* **19** pp 5816–5846

[18] Jones P D Lister D H Osborn T J Harpham C Salmon M and Morice C P 2012 Hemispheric and large-scale land surface air temperature variations: An extensive revision and an update to 2010 *J. Geophys. Res.* **117** D05127

[19] Kalnay E et al 1996 The NCEP / NCAR 40-year reanalysis project *Bull. Amer. Meteor. Soc.* **77** pp 437–471

[20] Compo G P et al 2011 The Twentieth Century Reanalysis Project *Quarterly J. Roy. Meteorol. Soc.* **137** 1–28

[21] Stickler A et al 2014 ERA-CLIM: Historical Surface and Upper-Air Data for Future Reanalyses *Bull. Amer. Meteor. Soc.* **95** 9 pp 1419–1430

[22] Kobayashi S et al 2015 The JRA-55 Reanalysis: General Specifications and Basic Characteristics *J. Met. Soc. Jap.* **93** 1 pp 5–48

[23] Hirahara S Ishii M and Fukuda Y 2014 Centennial-scale sea surface temperature analysis and its uncertainty *J. of Climate* **27** pp 57–75

[24] Levitus S Antonov J I Boyer T P Locarnini R A and Garcia H E 2009 Global ocean heat content 1955–2008 in light of recently revealed instrumentation problems *Geophys. Res. Lett.* **36** L07608

[25] Ishii M Kimoto M Sakamoto K and Iwasaki S I 2006 Steric sea level changes estimated from historical ocean subsurface temperature and salinity analyses *J. Oceanography* **62** 2 pp 155–170

[26] Levitus S 1982 *Climatological atlas of the world ocean* (Washington D.C.: US Government Printing Office 13 NOAA) p 163

[27] Taylor K E Stouffer R J and Meehl G A 2012 Overview of CMIP5 and the experiment design *Bull. Am. Meteor. Soc.* **93** pp 485–498

[28] Yu L Jin X and Weller R A 2008 *Multidecade Global Flux Datasets from the Objectively Analyzed Air-sea Fluxes (OAFlux) Project: Latent and sensible heat fluxes, ocean evaporation, and related surface meteorological variables* (Massachusetts: Woods Hole Oceanographic Institution OAFlux Project Technical Report OA-2008-01) p 64.

[29] Lee S-K et al 2011 What caused the significant increase in Atlantic Ocean heat content since the mid-20th century? *Geophys. Res. Lett.* **38** L17607

[30] Gulev S K and Latif M 2015 The origins of a climate oscillation *Nature* **521** pp 428–430
[31] Dvoryaninov G S Kubryakov A A Sizov A A Stanichny S V and Shapiro N B 2016 The North Atlantic Oscillation: A dominant factor in variations of oceanic circulation systems of the Atlantic Ocean *Doklady Earth Sciences* **466** 1 pp 100–104
[32] Domingues C M et al 2008 Improved estimates of upper-ocean warming and multi-decadal sea-level rise *Nature* **453** pp 1090–1094
[33] Whitmarsh F J Zika D and Czaja A 2015 Ocean heat uptake and the global surface temperature record *Grantham Institute Briefing Paper* **14**
[34] Polyakov I V Alexeev V A Bhatt U S Polyakova E I and Zhang X 2010 North Atlantic warming: patterns of long-term trend and multidecadal variability *Climate Dynamics* **34** 2-3 pp 439–457
[35] Liu W Xie S P and Lu J 2016 Tracking ocean heat uptake during the surface warming hiatus *Nat. Commun* 7 10926
[36] Tollefson J 2014 Climate change: The case of the missing heat *Nature* **505** 7483 pp 276–278
[37] Byshev V I Neiman V G Romanov Yu A and Serykh I V 2009 On the spatial nonuniformity of some parameters of global variations in the recent climate *Doklady Earth Sciences* **426** 1 pp 705–709

# Development of lake parametrization in the INMCM climate model

**V Bogomolov**[1,2], **V Stepanenko**[3,4] **and E Volodin**[5]

[1]Institute of Monitoring of Climatic and Ecological Systems, Siberian Branch of the Russian Academy of Sciences, Tomsk, 10/3, Academichesky ave., Russia
[2] V.E. Zuev Institute of Atmospheric Optics, Siberian Branch of the Russian Academy of Sciences, Tomsk, 1, Academician Zuev square, Russia
[3] Research Computing Center, Moscow State University, Moscow, Leninskiye Gory str., 1, bld.4, Russia
[4] Department of Geography, Moscow State University, Moscow,Leninskiye Gory str., 1, Russia
[5] Institute of Numerical Mathematics, Russian Academy of Sciences, Moscow, Gubkina str., 8, Russia

Bogomolov@scert.ru

**Abstract**. Land surface schemes (LSS or terrestrial models) are a crucial component of both Numerical Weather Prediction (NWP) systems and climate models. An important land- surface type is lakes. This paper presents the mechanism of incorporation of a model LAKE into a coupled general circulation model of the atmosphere and ocean INMCM4, with a space resolution $2^{o}$ to $1.5^{o}$ and 21 levels in height, with two-way interaction. A new map for 14 land types distribution was created using a digital map of inland waters for the entire globe. The digital map of water bodies includes the fraction of lake area on the land surface, and the average depth of water bodies, both on the grid of the climate model. This digital map is based on a dataset consisting of 14 000 freshwater lakes. In order to increase the time step in the LAKE model, the k-ε parameterization has been replaced by the parameterization of Henderson-Sellers. With the amended INMCM4 model, numerical experiments were carried out to simulate global climate during the second half of the XX century. The effects of the new lake parameterization on the surface temperature and heat fluxes are analyzed.

## 1. Introduction

The land surface scheme (LSS) is a key component of numerical weather prediction systems and climate models. The permafrost areas are particularly difficult to simulate within the framework of the LSS. Firstly, permafrost can only be truly reproduced if the subsurface phase transitions as well as snow cover processes are correctly reflected in the model. The complexity of the problem is demonstrated by the huge spread of modern permafrost areas in the CMIP5 models ($4\text{-}25 \cdot 10^{6}$ km$^{2}$, [11]). In this regard, we note that the LSS of INM RAS climate model is characterized by the maximal number of layers in the soil (23) among the CMIP5 models, allowing us to calculate in detail vertical heat transfer and the processes of freezing and thawing [12]. It is also important that the state of permafrost in LSSs is sensitive to small-scale features of the underlying surface, such as the presence of a layer of moss [13]. This indicates some prospects for further improving the quality of modeling of permafrost by including new effects on the underlying surface, such as a special model for snow in forest landscapes, improving the schemes of calculation of turbulent fluxes in the case of stable stratification [14].

It is important that the landscape of permafrost zones is characterized by a specific hydrological regime, expressed by an abundance of lakes and rivers. Lakes (mostly thermokarst) and wetlands occupy a significant part of Siberia (especially in Western Siberia). They significantly modify not only the thermodynamic interaction of the active layer with the atmosphere, but they are also an important source of greenhouse gases [15].

Lakes for a long time have not been recognized as important "actors" in the climatic system, as they occupy less than 4% of the land surface. However, the regional thermodynamic and dynamic effects of lakes on weather and climate are important for Canada, Finland, Western Siberia, and some other regions [1], [2], [3]. This motivated the inclusion of thermodynamic models of lakes in many weather systems and climate models [2], [1], [4], [5], [6]. Simultaneously with increasing knowledge on the significance of lakes for the current and future climates, several attempts have been made to develop lake models including not only thermodynamics and turbulence, but also biogeochemistry to simulate $CH_4$ and $CO_2$ in natural waters [8], [16], [17], [18].

In a coupled model of the general circulation of atmosphere and ocean, INMCM4, a participant of the CMIP5 experiments [19], the land cell can contain the following types of surface: vegetation, bare soil, snow and inland waters. The surface temperature, the sensible and latent heat fluxes from these four types are evaluated separately. The fluxes over an inland water body, as in the case of water droplets trapped by vegetation, are calculated under the assumption that the humidity of the air above this surface is saturated, but the layer of water does not have its own heat capacity. In addition, the soil under the different land surface types has the same vertical profiles of temperature, humidity, and ice concentration. It is quite a rough approximation for lakes, because the nature of the vertical heat exchange there is fundamentally different from that of the soil. Essentially, the same parameterization is used for wetlands. This leads to an inaccurate assessment of the role of lakes and wetlands in the formation of the regional climates of northern regions.

## 2. The components of a new lake parameterization for a climate model

To account for heat capacity of water bodies, turbulent heat transfer in lakes, and their contribution to the latent and sensible heat fluxes from the surface to the atmosphere, a special code inside the Earth system model INMCM4 has been developed. It includes a parameterization of lake thermo- and hydrodynamics, LAKE, [16] and detailed digital maps of the land waters surface fraction, and their depth (http://www.flake.igb-berlin.de/ep-data.shtml).

### *2.1. Lake depth, lake map (global data)*

A digital map of water bodies was created to include two 2D arrays on the grid of the climate model (Figure 1): the fraction of a cell area occupied by the water surface and the average depth of water bodies in the cell when the cell contain more than 5 percent of the lakes, while the Lake digital map includes all lakes from a database. This digital map was developed using a dataset that includes about 14,000 freshwater lakes with data on their depths [9].

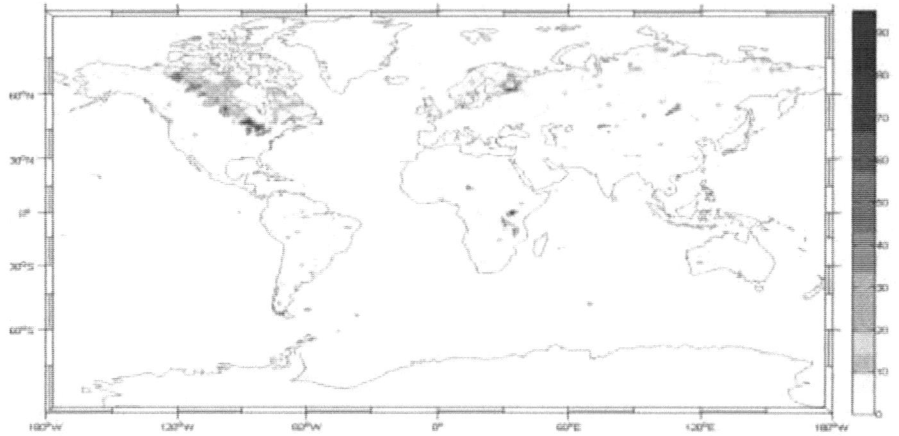

**Figure 1.** Map of water fraction in land cells of the INMCM4 model, %.

The previous mask of land surface types in INMCM4 included 13 types [20], the 13-th type corresponding to "wetlands", which included lakes inter alia. On the basis of the digital map of water bodies and the old land type mask, a new mask has been formed, comprising 14 types, where the 14-th type is lakes. At the same time, the fraction of lakes added to the 14-th type was excluded from the 13-th type. The fractions of other types are adjusted using a weighting function. In the old mask 1,018 cells included lakes, whilst in the new - 2422 cells (Figure 2). This new mask is generated using a developed suite of FORTRAN programs. This software package allows us to form a mask with the addition of a new type and correcting the fractions of the existing types using a weight function. It also meets any desired spatial resolution of the output data.

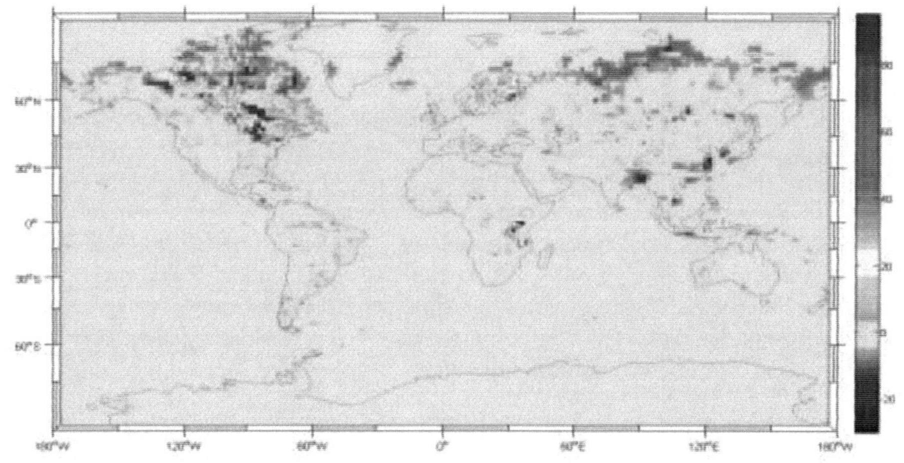

**Figure 2.** Difference between the lake fraction in the new mask and that in the old mask.

*2.2. Parameterization of vertical diffusivity, convective adjustment scheme in the case of unstable stratification for lakes with high water transparency*

In order to couple the models INMCM4 and LAKE, it was necessary to ensure in the LAKE model the same time step as in the climate model (1 hour). For this purpose, in the lake model, the time step-restricting k-ε turbulence closure [21] was replaced by parameterization of vertical diffusivity of Henderson-Sellers [10]. Before this replacement it had been shown that the difference in the surface heat fluxes and in the thermal regime of a lake caused by using these turbulent mixing schemes is small.

The numerical experiments with the lake model using the Henderson-Sellers parameterization have shown that mixing in water occurs only in cases of neutral and stable stratification, while in the case of unstable stratification, when the bottom temperature is higher than the surface temperature, the lake mixes very slowly (Figure 3). This situation is possible in lakes with high transparency and small depth (for example, 5 m), because the incoming solar radiation heats the bottom of the water body. For such cases of unstable density stratification, a bulk scheme of instantaneous mixing is introduced (the results using this scheme are presented in Figure 4).

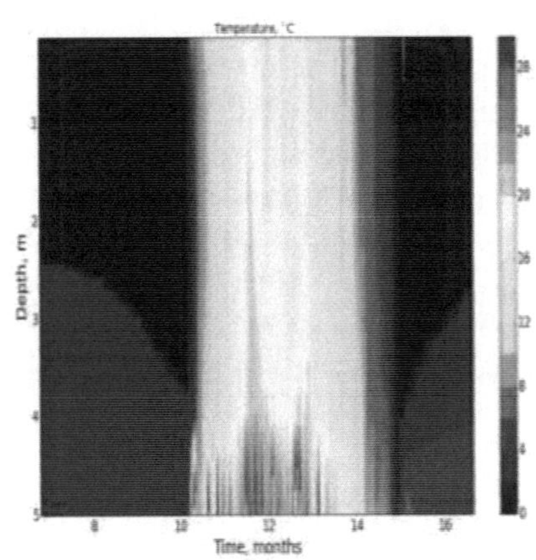


**Figure 3.** Time-depth distribution of temperature in a 5 m-deep water reservoir during the annual cycle without convective mixing scheme.

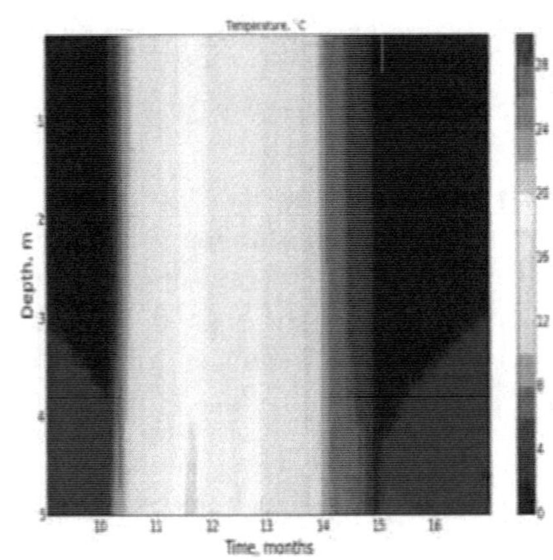


**Figure 4.** Time-depth distribution of temperature in a 5 m-deep water reservoir during the annual cycle with convective mixing scheme.

## 3. Validation of lake surface temperature from INMCM4 model vs. measurement data

The surface temperatures of lakes calculated by the LAKE parameterization with the atmospheric forcing from the INMCM4 model were validated vs. measurement data from the database https://portal.lternet.edu/nis/mapbrowse?packageid=knb-lter-ntl.10001.3. A comparison was carried out for lakes located in different climates and on different continents. From Table 1 it follows that the model satisfactorily reproduces the averaged temperature of the surface water with the meteorological forcing of the climate model.

**Table 1.** Modeled and observed mean surface temperatures of large lakes.

| Lake, country | The 5-year (1980-1985) averaged summer surface temperature, modelled, °C. | The 15-year (1986-2000) averaged summer satellite-derived surface temperature, °C. |
|---|---|---|
| Huron, Canada | 19.2 | 18.52 |
| Victoria, Tansania-Kenya-Uganda | 25.25 | 23.84 |
| Baikal, Russia | 14.83 | 12.74 |
| Ladoga, Russia | 15.49 | 14.61 |

The modified model INMCM4 was used to conduct climate simulations for the second half of the XX century. The difference of the monthly-averaged surface temperature and heat fluxes between the old lake parameterization of INMCM4 and the new parameterization (LAKE) was analyzed (Figures 5-8). A conspicuous feature seen in Figures 5-8 is the phase shift in the surface temperature and fluxes between the two model runs, with a warmer lake surface and larger heat fluxes during the late summer and autumn, when using the LAKE model. Due to the fact that the depth of daily active layer of the Earth, the heat-storage, is about 20 centimeters, while for a lake daily active layer, as a rule, is equal to mixed layer and is in the order of a few meters or more. The albedo of the land is equal to 0.1-0.3, and for a lake is 0.06. Thus lakes are more energy-intensive reservoirs with greater absorption of shortwave radiation. For this reason, the surface temperature of a reservoir, as compared to the land temperature, grows slower in summer and cooled slower in autumn. This explains differences in received flows of sensible and latent heat between the two parametrizations of inland waters. It corresponds well to the same well-known differences between the real lakes and the land surface due to deeper heat transfer in the lakes compared to that in soil.

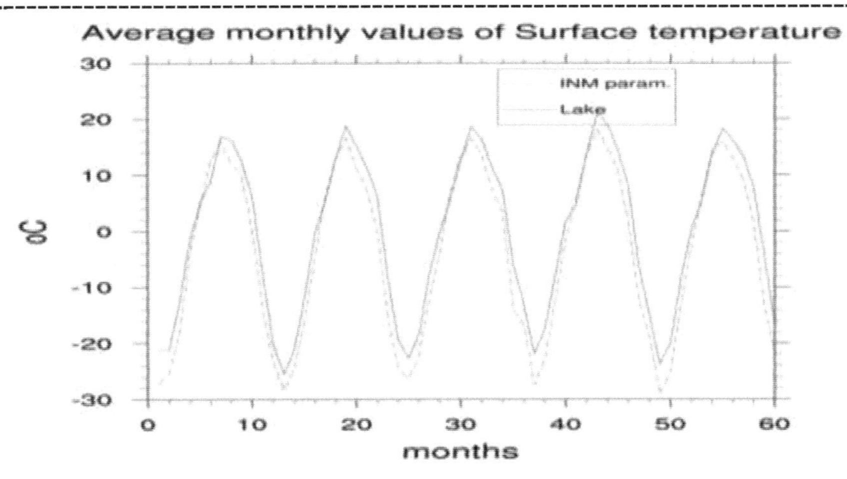


**Figure 5.** Monthly-averaged surface temperature for Lake Baikal, 60 months from 1980 to 1985.

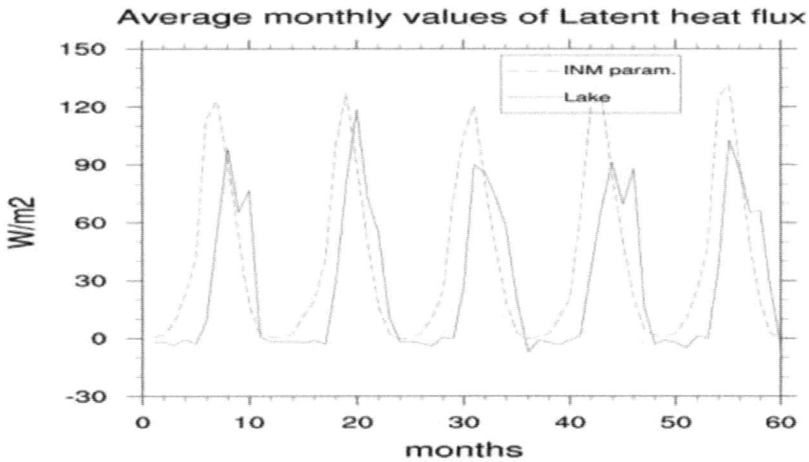


**Figure 6.** Monthly-averaged surface latent heat fluxes for Lake Baikal, 60 months from 1980 to 1985.

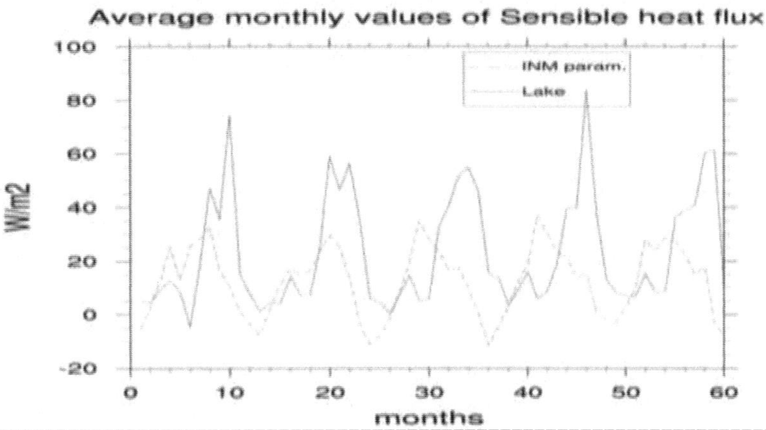


**Figure 7.** Monthly-averaged surface sensible heat fluxes for Lake Baikal, 60 months from 1980 to 1985.

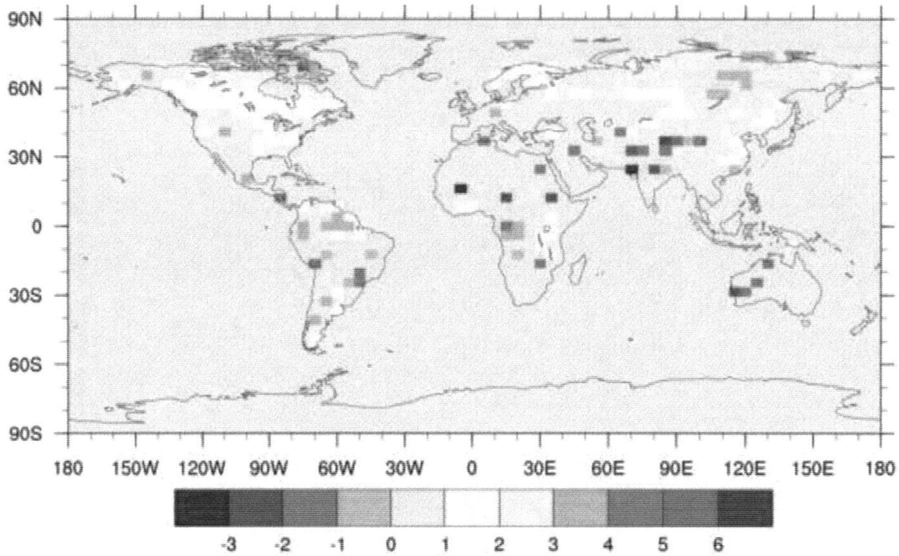


**Figure 8.** Difference between averaged annual temperatures of water bodies calculated with new and old lake parameterizations, 1981.

From this analysis it follows that a significant difference in the surface temperature between the two parameterizations holds after the temporal averaging. It is noteworthy that the use of the new lake parameterization decreases the surface water temperature in the desert areas, and increases in all the others. One of the likely causes of these differences is the soil type used in the climate model for the respective areas.

## Acknowledgments

Work on the new digital maps, preparation of the necessary observational data and global distribution of lakes, the inclusion of new physics in the lake model, validated by measurement data, as well as work on the integration of the models in the first phase (without a return interaction) was to RSF support projects № 16-19-10257. Work on the integration of the models in the second phase (reacting with reverse), experimenting with numerical climate model for the provision of computational resources INM was RSF support projects № 14-27-00126 in INM RAS.

## References

[1] Dutra E *et al* 2010 An offline study of the impact of lakes on the performance of the ECMWF surface scheme *Boreal Environment Research* **15** 100–112

[2] Martynov A *et al* 2012 Interactive lakes in the Canadian Regional Climate Model, version 5: the role of lakes in the regional climate of North America *Tellus* A **64** Available at: **http://www.tellusa.net/index.php/tellusa/article/view/16226/xml**

[3] Eerola K *et al* 2014 Impact of partly ice-free Lake Ladoga on temperature and cloudiness in an anticyclonic winter situation – a case study using a limited area model *Tellus* A **66** Available at: http://www.tellusa.net/index.php/tellusa/article/view/23929/xml

[4] Mironov D *et al* 2010 Implementation of the lake parameterisation scheme FLake into the numerical weather prediction model COSMO *Boreal Environment Research* **15**(2) 218–230

[5] Subin Z M *et al* 2012 Boreal lakes moderate seasonal and diurnal temperature variation and perturb atmospheric circulation: analyses in the Community Earth System Model 1 (CESM1) *Tellus* A **64** Available at: **http://www.tellusa.net/index.php/tellusa/article/view/15639/xml**

[6] Rontu L *et al* 2012 Data assimilation and parametrisation of lakes in HIRLAM *Tellus* A **64** Available at: **http://www.tellusa.net/index.php/tellusa/article/view/17611/xml**

[7] Volodin E M Dianskii N A and Gusev A V 2010 Simulating present-day climate with the INMCM4.0 coupled model of the atmospheric and oceanic general circulations *Izvestiya Atmospheric and Oceanic Physics* **46**(4) 414–431 Available at: **http://link.springer.com/10.1134/S000143381004002X**

[8] Stepanenko V M *et al* 2011 Numerical modeling of methane emissions from lakes in the permafrost zone *Izvestiya Atmospheric and Oceanic Physics* **47**(2) 252–264 Available at: **http://link.springer.com/10.1134/S0001433811020113**.

[9] Kourzeneva E Asensio H Martin E and Faroux S 2012 Global gridded dataset of lake coverage and lake depth for use in numerical weather prediction and climate modeling *Tellus* A **64** 15640 DOI: 10.3402/tellusa.v64i0.15640

[10] Henderson-Sellers B 1985 New formulation of eddy diffusion thermocline models *Applied Mathematical Modelling* **9**(6) 441–446 Available at: http://www.sciencedirect.com/science/article/pii/0307904X85901106

[11] Slater A G and Lawrence D M 2013 Diagnosing Present and Future Permafrost from Climate *Models Journal of Climate* **26**(15) 5608–5623

[12] Volodin E M Lykossov V N 1998 Parameterization of heat and moisture processes in soil-vegetation system 1 Description and calculations using local observational data *Izvestia Atmospheric and Oceanic Physics* **34**(4) 453–465

[13] Machul'skaya E E and Lykosov V N 2009 Mathematical modeling of the atmosphere-cryolitic zone interaction *Izvestiya Atmospheric and Oceanic Physics* **45**(6) 687–703 Available at: **http://link.springer.com/10.1134/S0001433809060024**

[14] Zilitinkevich S S *et al* 2013 A Hierarchy of Energy- and Flux-Budget (EFB) Turbulence Closure Models for Stably-Stratified Geophysical Flows *Boundary-Layer Meteorology* **146**(3) 341–373 Available at: **http://link.springer.com/10.1007/s10546-012-9768-8**

[15] Tranvik L J *et al* 2009 Lakes and reservoirs as regulators of carbon cycling and climate *Limnol. Oceanogr.* **54**(6) 2298–314

[16] Kessler M A, Plug L J and Walter Anthony K M 2012 Simulating the decadal- to millennial-scale dynamics of morphology and sequestered carbon mobilization of two thermokarst lakes in NW *Alaska Journal of Geophysical Research: Biogeosciences* **117**(G2) Available at: **http://doi.wiley.com/10.1029/2011JG001796**

[17] Tan Z Zhuang Q and Walter A K 2015 Modeling methane emissions from arctic lakes: Model development and site-level study *Journal of Advances in Modeling Earth Systems* 459–483 Available at: **http://doi.wiley.com/10.1002/2014MS000344**

[18] Stepanenko V, Mammarella I, Ojala A, Miettinen H, Lykosov V and Vesala T 2016 LAKE 2.0: a model for temperature, methane, carbon dioxide and oxygen dynamics in lakes *Geosci. Model Dev.* **9** 1977-2006 doi:10.5194/gmd-9-1977-2016

[19] *IPCC 2013 Climate change* 2013 (New York: Cambridge University Press) Available at: **http://www.climatechange2013.org/**

[20] Wilson M F and Henderson-Sellers A 1985 A global archive of land cover and soils data for use in general circulation climate models *J. Clim.* **5** 119–143

[21] Goudsmit G-H 2002 Application of $K-\varepsilon$ turbulence models to enclosed basins: The role of internal seiches *J. Geophys. Res.* **107** 3230 doi:10.1029/2001JC000954

# Kolmogorov goodness-of-fit test for *S*-symmetric distributions in climate and weather modeling

**Z Zenkova**[1] **and L Lanshakova**[2]

[1] Associate Professor, Department of Applied Mathematics and Cybernetics, National Research Tomsk State University, 36 Lenin Street, Tomsk 634030, Russia
[2] Postgraduate Student, Department of Applied Mathematics and Cybernetics, National Research Tomsk State University, 36 Lenin Street, Tomsk 634030, Russia

E-mail: thankoff@fpmk.tsu.ru

**Abstract.** Statistical data treatment is an essential part of climate and weather modeling. The Kolmogorov goodness-of-fit test is a widely applicable statistical method to determine the cumulative distribution function of a continuous random variable, e.g., a precipitation level, wind force, etc. In this paper, the authors consider a problem of goodness-of-fit testing involving additional information about *S*-symmetry of the cumulative distribution function and its influence on the Kolmogorov statistic distributions. A definition of *S*-symmetry is given; it is a generalized classical definition of distribution symmetry. It is proved that any continuous increasing cumulative distribution function is *S*-symmetric. A uniform distribution is considered as an example of an *S*-symmetric distribution. A modification of the Kolmogorov statistic using additional information about the new type of symmetry is proposed. The exact and asymptotic distributions under the null and the alternative hypothesis of the modified statistics are described. The authors also provide an example which proves that the modified test is more powerful than the non-modified one. The new test is used to check the hypothesis of a uniform distribution of the average sum of precipitation.

## 1. Introduction

Climate and weather modeling is a complex task which involves a wide range of different methods and techniques [1-2], one of them is a statistical data treatment [3-5]. It allows description, analysis and forecasting of weather processes, helps detect and prove scientifically implicit and subtle climate changes.

Despite plenty of observations and collected information evidence in the field of weather and climate modeling, inaccurate results are still a problem [6-8], because finding exact values is statistically impossible. So, a lot of research aims to improve statistical procedures. A short review of the methods is given [9], which can improve an accuracy of statistical treatment using additional and auxiliary information, especially for incomplete and censored data [10].

In this paper, the authors consider a new definition of *S*-symmetry of cumulative distribution function (c.d.f.) and use it in the modification of Kolmogorov goodness-of-fit test. The distributions of modified statistics are provided for null and alternative hypotheses for asymptotic and non-asymptotic cases. An example of power increase is given.

## 2. Kolmogorov goodness-of-fit test for $S$ -symmetric distribution

The classical test [11] checks whether the unknown c.d.f. $F(x)$ of a sample $X=(X_1,X_2,...,X_N)$ is equal to a hypothetical c.d.f. $G(x)$. It is based on statistics

$$d_N=\sup_{x\in\mathbf{R}}\left|F_N(x)-G(x)\right|,$$

where $F_N(x)$ is the empirical distribution function (e.d.f.) [12]. The null distribution of $d_N$ is equal to [11-12]

$$P(d_N\le y)=N!\det\left[\frac{\left(\max\{0;\left[\min(1;b_i)-\min(1;a_j)\right]\}\right)^{j-i+1}}{(j-i+1)!}\right]_{i,j=\overline{1,N}}, \qquad (1)$$

where for $i=\overline{1,N}$, $j=\overline{1,k}$

$$a_i=\min\left[1;\max\left(0;\frac{i}{N}-y\right)\right],\ b_i=\max\left[0;\min\left(1;\frac{i-1}{N}+y\right)\right].$$

The asymptotic null distribution of $d_N$ can be calculated for $z>0$ by means of the formula:

$$\lim_{N\to\infty}\mathrm{P}\left(\sqrt{N}d_N\le z\right)=K(z)=\sum_{k=-\infty}^{+\infty}(-1)^k\exp\left(-2k^2z^2\right), \qquad (2)$$

where $K(z)$ is Kolmogorov distribution [13].

In this paper, the authors consider the hypothesis:

$$H_0^S:\ F(x)=G(x),\ F,G\in\Xi^S,\ \forall x\in\mathbf{R},$$

against

$$H_1^S:\ F(x)=V(G(x)),\ F,V,G\in\Xi^S,\ \forall x\in\mathbf{R},$$

where $\Xi^S$ is a set of $S$-symmetric c.d.f.

**Definition**. If for $c_1<c_2<...<c_k$, $i=\overline{1,k}$, $k>1$,

$$p_i=F(c_i)=F(c_i+0),$$

$$p_0=F(c_0)=0,\ p_{k+1}=F(c_{k+1})=1,$$

for $j=\overline{1,k}$ and $x\le c_{j+1}$ c.d.f. $F(x)$ satisfies the conditions:

$$F\left(\max\{x,S_j(x)+0\}\right)=p_{j+1}-\frac{p_{j+1}-p_j}{p_j}F\left(\min\{x,S_j(x)+0\}\right), \qquad (3)$$

where $S_j(x)$ are continuous monotonically decreasing functions,

$$\left(S_j\right)^{-1}(x)=S_j(x), \qquad (4)$$

$x\le c_{j+1}$, $j=\overline{1,k}$, $S_j(c_i)=c_i$, $i=\overline{1,k}$, $S_j(c_{j+1})=c_0$, then c.d.f. $F(x)$ is **$S$-symmetric c.d.f.**

For $k=1$, $p_1=F(c_1)=0.5$, $S_1(x)=2c_1-x$ we can obtain classical symmetry of c.d.f. around $c_1$:

$$F(x)=1-F(2c_1-x).$$

**Theorem 1**. If $F(x)$ is continuous increasing c.d.f. then it is $S$-symmetric c.d.f. with

$$S_j(x)=\begin{cases}F^{-1}\left(p_{j+1}-\dfrac{p_{j+1}-p_j}{p_j}F(x)\right), & x\le c_j,\\[2ex] F^{-1}\left(p_j\cdot\dfrac{p_{j+1}-F(x)}{p_{j+1}-p_j}\right), & c_j<x\le c_{j+1},\end{cases} \qquad (5)$$

$p_j = F(c_j)$, $j=\overline{1,k}$, where values $p_j$ and $c_j$ are known, $p_0 = 0$, $p_{k+1} = F(c_{k+1}) = 1$, $F^{-1}(x)$ is the inverse function for $F(x)$, i.e. $F^{-1}(F(x)) = x$.

**Proof.** For $j=\overline{1,k}$ let us find $S_j(x)$ which satisfies the conditions (3). If $x \le c_j$ then (3) is

$$F(S_j(x)) = p_{j+1} - \frac{p_{j+1}-p_j}{p_j} F(x),$$

because $F(x)$ is continuous and increased, for $x \in (c_j; c_{j+1}]$ equation (3) is

$$F(x) = p_{j+1} - \frac{p_{j+1}-p_j}{p_j} F(S_j(x)),$$

therefore

$$F(S_j(x)) = p_j \cdot \frac{p_{j+1} - F(x)}{p_{j+1}-p_j}.$$

It is well-known, that if $F(x)$ is continuous and increased then its inverse function $F^{-1}(x)$ exists, so (5) is obvious. Let us prove that for $j=\overline{1,k}$ $S_j(x)$ are decreasing functions.

For any $j=\overline{1,k}$, $x_1 < x_2 \le c_j$

$$S_j(x_1) = F^{-1}\left(p_{j+1} - \frac{p_{j+1}-p_j}{p_j} F(x_1)\right) > F^{-1}\left(p_{j+1} - \frac{p_{j+1}-p_j}{p_j} F(x_2)\right) = S_j(x_2)$$

because $F(x)$ and $F^{-1}(x)$ are both increasing functions and

$$p_{j+1} - \frac{p_{j+1}-p_j}{p_j} F(x_1) > p_{j+1} - \frac{p_{j+1}-p_j}{p_j} F(x_2).$$

If $c_0 \le x_1 \le c_j < x_2 \le c_{j+1}$ then

$$S_j(x_1) = F^{-1}\left(p_{j+1} - \frac{p_{j+1}-p_j}{p_j} F(x_1)\right) > F^{-1}\left(p_j \cdot \frac{p_{j+1} - F(x_2)}{p_{j+1}-p_j}\right) = S_j(x_2),$$

because $0 = F(c_0) \le F(x_1) \le F(c_j) = p_j < F(x_2) \le F(c_{j+1}) = p_{j+1}$ and so

$$p_{j+1} - \frac{p_{j+1}-p_j}{p_j} F(x_1) \in \left[p_{j+1} - \frac{p_{j+1}-p_j}{p_j} \cdot p_j;\ p_{j+1} - \frac{p_{j+1}-p_j}{p_j} \cdot 0\right] = [p_j;\ p_{j+1}]$$

$$p_j \cdot \frac{p_{j+1} - F(x_2)}{p_{j+1}-p_j} \in \left[p_j \cdot \frac{p_{j+1}-p_{j+1}}{p_{j+1}-p_j};\ p_j \cdot \frac{p_{j+1}-p_j}{p_{j+1}-p_j}\right) = [0;\ p_j)$$

If $c_j < x_1 < x_2 \le c_{j+1}$ then

$$\frac{p_{j+1} - F(x_1)}{p_{j+1}-p_j} > \frac{p_{j+1} - F(x_2)}{p_{j+1}-p_j},$$

thus

$$S_j(x_1) = F^{-1}\left(p_j \cdot \frac{p_{j+1} - F(x_1)}{p_{j+1}-p_j}\right) > F^{-1}\left(p_j \cdot \frac{p_{j+1} - F(x_2)}{p_{j+1}-p_j}\right) = S_j(x_2),$$

because $F(x)$ is increasing function. Therefore, for $j=\overline{1,k}$ $S_j(x)$ are decreasing functions.

As for $j=\overline{1,k}$ $p_j = F(c_j)$ then it is obvious that $S_j(c_j)=c_j$. Let us prove that $S_j(c_{j+1})=c_0$. Indeed, as $F(c_{j+1})= p_{j+1}$ then

$$S_j(c_{j+1})=F^{-1}\left(p_j \cdot \frac{p_{j+1}-F(c_{j+1})}{p_{j+1}-p_j}\right)=F^{-1}\left(p_j \cdot \frac{p_{j+1}-p_{j+1}}{p_{j+1}-p_j}\right)=F^{-1}(0)=c_0 .$$

For proving (3) we should find the inverse functions for $S_j(x)$, $j=\overline{1,k}$. If $x \le c_j$ then

$$y=S_j(x)=F^{-1}\left(p_{j+1}-\frac{p_{j+1}-p_j}{p_j}F(x)\right),$$

thus

$$F(y)=p_{j+1}-\frac{p_{j+1}-p_j}{p_j}F(x),$$

which allows obtaining

$$x=S_j^{-1}(y)=F^{-1}\left(p_j \cdot \frac{p_{j+1}-F(y)}{p_{j+1}-p_j}\right)=S_j(y),$$

because $S_j(\cdot)$ is decreasing and so for $x \le c_j$ $c_j < y = S_j(x) \le c_{j+1}$.

Thus, the theorem is proved.

To test the hypothesis $H_0^S$ the modified Kolmogorov statistic can be used

$$d_N^S = p^* \cdot \sup_{x\in\mathbf{R}}\left|F_N^S(x)-G^*(x)\right|, \tag{6}$$

where

$$p^* = p_1 \cdot \prod_{j=1}^{k} \max\left(1; \frac{p_{j+1}-p_j}{p_{j+1}}\right), \quad G^*(x)=\min\left(1; p_1^{-1} \cdot G(x)\right),$$

$F_N^S(x)$ is e.d.f. based on $k$ times symmetrized sample $X^*=\left(X_1^*,...,X_N^*\right)$. Here, for $i=\overline{1,N}$ $X_i^*=X_i^{(k)}$,

$$X_i^{(j)}=\min\left[X_i^{(j-1)}; S_{k-j+1}\left(X_i^{j-1}\right)\right],$$

$j=\overline{1,k}$, $X_i^{(0)}=X_i$. It is easy to prove that random variable $X_i^*$ has c.d.f. $F^*(x)=\min\left\{p_1^{-1}\cdot F(x), 1\right\}$.

Note that e.d.f. $F_N^S(x)$ is unbiased estimation of *S*-symmetric c.d.f. $F(x)$ with variance

$$\operatorname{var} F_N^S(x)=\operatorname{var} U^{(k)}(x),$$

where $\operatorname{var}U^{(k)}(x)$ is calculated by *k* times repetitions of the formula

$$\operatorname{var}U^{(j)}(x)=\begin{cases}\operatorname{var}U^{(j-1)}(x), & x \le c_j;\\ \left(p_{j+1}-p_j\right)^2 p_k^{-2}\cdot \operatorname{var}U^{(j-1)}\left(S_j(x)+0\right), & x\in\left(c_j;c_{j+1}\right]\end{cases}$$

wherein $\operatorname{var}U^{(0)}(x)=N^{-1}F(x)\left(p_1-F(x)\right)$ and $p_{k+1}=1$.

The alternative distribution of the modified statistic $d_N^S$ can be calculated for $F(x)=V\left(G(x)\right)$, $y>0$, by the formula:

$$P\left(d_N^S \le y\right) = N!\det\left[\frac{\left(\max\left\{0;\left[\min\left(1;\frac{V(p_1 b_i)}{p_1}\right) - \min\left(1;\frac{V(p_1 a_j)}{p_1}\right)\right]\right\}\right)^{j-i+1}}{(j-i+1)!}\right]_{i,j=\overline{1,N},} \tag{7}$$

where for $i=\overline{1,N}$, $j=\overline{1,k}$

$$a_i = \min\left[1;\max\left(0;\frac{i}{N}-\frac{y_1}{p_1}\right)\right],\ b_i = \max\left[0;\min\left(1;\frac{i-1}{N}+\frac{y_1}{p_1}\right)\right],$$

$$y_j = \min\left(y_{j+1};\frac{p_j y_{j+1}}{p_{j+1}-p_j}\right),\quad y_{k+1} = y.$$

If in (7) $V(t) = t$ then it allows finding the null distribution of $d_N^S$ as

$$P\left(d_N^S \le y\right) = N!\det\left[\frac{\left(\max\left\{0;\left[\min(1;b_i) - \min(1;a_j)\right]\right\}\right)^{j-i+1}}{(j-i+1)!}\right]_{i,j=\overline{1,N},}$$

where for $i=\overline{1,N}$ $a_i$ and $b_i$ depend on $y_1$ which is obtained by $k$ times repeating of the formula $y_j = \min\left(y_{j+1};\frac{p_j y_{j+1}}{p_{j+1}-p_j}\right)$. Thus

$$y_1 = y_2 \min\left(1;\frac{p_1}{p_2-p_1}\right) = y_3 \min\left(1;\frac{p_2}{p_3-p_2}\right)\cdot\min\left(1;\frac{p_1}{p_2-p_1}\right) = \ldots =$$

$$= y\prod_{j=1}^{k}\min\left(1;\frac{p_j}{p_{j+1}-p_j}\right) = \frac{y\cdot p_1}{p_1\prod_{j=1}^{k}\max\left(1;\frac{p_{j+1}-p_j}{p_j}\right)} = \frac{y\cdot p_1}{p^*}.$$

So for $i=\overline{1,N}$

$$a_i = \min\left[1;\max\left(0;\frac{i}{N}-\frac{y}{p^*}\right)\right] \text{ and } b_i = \max\left[0;\min\left(1;\frac{i-1}{N}+\frac{y}{p^*}\right)\right].$$

Therefore, comparing with (1) we can conclude that for null hypothesis

$$P\left(d_N^S \le y\right) = P\left(d_N \le y\cdot\left(p^*\right)^{-1}\right).$$

As a result, the asymptotic null distribution of modified statistic (6) can be calculated for $z>0$ as

$$\lim_{N\to\infty} \mathrm{P}\left(\sqrt{N}d_N^S \le z\right) = K\left(\left(p^*\right)^{-1}\cdot z\right).$$

Here function $K(\cdot)$ is Kolmogorov distribution as in formula (2).

## 3. *S*-symmetry of the uniform distribution

Let us consider the uniform distribution $U(x)$.

**Theorem 3**. The uniform c.d.f. $U(x)$ is satisfied by (3) for $k = 1$, if for arbitrary $c_1 = p_1$, $0 < p_1 < 1$,

$$S_1(x) = \begin{cases} 1 - \dfrac{1-p_1}{p_1} \cdot x, & 0 \le x \le c_1, \\ \dfrac{1-x}{1-p_1} \cdot p_1, & c_1 < x \le 1. \end{cases}$$

**Proof.** Obviously, that function $S_1(x)$ is continuous and monotonically decreasing with $p_0 = F(c_0 = 0) = 0$, $p_2 = F(c_2 = 1) = 1$. Let $c_1 = p_1$ is arbitrary and $0 < c_1 < 1$ then in (3) we can obtain

$$F(\max\{x, S_1(x) + 0\}) = 1 - \frac{1-p_1}{p_1} F(\min\{x, S_1(x) + 0\}).$$

If $0 \le x \le c_1$ then

$$\max\{x, S_1(x) + 0\} = \max\left\{x, 1 - \frac{1-p_1}{p_1} \cdot x\right\} = 1 - \frac{1-p_1}{p_1} \cdot x$$

but

$$\min\{x, S_1(x) + 0\} = \min\left\{x, 1 - \frac{1-p_1}{p_1} \cdot x\right\} = x$$

because $0 < p_1 < 1$, therefore it satisfies (3). For $c_1 < x \le 1$ the proof is similarly. The result can be easily generalized for the uniform distribution $U_{[a,b]}(x)$, $x \in [a,b]$.

Note that classical symmetry around 0.5 can be obtained if $c_1 = p_1 = 0.5$.

Thus, the theorem is proved.

For $x \in \mathbf{[0,1]}$ let us consider a hypothesis

$$H_0^S: \ F(x) = U(x) = x$$

against

$$H_1^S: \ F(x) = U_1(x),$$

where for integer $m > 0$

$$U_1(x) = \begin{cases} p_1 \left(\dfrac{x}{p_1}\right)^m, & x \in [0, p_1], \\ 1 - (1-p_1)\left(\dfrac{1-x}{1-p_1}\right)^m, & x \in (p_1, 1]. \end{cases}$$

Function $U_1(x)$ is satisfied by (3) for the same $S_1(x)$ and for $c_1 = p_1 = 0.5$ we have classical symmetry again.

The plots of functions $U(x)$, $U_1(x)$ for $m = 3$, $c_1 = p_1 = 0.7$ and $U_1(x)$ for $m = 3$, $c_1 = p_1 = 0.5$ are demonstrated in Figure 1.

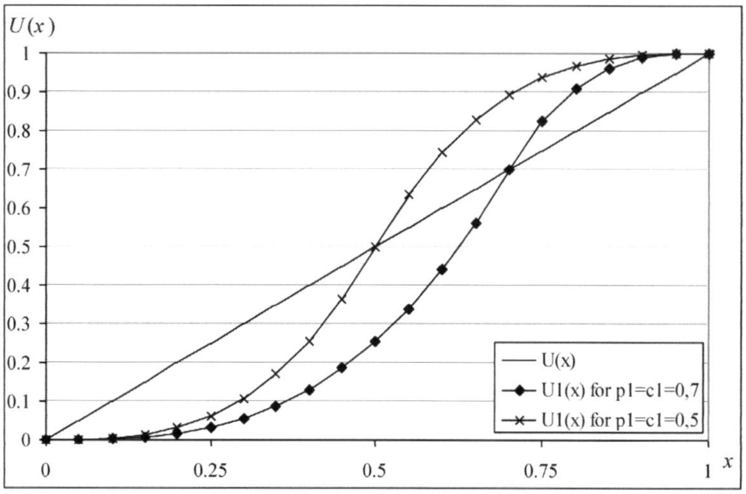


**Figure 1.** Functions $U(x)$, $U_1(x)$ for $m = 3$, $c_1 = 0.7$ and $U_1(x)$ for $c_1 = 0.5$.

The dependence of powers of modified ($Ws$) and non-modified ($W$) tests on a sample size $N$ with significance level $\alpha = 0.05$, $m = 3$, $p_1 = 0.7$ are shown in figure 2. It is obvious that the modified test is more powerful than the classical one. The result is easily generalized for the uniform distribution $U_{[a,b]}(x)$, $x \in [a,b]$.

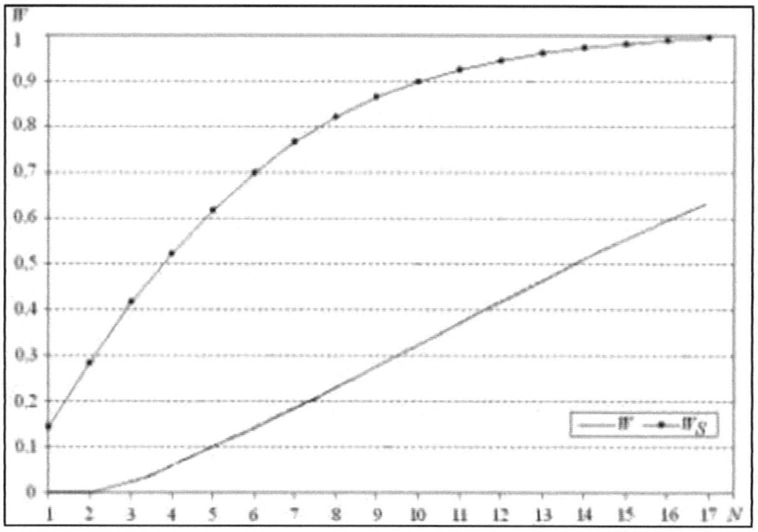


**Figure 2.** The powers of Kolmogorov goodness-of-fit tests: modified ($Ws$) and non-modified ($W$) with $\alpha = 0.05$, $m = 3$, $p_1 = 0.7$.

## 4. Example

Let us use a new technique to check the hypothesis that an average sum of precipitation' distribution is uniform. The data was collected during 1961-1990 in Tomsk, Russia [14] and shown in table 1.

**Table 1.** Average sums of precipitation within a year during 1961-1990 in Tomsk, Russia.

| Month | Jan. | Feb. | Mar. | Apr. | May | June | July | Aug. | Sept. | Oct. | Nov. | Dec. |
|---|---|---|---|---|---|---|---|---|---|---|---|---|
| **Average sums of precipitation, mm.** | 21 | 22 | 27 | 32 | 41 | 43 | 46 | 51 | 52 | 60 | 71 | 73 |

For $x \in [10{,}90]$ let us consider the hypothesis

$$H_0^S : F(x) = U_{[10,90]}(x) = \frac{x-10}{80}.$$

The classical Kolmogorov test allows obtaining the statistic $d_N = 0.225$. The critical value is $d_{N,\alpha} = 0.375$ for $N$=12 and significance level is $\alpha = 0.05$, so we can make a conclusion that the null hypothesis $H_0^S$ is true, because $d_N < d_{N,\alpha}$.

It was shown above that the uniform distribution satisfies the condition (3), hence, the modified test can be used to check $H_0^S$. For simplicity, let us consider the classical symmetry about $c_1 = 50$ with $p_1 = 0.5$, so symmetrized sample is obtained using the formula

$$X_i^* = X_i^{(1)} = \min[X_i; 100 - X_i],\ i = \overline{1,N},$$

where

$$X^* = (21, 22, 27, 32, 41, 43, 46, 49, 48, 40, 29, 27),$$

and the resulting modified statistic $d_N^S = 0.2125$.

According to formula (8), for significance level $\alpha = 0.05$ the critical value is

$$d_{N,\alpha}^S = \frac{0.375}{2} = 0.1875,$$

therefore, the hypothesis $H_0^S$ is rejected, because $d_N^S > d_{N,\alpha}^S$.

Since the power of the modified test is higher than that of the non-modified one, the results of its application are more adequate and reliable. So we should reject the hypothesis that the distribution of the average sum of precipitation is uniform in [10, 90] mm despite the classical test admits it.

## 5. Conclusions

The paper gives new definition of *S*-symmetric cumulative distribution function and considers the way of involving the additional information about *S*-symmetry to modify Kolmogorov goodness-of-fit test. It presents the precise null and alternative distributions and asymptotical null distribution of the modified statistics. The demonstrated example proves that the modification of the statistic helps increase the power of the new test. The authors apply the new method to check the hypothesis that an average sum of precipitation distribution is uniform and reject it despite the classical goodness-of-fit Kolmogorov test admits the hypothesis.

Therefore, using additional information about *S*-symmetry allows construction of more effective and powerful statistical procedures, which can be used in climate and weather modeling.

## References

[1] Stocker T 2011 *Introduction to Climate Modeling* (Berlin: Springer) p 179
[2] McGuffie K, Henderson-Sellers A 2005 *A Climate Modeling Primer* (New York: Wiley) p 296
[3] Hennemuth B, Bender S, Bülow K, Dreier N, Keup-Thiel E, Krüger O, Mudersbach C, Radermacher C, Schoetter R 2013 *CSC Rep.* **13**
[4] von Storch H, Zwiers F 1999 *Statistical Analysis in Climate Research* (Cambridge: Cambridge University Press) p 484
[5] Haltiner G, Williams R 1980 *Numerical Prediction and Dynamic Meteorology* (New York:

Wiley) p 477
[6] Lupo A, Market P S 2002 *Weath. Forecast.* **17** 891–897
[7] Martner B, Politovich M 1999 *Nat. Weath. Dig.* **23** 9–20
[8] Zhu A, Halton P 2014 *J. Climat. Weath. Forecast.* **2** 110
[9] Zenkova Z 2015 *AIP Conf. Proc.* vol 1688, 040002
[10] Klein J, Moeschberger M 2003 *Survival Analysis: Techniques for Censored and Truncated Data* (New York: Springer) p 551
[11] Kolmogorov A 1933 *Gorn. Inst. Ital. Att.* **4** 83–91
[12] D'Agostino R, Stephens M 1986 *Goodness-of-fit-techniques* (New York: Dekker) p 576
[13] Hollander M, Wolfe D 1973 *Nonparametric statistical methods* (New York: Wiley)
[14] Hydrometeorological Centre of Russia. *Available at:* http://meteoinfo.ru/?option=com_content&view=article&id=1736

# Mesoscale high-resolution modeling of extreme wind speeds over western water areas of the Russian Arctic

**Vladimir S Platonov and Alexander V Kislov**

Lomonosov Moscow State University, Faculty of Geography, Department of Meteorology and Climatology. 119991, Moscow, Leninskie Gory MGU, 1, GSP-1

vplatonov86@gmail.com, avkislov@mail.ru

**Abstract**. A statistical analysis of extreme weather events over coastal areas of the Russian Arctic based on observational data has revealed many interesting features of wind velocity distributions. It has been shown that the extremes contain data belonging to two different statistical populations. Each of them is reliably described by a Weibull distribution. According to the standard terminology, these sets of extremes are named 'black swans' and 'dragons'. The 'dragons' are responsible for most extremes, surpassing the 'black swans' by 10 – 30 %. Since the data of the global climate model INM-CM4 do not contain 'dragons', the wind speed extremes are investigated on the mesoscale using the COSMO-CLM model. The modelling results reveal no differences between the 'swans' and 'dragons' situations. It could be associated with the poor sample data used. However, according to many case studies and modeling results we assume that it is caused by a rare superposition of large-scale synoptic factors and many local meso- and microscale factors (surface, coastline configuration, etc.). Further studies of extreme wind speeds in the Arctic, such as 'black swans' and 'dragons', are necessary to focus on non-hydrostatic high-resolution atmospheric modelling using downscaling techniques.

## 1. Introduction

Investigation and modelling of the extreme wind velocities conditions in Arctic is presently one of the most important tasks for the purposes of ensuring the safety of infrastructure (the increasing exploration Arctic offshore oil and gas fields, development of navigation over the Northern Sea Route) and accompanying port facilities. In the maritime sectors, the extremes of low-level winds can generate huge oceanic waves and storm surges that consequently may lead to the damage of marine structures (ships, drilling platforms) and coastal erosion. This is especially the case during the cold period in the Arctic, where regular events of intense wind velocities are typically observed. The quantitative analyses and modelling of spatial variation of extreme wind patterns are important for effective wild fire management and sustainable long-term urban development on fire-prone landscapes. It is therefore worthwhile to properly assess the distribution of wind extremes and their origin.

The main features of wind conditions over Arctic basin are described in [1]. Particularly, the duration of periods when the wind speed doesn't exceed 15 m/s, is about 3 – 6 days in winter months. Then, the stormy weather much of time is prevailing over the Arctic basin, therefore it is very important to develop methods of an accuracy extreme wind conditions forecasting.

Analysis and forecast of the main hydrometeorological fields are frequently not fully adequate to the observations. In [2] it were compared different reanalyzes at meteorological stations north of 60° N and shown that over the most points errors are considerable and correlation coefficients are not significant. These conclusions are confirming a necessity and relevance of regional atmospheric modelling over the Arctic region using high-resolution techniques for purposes of reproducing extreme wind patterns.

## 2. Data and methods

### *2.1. Study area and data description*

Wind velocity at the boundary layer has strong spatio-temporal variability, therefore its correct simulation is still associated with significant errors. It is especially relevant to Arctic seashores with complicate surface and to poor data areas. There are many papers dedicated to the wind extremes' climate analysis as well as to an estimation and reproducing of wind speed extremes in the long-term datasets. Thus, [3] considered the 23-year dataset of wind speed and sea waves and concluded that wind extremes are underestimated by method of a sampling peaks above a certain threshold. [4, 5] estimated the NARR reanalysis over the USA Arctic region and inferred that $95^{th}$ percentile increased from 7 m/s in 1979 to 10.5 m/s in 2009.

Many studies [6, 7, 2, 8] have shown a definite advantage of mesoscale atmospheric models in the reproduction of wind velocity pattern against the most modern reanalyses. Cited papers noted that decreasing of horizontal resolution of the regional atmospheric models (e.g., WRF or HIRHAM [9]) brings the wind speed spectra closer to the well-known "-5/3" law, however there are significant underestimations of the wind speed on mesoscale comparing to the observations. Nevertheless, the mesoscale modelling demonstrates more adequate description of the extreme wind field patterns comparing to reanalyzes definitely.

This study was performed over the Arctic region from the Kola Peninsula until Chukotka Peninsula including both coastal area (predominantly) and inland territory. Strong wind speed events are often noted in the region in the cold time of year during the passage of meteorological synoptic storms. Wind speeds of more than 30 m/s are observed during this time over the marine surface, inducing high waves of more than 4 m. A dataset of observed hourly 10-minute mean wind speed data from stations was obtained, with the record period varying from station to station. For the present study, we used the period 1966-2013, which was covered by data of 40 Russian stations.

### *2.2. Statistical methods and data analysis features*

Research in the statistical analysis of extreme values has flourished over the past several decades: new probability models, inference and data analysis techniques have been introduced; and new application areas have been explored [10 – 12].

Extreme value analysis of wind speeds ($U$) is generally performed through implementation of the following idea. Beginning with a parent distribution whose cumulative distribution function (cdf) is $F(U)$, the distribution is sampled $n$ times, and the maximum value of the $n$ samples is obtained. This maximum value has a cumulative distribution function of its own of simply $F^n(U)$. This relationship leads to the extreme value theory noted by Fisher and Tippett [13] that for sufficiently long sequences of independent and identically distributed random variables, the maxima of these sequences can be fit to one of three limiting distributions. This result has been quantitatively refined by Gnedenko [14]. One representative of these three limiting distributions is the Weibull distribution, which has traditionally been used for the statistical modelling of wind extremes (see [15]).

In another approach, the Pareto distribution is applied to the peaks of independent storms that exceed a sufficiently high threshold (see [15] and [16]).

Statistical method of the extreme value analysis of wind speeds is important because it allows us to detect their statistical model. Note that the same statistical distribution suggests a common originating mechanism. We plan to use such idea to interpret of the extreme wind records.

It is a condition of extreme value analysis that the extremes selected for examination have to be independent. Annual (or seasonal) maximum wind speeds chosen from each year are statistically independent. However, when several data points are taken from each season, there may well be several clustered maximum speeds from a single storm. Such events are unlikely to be statistically independent. Various strategies are invoked to remove dependent events before proceeding with a statistical analysis. A simple method is to require a minimum time separation or "deadtime" between selected events. Working with Arctic wind climate, we use the autocorrelation coefficient $r(\tau)$ to establish a "deadtime" between consequent wind fluctuations. Its value is a measure of the correlation of neighbouring wind events. It was shown to be less than 0.05 for $\tau$ equal to 48 or 72 hours. Therefore, we use a "deadtime" of 72 hours. The same values (48 – 60 h) were used earlier [17 – 19].

There were revealed many interesting features of wind velocity statistical distribution. In figure 1 we plot several empirical cdfs on the bases of station measurements. Configuration of empirical points in the form of columns is determined by the fact that the data are quantized due to specified accuracy of measurement. Pictures are the "Weibull Plots", which are a specific nonlinear transformation of the data, and a straight line is recovered if the sample is Weibull. We can see that at all sites, we found that the empirical cdfs consistently deviate from the theoretical line starting with certain large threshold values ($U_{th}$). This deviation means that the standard Weibull model doesn't describe the empirical cdf, starting from a certain threshold $U > U_{th}$. In this way, the greatest extremes namely, which are the most important, should to be described by some other law, not by the standard Weibull distribution.

Note additionally that application of a sufficiently high threshold and, consequently, detection of especially high wind speeds allows us to describe for their approximation the peaks-over-threshold modelling approach, using the Pareto distribution. It has a cumulative distribution function

$$\Phi(U) = 1 - (U_{th}/U)^{\gamma} \quad (1)$$

It is worth mentioning here that the threshold value is not assigned a priori (as is usually done [16]), but is explicitly estimated previously.

Generally, the implemented analytical approach allows us to detect that the majority of extreme wind speed events (below the threshold value) adhere to the Weibull distribution. The same statistical distribution of population could be considered a result of the same organization principle, and this suggests a common generating mechanism for each representative of this population. This idea allows us to understand that a large extreme is not distinguished from its small siblings apart from its large power. The occurrence of large extremes looks like the appearance of a few black swans in a flock of white swans. This terminology was introduced by N N Taleb [20], as a metaphor to describe an event that comes as a surprise.

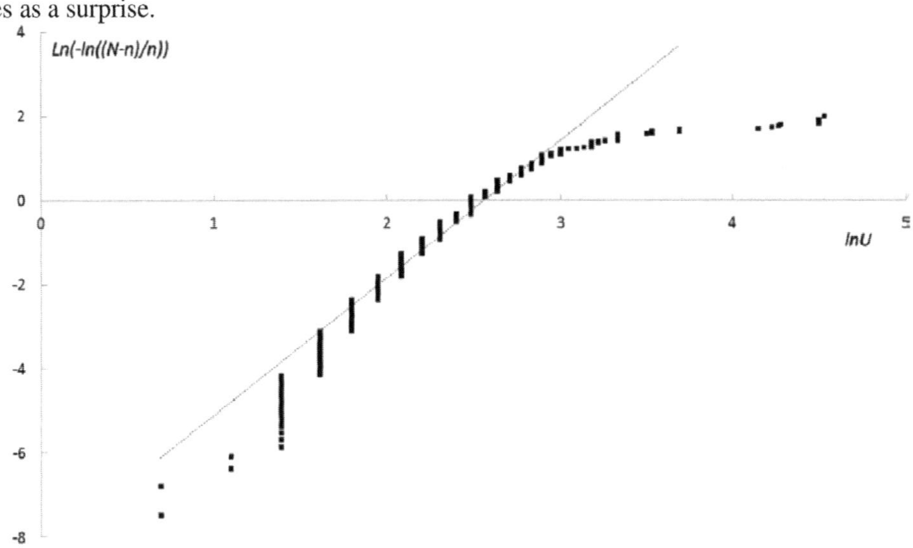


**Figure 1.** Empirical cumulative distribution functions of wind speed maxima (station observations, cold period) for 72 hours' time step records straightening on the coordinate axis of the Weibull distribution, and linear regression line corresponding to the Weibull function for station Teriberka.

However, there are extreme wind speed events located above the threshold value that are much more pronounced than predicted by the extrapolation of 'black swans' law distributions in their tail. They adhere again to the Weibull distribution. Such events were termed 'kings' (taking into account the special position of the fortune of kings, which appear to exist beyond the Zipf law distribution of the wealth of their subjects [21]) or 'dragons' (to stress that we address a completely different, beyond the normal, type of animal). D. Sornette [22] introduced the concept of 'dragon-kings' to refer to such extreme events. The same statistical distribution suggests a common generating mechanism different from that responsible for extrema at $U \leq U_{th}$.

It is not clear to what extent such excessively metaphoric terminology is required for our case because it was originally introduced to describe unique extraordinary events. However, it allows us to mark events that adhere to different Weibull distributions. Therefore, we will use below these terms: 'swans' or 'black swans' (hereafter BSs) and 'dragons' (hereafter the Ds).

*2.3. Global climate models estimation*

The next step of our analysis was to investigate to what extent the above-mentioned peculiarities of wind extremes are simulated by climate models. We analysed a dataset of wind simulation of the INM-CM4 climate model [23–25]. The establishment of the correspondence between wind simulation products and near-surface observations could help us to assess the quality of modelling products and their capability to reproduce the wind extremes. Apart from that, it is important to advance our understanding of the origin of the BSs and the Ds.

According to cdfs on the bases of the INM-CM4 simulations we concluded that the Weibull distribution is a good approximation of the modelled wind speed extremes. We found very small deviation in cdfs from the theoretical line starting with certain large threshold values. Keeping in our mind that noticeable deviation was a typical feature of empirical cdfs and using our terminology, we can conclude that the INM-CM4 model wind speed extremes are the BSs and there are no the Ds.

This conclusion is supported not only by the specific location of points along a theoretical line but also by the fact that modelled wind speed extremes themselves are close to observation data adhering the BSs besides the Zimnegorsky Mayak data and the Teriberka data, where observed *U(0.99)* are almost half times greater than modelled values. Probably, this is due to inadequate distribution of land and sea in the INM-CM4. Their geographical peculiarities are the same (the maxima are at the coastal area).

*2.4. Regional model and experiments description*

The discovered phenomenon of the absence of the representatives of the Ds in modelling data is very important. It means, the global climate models are not capable to reproduce many processes and effects responsible for the extreme wind cases referred to Ds type formation. Therefore, an investigation and analysis of BSs and Ds phenomena is appropriate to carry out on the mesoscale. Since the mesoscale hydrodynamic atmospheric modelling with high resolution is the widespread method of an investigation of detailed structure of meteorological fields including wind speed, we have tried to estimate possible genetic differences appearing in Ds and BSs cases using the regional model. We have sorted out from many numbers of these cases a few ones to execute the model experiments for the following dates: 15 – 17.12.1997, 29 – 30.10.2000, 26.01.2002, 12.12.2013 (see Table 1). The period of each experiment was about 7 days with the extreme event in the middle of period.

In this study, we have used the climate version of COSMO model (COSMO-CLM). It is well-known non-hydrostatic regional atmospheric model developed by German Weather Service (DWD) and CLM-Community (see CLM Community site http://www.clm-community.eu/, [26, 27]). COSMO-CLM model is based on the primitive Navier-Stokes equations describing the dynamics of compressible fluid in the moist atmosphere. Model equations are solving on the rotational grid 'latitude-longitude' ($\lambda$, $\varphi$) with pole tilt. It helps to minimize the problem of meridians convergence over the pole. The numerical scheme is realized on Arakawa C-grid, the vertical coordinate is hybrid Gal-Chen coordinate.

**Table 1.** Dates of extreme wind speed cases used for performed experiments.

| 'BSs' | 'Ds' |
|---|---|
| 15.12.1997 | 17.12.1997 |
| 29 – 30.10.2000 | 12.12.2013 |
| 26.01.2002 | |

Model runs were performed for the unified large domain with spatial resolution of 0.12°, covered the Barents Sea, part of Kara Sea, northern European territory of Russia and the surrounding water areas. Driving conditions came from ERA-Interim reanalysis (~0.75° resolution). After that, the downscaling technology was performed for the different 'small' domains (resolution about ~2.8 km), inside the 'large' domain (see figure 2). Information about parameters of experiments is shown in the Table 2. Standard configuration of COSMO-CLM model (version 5.0) was applied: Runge-Kutta integration scheme with 5th advection order; 40 vertical levels; prognostic TKE-based scheme for turbulence. More detailed description of COSMO-CLM model parameterizations could be found in the COSMO Model documentation (http://www.cosmo-model.org/content/model/documentation/core/default.htm).

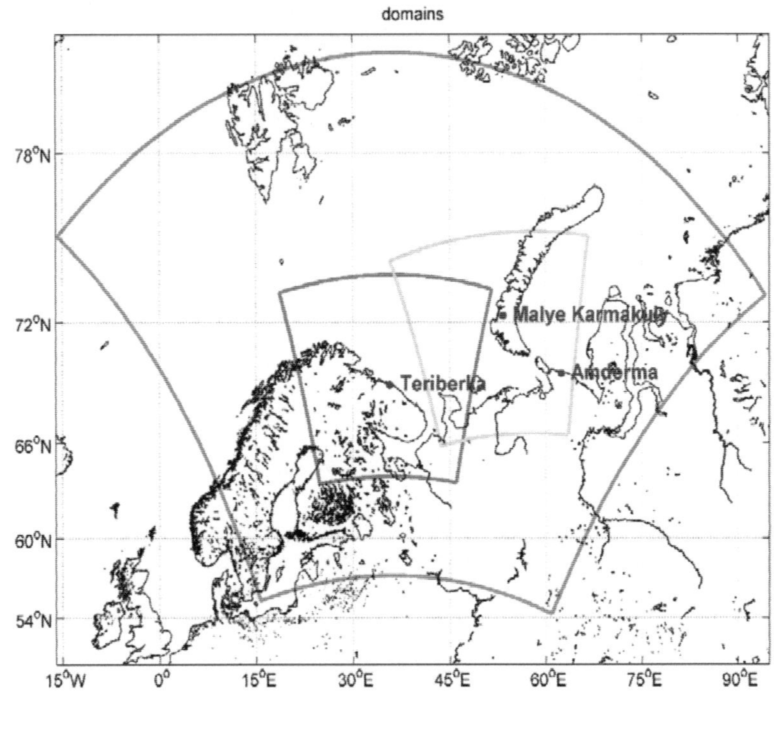


**Figure 2.** Model domains used in experiments and meteorological stations used for verification.

**Table 2.** Main parameters of the performed experiments.

| Parameters of experiments | Two model domains using downscaling | |
|---|---|---|
| Experiment's duration | Approx. 7 days (extremes ~in the middle) | |
| Horizontal resolution | 0.165° (~18 km) | 0.025° (~2.8 km) |

| | | |
|---|---|---|
| Domain size (number of points) | 164*146*40 | 380*400*40<br>326*364*40 |
| Time step, s | 100 | 40 |
| Initial and boundary conditions | ERA-Interim (~0.75°), every 6 h | COSMO-CLM 18 km, every 3 h |

## 3. Modelling results and discussion

### *3.1. Modelling results*

Considering the main modelling results for extreme wind speed cases, it should be noted that COSMO-CLM model reproduced the common atmospheric patterns and the dynamics of synoptic-scale processes very well for all cases. It was revealed by comparison modelling results with the weather charts.

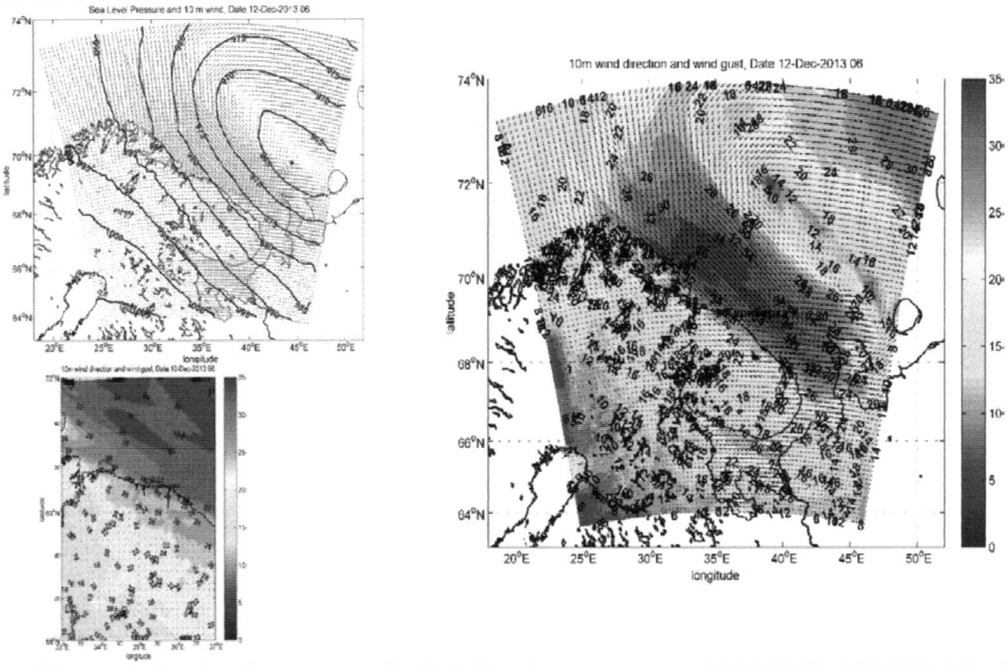


**Figure 3.** Sea level pressure and wind direction patterns on 12.12.2013, 06 UTC (hPa, left top), wind direction and gusts (m/s, right top), detailed map (bottom).

Further, we'll consider the reproducing of atmospheric circulation during different situations. In case of 29 – 30.11.2000 there are occurred an increasing of wind speed over the southern edge of Novaya Zemlya Island and in the Kara Strait. On the northeastern periphery of the moderate cyclone moved over the south of the Barents Sea. The strong northeastern winds were quasi-parallel to the strait, and initiated katabatic accelerations and wind gusts over the south-western coast of Novaya Zemlya. The mean wind velocities in the strait came to 20 m/s, gusts were up to 24 – 26 m/s. The wind pattern over the seashores is spotted and striped and associated with the mesoscale effects, such as a hydrodynamic impact of land and increasing of velocity over the strait, obviously.

The next situation, on 12.12.2013 (see figure 3) was characterized by the maxima wind speed among the examined and one of the best reproduced by model. The strong cyclone with the 965 hPa isobar minima at the center, displaced from the NW to the SE of the Barents Sea, and highly condensed isobars located quasi parallel to the coastline. This combination of factors and the long maintenance of this situation could lead to formation of outstanding extreme wind speed values, exceeded 30 m/s in averages and 35 m/s in wind gusts. Wherein, the maxima wind area continued for many hundreds kilometers to the sea. More detailed map demonstrates (see figure 3) very significant wind shear near the coast: from 12 to 20 m/s in average velocities, from 26 to 32 m/s in gusts.

The next investigated case refers to 26.01.2002. The intense cyclone shifted over the north of the European Russia. Coupled with a strong pressure gradient it led to increasing of wind speed up to 22 – 24 m/s on the Barents Sea coast and up to 30 m/s on the White Sea coast. This case was reproduced by model with many significant errors: the correlation coefficient was 0.64, mean and RMS errors exceeded 4 m/s as on coarse as on high resolutions.

The last considered case is perhaps the most interesting, because there were two consequent extreme situations during three days, relating to the two different types: BSs and Ds. There are 15.12.1997 and 17.12.1997, respectively (see figure 4). The strongest anticyclone on the south, extreme pressure gradient and isobars quasi-parallel to the coastline – these factors combined to the super extreme situation. It was enhanced by its long duration, evolving, and moving the cyclone to the east. As a result, the wind speed maxima was observed over the narrow coastal band few tens of kilometers wide. The average wind speeds exceeded 30 m/s. This case was sufficiently reproduced by model with the both resolutions.

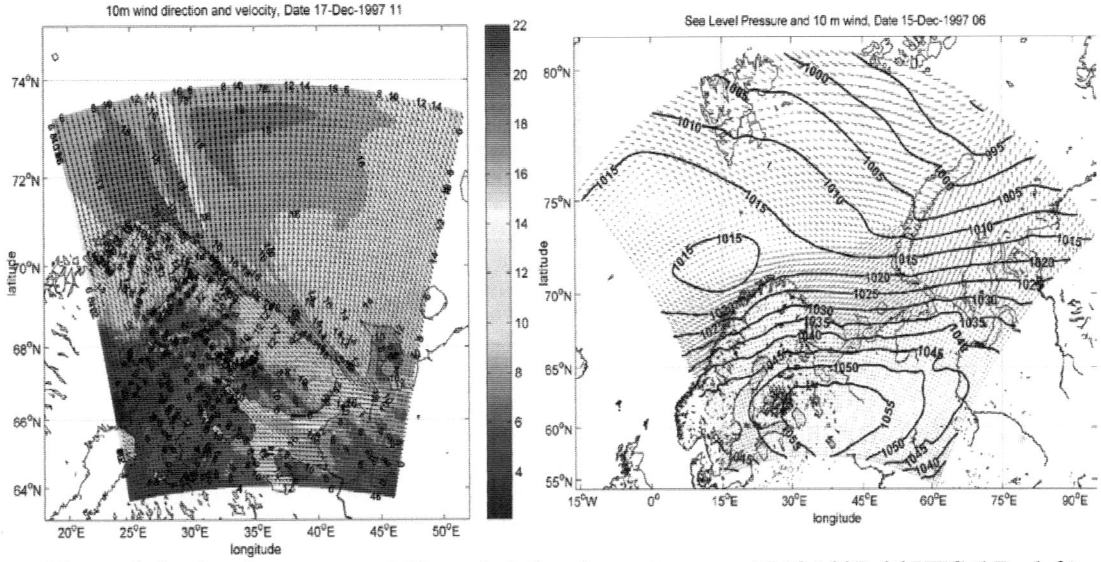


**Figure 4.** Sea level pressure and 10 m wind direction patterns on 15.12.1997, 06 UTC (hPa, left), wind speed and direction on 17.12.1997, 11 UTC (m/s, right).

### *3.2. Error statistics analysis and discussion*

We have done an elementary verification according to the simple method of comparison observational data with modelling results. It is the 'nearest neighbor' method without any interpolations, weighting coefficients etc. Many errors statistics for each case study, namely correlation coefficients, mean and median errors, RMSE, standard deviations are listed in the Table 3. We can see there are no significant differences in errors between coarse and high resolutions, except the mean errors. It could be associated with the used estimation method. In general, the COSMO-CLM model reproduces the synoptic-scale dynamics and general synoptic-scale wind velocity patterns well as both with the coarse (18 km), and high resolutions. Model with 2.8 km resolution succeed to reproduce detailed spotty wind pattern, caused by local orography or/and dynamic factors.

**Table 3.** Error statistics for performed experiments (meteostaion Teriberka).

| Experiment /statistics | Correlation coefficient | Mean error | Median error | RMSE | STD |
|---|---|---|---|---|---|
| 2013 18 km | 0.90 | -1.49 | -1.00 | 3.63 | 3.34 |

| | | | | | |
|---|---|---|---|---|---|
| 1997 18 km | 0.86 | -1.43 | -0.87 | 3.95 | 3.71 |
| 2002 18 km | 0.67 | -0.31 | -0.60 | 4.20 | 4.23 |
| 2013 2,8 km | 0.87 | -2.88 | -2.27 | 4.73 | 3.80 |
| 1997 2,8 km | 0.88 | -3.07 | -2.45 | 4.71 | 3.60 |
| 2002 2,8 km | 0.82 | -3.31 | -2.76 | 5.00 | 3.78 |

There are many features by interpretation and estimation these statistics results. COSMO-CLM model with 2.8 km resolution succeed to reproduce the detailed extreme wind speed pattern, caused by local orography and dynamical factors, usually. On the one hand, the model underestimates the real observed average wind speeds and gusts on the seacoast by 4 – 5 m/s systematically. On the other hand, these extreme velocities of air particles (20 m/s) doesn't make a certain physical sense to estimate model errors at the certain station point. We have to take into account that during 10 minutes of the average wind speed measurement the air particle could be moved on many kilometers or a few tens kilometers from sea to the land, i.e. many model grid points. Therefore, we should in future estimate wind velocity values for some surrounding area, according to the distance, corresponding to wind velocities. Taking into account these reasons, we can ascertain, that COSMO-CLM model reproduces wind velocity pattern quite adequately, but using the resolution 5 km and less, only.

## 4. Conclusions

Extreme value analysis has been implemented to estimate the statistical properties of extreme wind speed over the European and Siberian parts of Arctic region from the Kola Peninsula to the Chukotka Peninsula.

It was shown that for all stations a volume of observed samples of extreme wind speed are composed of two sets of variables. All samples of each population have the same statistical properties but one population is sharply different from another. So different origin of strong wind events adhering to two groups can be concluded. Using metaphoric terminology, we marked these events as the 'black swans' (BSs) and the 'dragons' (Ds). However, the global climate model extreme wind speeds dataset (INM-CM4 data) consist of only the BSs. Therefore, we have applied the mesoscale nonhydrostatic climate model COSMO-CLM to reproducing and investigation the physical mechanisms of BSs and Ds formation.

Analyze has shown that model reproduces the synoptic-scale dynamics and general synoptic-scale wind velocity patterns well as both with the 18 km, and 2.8 km resolutions and, in general, simulates wind speed quite adequately.

With respect to revealing many differences between 'BSs' and 'Ds' situations, there were found out no clear distinctions. It could be associated with a poor sample of cases. Based on examined cases, we can assume it caused by the rare overlay of large-scale synoptic factors (strong pressure gradient, orientation of the isobars with respect to coastline) and many local meso- and microscale factors (surface, coastline configuration etc.). In the future, further studies of the extreme wind speeds genesis in the Arctic, such as the 'black swans' and 'dragons', urged to focus on nonhydrostatic high-resolution modeling using downscaling techniques.

## Acknowledgements

Funding for this research was provided by grants from the Russian Science Foundation (Grant No. 11.G34.31.007) and Grant from the Academic Council of Faculty of Geography, Lomonosov Moscow State University.

## References

[1] *Reference data on the wind and waves regime of the Barents, Okhotsk Sea and Caspian Sea* (In Russian). Ed. by Lopatukhin L, Bukhanovsky A, Degtyarev A, Rozhkov V 2003 SPb Russian Maritime Shipping Register p 213

[2] Lindsay R, Wensnahan M, Schweiger A, Zhang J Evaluation of seven different atmospheric reanalysis products in the arctic. 2014 *J. Climate* **27** 2588 – 606
[3] Vinoth J and Young I Global estimates of extreme wind speed and wave height. 2011 *J. of Climate* **24** 1647–65
[4] Stegall S, Zhang J Wind field climatology, changes, and extremes in the Chukchi–Beaufort Seas and Alaska North Slope during 1979–2009. 2012 *J. of Climate* **25** 8075–89
[5] Hundecha Y, St-Hilaire A, Ouarda T, El-Adlouni S, Gachon P Nonstationary extreme value analysis for the assessment of changes in extreme annual wind speed over the Gulf of St. Lawrence, Canada. 2008 *J. of Appl. Met. And Clim.* **47** 2745–59
[6] Frehlich R and Sharman R The use of structure functions and spectra from numerical model output to determine effective model resolution. 2008 *Mon. Wea. Rev.* **136** 1537–53
[7] Larsen X, Ott S, Badger J, Hahmann A, Mann J Recipes for correcting the impact of effective mesoscale resolution on the estimation of extreme winds. 2012 *J. of Appl. Met. And Clim.* **51** 521–33
[8] Skamarock W Evaluating mesoscale NWP models using kinetic energy spectra. 2004 *Mon. Wea. Rev.* **132** 3019–32
[9] Bøssing Christensen O, Drews M, Hesselbjerg Christensen J, Dethloff K, Ketelsen K, Hebestadt I, Rinke A The HIRHAM Regional Climate Model. Version 5 (beta). 2007 *Danish Climate Centre* Danish Meteorological Institute p 22
[10] Beirlant J, Goegebeur Y, Segers J, Teugels J, De Waal D, Ferro C *Statistics of extremes: theory and applications.* 2004 (Wiley Series in Probability and Statistics, John Wiley & Sons Ltd., Chichester)
[11] Coles S *An introduction to statistical modeling of extreme values.* 2001 (Springer Series in Statistics, Springer-Verlag)
[12] de Haan L and Ferreira A *Extreme value theory: an introduction.* 2006 (Springer Series in Operations Research and Financial Engineering, Springer)
[13] Fisher R and Tippett L Limiting forms of the frequency distribution of the largest or smallest members of a sample. 1928 *Proceedings of the Cambridge Philosophical Society* **24** 180-90
[14] Gnedenko B Sur la distribution limite du terme maximum d'unesérieealéatoire. 1943 *Annals of Maths.* **44** 423-53 (In French)
[15] Palutikof J, Brabson B, Lister D, Adcock S A review of methods to calculate extreme wind speeds. 1999 *Met. Appli.* **6** 119-32
[16] Brabson B and Palutikof J Tests of the generalized Pareto distribution for predicting extreme wind speeds. 2000 *J. of App. Met. and Clim.* **39**, 1627-40
[17] Coles S and Walshaw D Directional modelling of extreme wind speeds. 1994 *J. of the Roy. Stat. Soc. Series C (App. Stat.)* **43** 139-57
[18] Cook N *The designer's guide to wind loading of building structures. Part 1: Background, damage survey, wind data and structural classification.* 1985 (Building Research Establishment. Garston and Butterworths, London)
[19] Gusella V Estimation of extreme winds from short-term records. 1991 *J. of Struc. Engin.* **117** 375-90
[20] Taleb N *The black swan: The impact of the highly improbable fragility.* 2010 (New York, Random House) p 300
[21] Laherrère J and Sornette D Stretched exponential distributions in nature and economy: fat tails with characteristic scales. 1999 *The Eur. Phys. J. B* **2** 525-39
[22] Sornette D Dragon-Kings, Black Swans and the prediction of crises. 2009 *Int. J. Terr. Sci. and Engin.* **2** 1–18
[23] Volodin E, Diansky N, Gusev A Simulating present-day climate with the INMCM4.0 coupled model of the atmospheric and oceanic general circulations. 2010 *Izv. Atmos. Ocean. Phys.* **46** (4) 414–431

[24] Kislov A, Matveeva T An extreme value analysis of wind speed over the European and Siberian parts of Arctic region. 2016 *Atm. and Clim. Sci.* **6** 205-23

[25] Kislov A, Matveeva T, Platonov V Wind speed extremes in Arctic area. 2015 (In Russian) *Fund. and Appl. Clim.* 2015 **2** 63–80

[26] Böhm U, Kuecken M, Ahrens W, Block A, Hauffe D, Keuler K, Rockel B, Will A CLM – the climate version of LM: brief description and long-term applications. 2006 *COSMO Newsletters* **6** 225–35

[27] Rockel B, Will A, Hense A The regional climate model COSMO-CLM (CCLM). 2008 *Met. Zeit.* **17** **4** 347–48

# Monitoring of high-altitude terrestrial ecosystems in the Altai Mountains

**E E Timoshok, E N Timoshok, S A Nikolaeva, D A Savchuk, E O Filimonova, S N Skorokhodov, A Yu Bocharov**

Lab. of Dynamics and Stability of Ecosystems, Institute of Monitoring of Climatic and Ecological Systems of the Siberian Branch of RAS, 10/3 Prosp.Academichesky, Tomsk, 634055, Russia
E–mail: timoshokee@mail.ru

**Abstract**. The Aktru mountain glacier basin (the North-Chuya Ridge, Altai Mountains) is a region of highly important ecosystems. We have been performing a monitoring of the autotrophic component of the basin ecosystems for the last 16 years. The primary indicator species with the most clearly defined response to climatic changes are Siberian stone pine and Siberian larch with their individuals and populations. The ecosystem level of the monitoring includes that of old forests, ecotone ecosystems, and ecosystems on the new moraines. The old forests have remained stable for about 1000 years. The reasons for this stability are the long lifespan and the long generative stage of stone pine and larch, their ability to produce several growth forms, optimal ecological conditions of the basin for these species and high α-diversity of the old forests. The treeline has moved up by 100–200 m for the last 150 years and by 40–90 m for the last 40 years, mostly because of an invasion of stone pine to the ecotone. The primary successions on the moraines are also relatively stable, although at present only stone pine has been involved in the successions. No regeneration of larch has been observed for the last 16 years in the entire basin.

## 1. Introduction

The modern scientific community shows great interest in the studies of mountains under modern global climate changes. The reasons for this interest are numerous. These include: the assumption of early response of mountain ecosystems to climatic changes, high biodiversity of ecosystems and species in mountain regions caused by the existence of altitudinal belts, the importance of mountains as sources of water, mineral, and biological resources necessary for the population wellbeing, and many other reasons, such as the high cultural value of mountains [1,2].

Monitoring of different components of mountain environments as a whole system is one of the most important goals of modern science [3]. The efforts of some networks (GLOCHAMORE; GLORIA, MIREN) and numerous independent observatories are focused on various kinds of the monitoring. The research program of the GLOCHAMORE network is the widest and most diverse in comparison with the other networks. However, the monitoring of wildlife within the framework of the program is limited. An "optimal" level of monitoring, which is the widest of the three suggested levels of monitoring, includes only 7 indices used for the wildlife monitoring, namely: species composition, cover/abundance, vegetation type, moss layer cover, the presence and the quantity of dung (for indirect assessment of the abundance of grazing animals), the presence of some local species of invertebrates and the treeline altitude measured through satellite imagery deciphering [3]. The GLORIA and MIREN networks were designed for biospheric monitoring, but the only indices directly related to wildlife are the species composition and the abundance of species [4].

Our investigations are based on an assumption of a key role of the autotrophic component of ecosystems, an integrated approach, and an idea of multiple levels of wildlife organization. We select the autotrophic component of ecosystems (plant communities) as the main object of monitoring. The autotrophic component is directly affected by the climate and, at the same time, it affects the soil and animal population dynamics. The stability of the autotrophic component is necessary for the ecosystem's stability as a whole, and its change automatically changes the ecosystem as a whole. The essence of the applied integrated approach is simultaneous monitoring of a number of indices with the purpose of a comprehensive estimation of the changes in our object. The integrated approach allows us to obtain information that cannot be obtained by monitoring of individual parameters.

## 2. Study area

We have been monitoring high-altitude areas of the Altai Mountains for 16 years. We have a group of observatories for the monitoring of terrestrial ecosystems in 5 mountain areas. There are two observatories at the North-Chuya Ridge (Aktru, Korumdu): two at the South-Chuya Ridge (Akkol, Jazator) and one at the Katun Ridge (Akkem).

Our primary observatory is the Aktru mountain glacier basin. In this basin we perform regular annual studies of the ecosystems. The basin was selected because of its highly representative climatic, glaciological [5], and floristic peculiarities [6]. The basin has been used for glaciological (since 1952) and climatic (since 1956) monitoring by Tomsk State University scientists, and was chosen by the UNESCO Executive Committee as one of 8 glaciological and meteorological representative areas of the former USSR during the International Geophysical Year (1957–1958). We have been performing monitoring of the ecosystem in the basin since 1999.

The basin (50°05' N, 87°45' E, 2150–5040 m a.s.l.) is situated in the highest part of the North-Chuya Ridge, at the mountain junction Bish-Iirdu. The current relief of the basin is complex, deeply disjointed and typical for high mountains [7]. The slopes of the Aktru River valley are steep and asymmetrical. The valley glaciers of Maly Aktru, Levy Aktru, and Pravy Aktru and their moraines are located at the upper sources of the Aktru River. The new moraines have been deglaciated since the mid-19th century up to the present day.

The Aktru basin has severe high-mountain climate. The basin has a high solar radiation level (975–1045 W/ $m^2$), low temperatures (the mean annual temperature is –5,2°C, and the mean summer temperature is +8,7°C), significant precipitation (the mean annual precipitation is 563 mm), and high relative air humidity (about 67%) [8, 9]. The winds are usually south or south-west throughout the year, and the highest wind speed is in winter (up to 20–24 m/s). The precipitation is inhomogeneous, with the greatest precipitation amount in summer. The winter winds greatly affect the snow cover. Our winter observations indicate that some parts of the valley slopes (especially in the treeless areas) may lack snow, which is probably removed to the lower parts of the slope by the wind and due to gravity.

## 3. Materials and methods

The monitoring has been made at levels of the organism, population, and ecosystem. The main objects of the monitoring are old forests (2100–2300 m a.s.l.), phytocenoses of forest-tundra ecotone (2200–2500 m a.s.l.) and new moraines of the Maly Aktru glacier (2200–2300 m a.s.l.) at the ecosystem level, and indicator species of Siberian stone pine (*Pinus sibirica* Du Tour) and Siberian larch (*Larix sibirica* Ledeb.) at the organism and population levels. These species are of high importance for the ecosystem. Since they are dominant in the vegetation communities, they have a very long lifespan (up to 800 years) and clearly distinguishable generations. They are also responsive and sensitive to climate changes at the organism (radial increment and growth form) and population (tree age structure) levels.

These objects were studied on ecological and cenotic transects along the elevation gradient from 2100 to 2500 m a.s.l. from old forests on slopes of various orientations and steepness to high-altitude tundra and from these forests through the new moraines of the Maly Aktru glacier to its terminus. We studied 70 sample areas. We measured all trees, seedlings, and saplings (diameters and heights) and took cores from 500 sample trees. The cores were the basis of the age structure and ring width chronologies. The ring widths were measured by LINTAB with the software TSAP [10] with an accuracy of 0.01 mm. Individual ring series were dated using a combination of cross-correlation analysis [11] and graphic dating [12] in the TSAP [10] and COFECHA programs [11]. Then the series were generalized into chronologies by the standardization routine (minimization of the age trend and the autocorrelation component in the chronology) by the CRONOL program [11]. To find the correlation between the tree ring width chronologies and the climate variables, the monthly temperature and precipitation values of the Aktru weather station were used. Siberian stone pine and Siberian larch growth forms were identified by I. Serebryakova [13] and A. Chistyakova [14]. In all the sample areas, comprehensive phytosociological descriptions were made in Russia using a standard method [15]. The seedlings and saplings less than 50 years old of the Siberian stone pine and Siberian larch were called young individuals.

## 4. Results and discussion

The most interesting object of monitoring are the *old forests*, which are usually dominated (up to 80–90% of all the forest trees) by Siberian stone pine. These forests provide the best data on the contemporary and retrospective states of high-altitude forest ecosystems at the organism, population, and ecosystem levels.

Long-term changes in tree ring widths indicate general trends of increasing growth in the mid-18th and late 19th centuries. A general decrease in the growth of Siberian stone pine and Siberian larch was found for the late 17th and early 18th centuries, in the middle of the first half of the 19th century, and the mid-20th century. Major peaks of growth depression of Siberian stone pine were in 1544, 1599, 1630, 1662, 1706, 1798, and 1829, and of Siberian larch, in 1497, 1580, 1688, 1822, and 1961. The most prominent and long periods of increasing growth indexes took place in 1544–1575, 1704–1776, 1828–1902, and the last decade of the 20th century for Siberian stone pine and in 1499–1562, 1580–1616, 1689–1726, and 1824–1888 for Siberian larch. Major periods of decreasing growth of Siberian stone pine were identified in 1576–1600, 1683–1700, 1776–1826, and less significant ones, in 1902–1913. As for Siberian larch, the same periods occurred in 1562–1594, 1616–1689, 1800–1824, and 1890–1963. The amplitude of radial growth fluctuations (Figure 1a) in Siberian larch (which prefers arid habitats and deciduous) is higher than those in Siberian stone pine (which prefers humid habitats and evergreen), i.e. the curve of the radial growth is smoother for Siberian stone pine. The behavior of the radial growth curves is generally similar for both species.

The highest correlation between the radial growth and the climatic variables (according to the Aktru weather station data) was found for Siberian larch (the correlation coefficient with the May temperature is +0.55; with the June and July temperatures, +0.50, and with the June precipitation, 0.34). This significant correlation between the spring and summer temperatures and the radial growth in the Siberian larch allowed us to reconstruct these temperatures for the entire mountain glacier basin (Figure 1d). The variability of mean annual spring and summer temperatures varied from 4.7 to 9.7°C. Major periods of temperature increase were in the mid-16th, early 17th, mid-18th, and late 19th centuries. Major periods of temperature decrease were in the late 16th, late 17th, and early 19th centuries.

The age structure of the tree storey of old forests consists of three stone pine and larch age generations. The forests also contain single stone pine and larch trees from the previous generations; they are 550–600 or even 730 and 830 years old, they settled in 1150–1460 and have survived since the Medieval Warming, which ended in the 90s of the 15th century in the Altai Mountains [16]. The first generation of Siberian stone pine is the oldest one; it was formed in 1490–1630. The first generation of larch was formed in 1440–1580. The second generations of stone pine and larch were formed in 1660–1790 and 1620–1740, respectively, the third generations, in 1840–1960 and 1900–1950, respectively (Figure 1). The range between the oldest and youngest trees is 140 years for the first generation in both species, 120 years for the second generation in stone pine and 130 years in larch, and 120 years for the third generation in stone pine and 150 years in larch. The third generation is the youngest. It began to form at the end of the Little Ice Age (Figure 1c). The least number of larch trees that settled down between the generations was detected in the late 16th–early 17th centuries and in the late 18th century (1750–1780). The worst colonization of stone pine occurred in the mid-17th century and in the first half of the 19th century. These decreases in regeneration are clearly connected with intra-century coolings (Figure 1d) which occurred during a long-term decrease in the temperatures, i.e. in the Little Ice Age [16]. The times of formation of the stone pine and larch

IOP Conf. Series: Earth and Environmental Science **48** (2016) 012008 doi:10.1088/1755-1315/48/1/012008

generations are different and separated by 40–50 years (Figure 1b,c) due to different reactions of these species to decreases in air temperature during the coolings (Figure 1d).

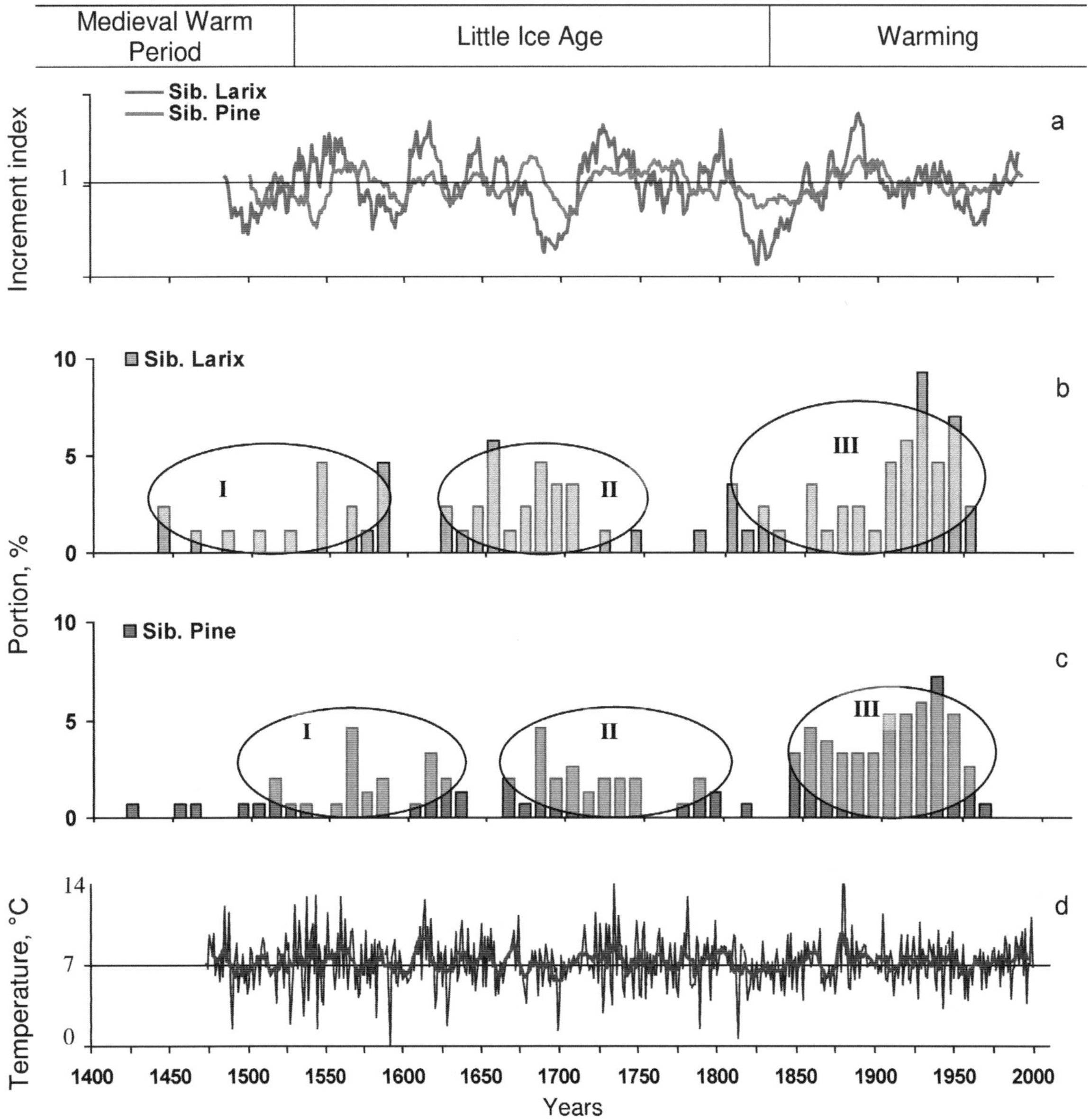


**Figure 1.** Radial increment with 7–year smoothing in Siberian larch and Siberian stone pine (a), age structure of the tree storey of old forests (b, c), and reconstructed annual and smoothed (7–year) mean spring-summer temperature (d) in Aktru mountain glacier basin (North-Chuya Ridge, Altai Mountains).

The modern, fourth, generation of Siberian stone pine in the old forests is represented by rare young individuals from 1–3 to 50 years of age. The same generation of Siberian larch consists of single individuals about 40–50 years of age.

Modern warming has been registered in the Altai region since the 1980s. During this warming, we observe an increase only in the Siberian stone pine regeneration, but no regeneration of the Siberian larch.

In the old forests studied, six growth forms (single-stemmed tree, several-stemmed tree, grove-forming tree, tree with a multi-storey root system, "living" windfallen tree, and "living" stump) were identified in Siberian stone pine. The single- and multi-top single-stemmed trees prevail (over 90% of the total) in the tree storey. Such Siberian stone pine trees can be more than 450–600 years old (Figure 2a–c) and can bear cones during 500 years [17]. Trees with a multi-storey root system are formed when the tree butts are periodically buried down by rock debris, "living" stumps (Figure 2d) are formed when the trunks of old trees break, a grove-forming tree is formed when lower branches establish, and a "living" windfallen tree (Figure 2e) appears when the trunks of old trees fall down and several living adventitious roots remain in the soil [18]. Individuals of the growth forms can live for 300–700 years and produce cones during many recent decades. High solar radiation, sufficient precipitation, and a relatively low wind speed are the factors contributing to these growth forms. Grove-forming trees, "living" windfallen trees, and "living" stumps are found only in old forests. Single-stemmed trees and several-stemmed trees are identified only in Siberian larch.

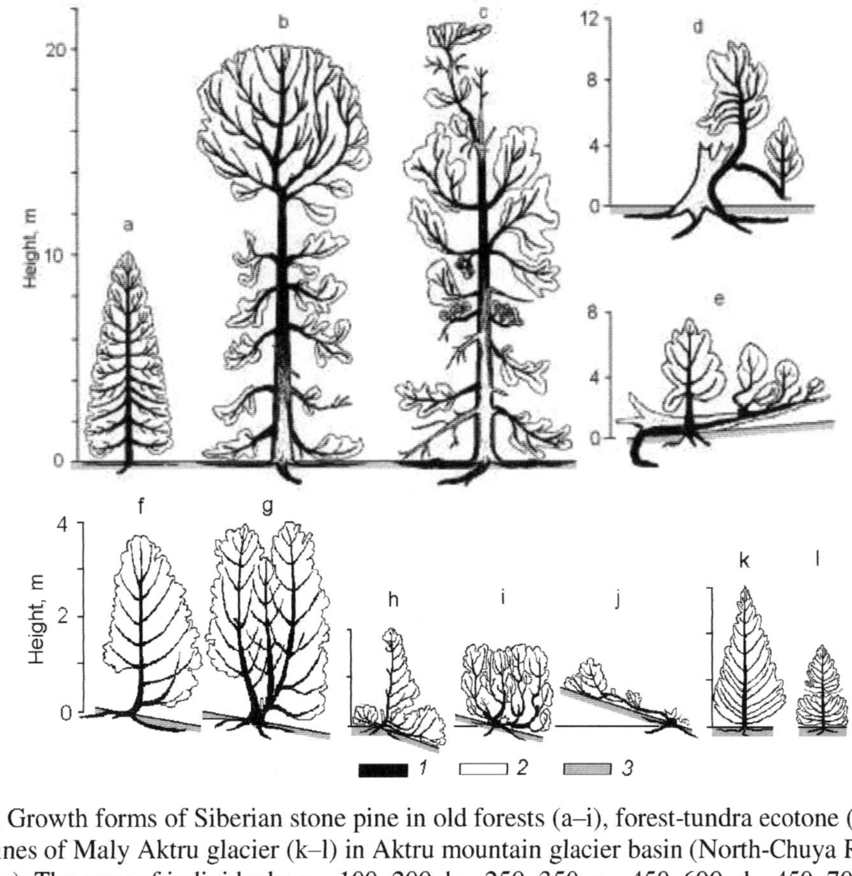


**Figure 2.** Growth forms of Siberian stone pine in old forests (a–i), forest-tundra ecotone (f–j) and on new moraines of Maly Aktru glacier (k–l) in Aktru mountain glacier basin (North-Chuya Ridge, Altai Mountains). The ages of individuals: a – 100–200, b – 250–350, c – 450–600, d – 450–700, e – 300–700, f, g – 50–125, h, i, j – 40–100, k – 50–100, l – 40–60 years. *1* –living and *2* –dead sections of trunks and branches; *3* –substrate surface.

Investigation of the shrub and herb storeys of old forests allowed us to determine high α–diversity of vessel plants: the density of species measured on sampling areas of 100 $m^2$ and consists of 36 species on average; the general species richness of old forests is 104 species, which is equal to 50% of the entire vessel plant flora of the stone pine forest of the entire Altai Mountains, thoroughly investigated by A.V. Kuminova [19]. Our long-time (16-year) monitoring of these storeys shows their significant stability under the impact of modern global changes.

The high stability of the old forests under the climate change impacts is mostly due to the stability of larch and stone pine populations. The ecological conditions of the North-Chuya Ridge are optimal for both species in precipitation, dampening, and richness of the soils [20], which contributes to diversity of growth forms of primary trees and high α-diversity of vessel plants of the shrub and herb storeys of these forests.

The basin also contains pronounced early response areas which are *forest-tundra ecotones* (2200–2500 m a.s.l) where the reaction of terrestrial ecosystems to global and regional climate changes is maximal. An invasion of stone pine and larch into the forest-tundra ecotones began after the start of a warming in the 60s of the19th century. This warming was followed by the strongest cooling in the last millennium, which occurred in the first half of the 19th century. This cooling greatly destroyed a previous generation of trees which existed earlier in the ecotone. The dead adult trees were found by M.V. Tronov [2 1] in 1936.

At present, the most abundant tree species of the ecotone is stone pine (60–80% of all trees), while the presence of larch is significantly lower (20–20%). Here the age structures of the tree populations are totally different from those of the old forest below the ecotone. There is only one adult generation of Siberian stone pine and Siberian larch in the ecotone. The age of the trees is 50–130 years (Figure 3). The generation began to form in the 60s of the 19th century and continued during the subsequent warming [16]. The trees of the second generation of stone pine are young and numerous, and the age of individuals is 4–50 years. The second larch generation is represented by individuals 36–49 years old. An invasion of stone pine to the lower and higher sections of the ecotone was simultaneous and occurred in 1860–1960. An invasion of larch to the lower section of the ecotone occurred in 1870–1960, and to the upper section, 20 years later (since 1900). A fast and effective invasion of stone pine began in 1920s and continues up to the present day. An invasion of larch was insignificant during the 20th century and stopped in the 1980s (Figure 3).

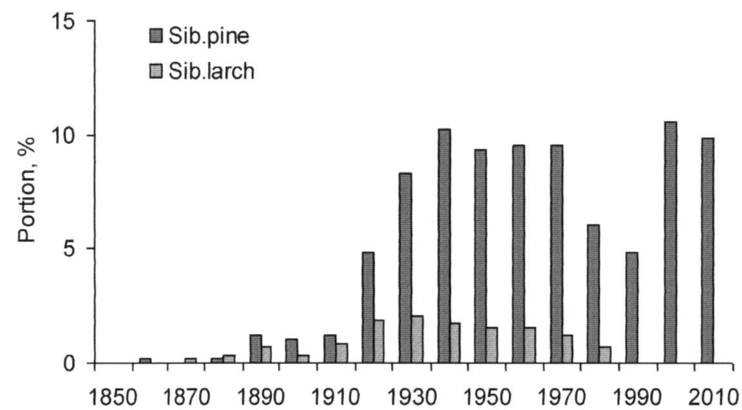


**Figure 3.** Age structure of the tree storey in forest-tundra ecotone of the Aktru mountain glacier basin (North-Chuiya Ridge, Altai Mountains).

The treeline moved up by 60–210 meters since the beginning of the warming in the 60s of the 19th century. It moved up by 40–120 meters (up to 2390 m a.s.l.) from the 1960s to mid-1970s and by 20–90 m more (the stone pine line to 2480 m, and the larch line to 2420 m a.s.l.) for the last 40 years. Our monitoring data indicate that the invasion of stone pine to the ecotone is continuing.

In the forest-tundra ecotone, six growth forms (single-stemmed tree, several-stemmed tree, multi-stemmed tree, skirted tree, shrub-like tree, and elfin wood) were identified in Siberian stone pine. Single- and multi-top single-stemmed and multi-stemmed trees prevail (70–80% of the total) in the tree storey. Their age is rarely more than 125 years. Such trees have often an asymmetric crown (Figure 2f–g). The skirted tree, the shrub-like tree, and the elfin wood (Figure 2h–j) are found only in the ecotone rather than in the old forest, they are 40–100 years old and do not produce cones. These forms are found throughout the ecotone, and the elfin wood is formed only in the upper section of the ecotone. They are formed when the tree crown and the trunk are damaged mechanically. Climatic (strong winter winds and thin snow covers) and geomorphological (rockfalls and avalanches) factors contribute to these growth forms, as in the other high-altitude areas [22, 23]. Three growth forms (single-stemmed tree, several-stemmed tree, and "living" stump) were identified in Siberian larch. The Siberian stone pine and Siberian larch individuals keep steadily their growth forms at least for 15 years, according to our monitoring data.

The monitoring of the shrub and herb–low shrub storeys of the ecotone plant communities allowed us to find high α-diversity of the vessel plants: the species density in the ecotone is about 41 species per 100 $m^2$, and the general species richness is 222 species. The species include about 36% of all periglacial flora of the Altai–Sayan mountain area [24]. Our 16–year long observations show high stability of the species composition and the structure of these storeys during the entire observation period.

Some *new moraines* of the valley Maly Aktru Glacier (2200–2300 m a.s.l.) are of special interest, since the developing ecosystems there may also provide valuable data on the early response of ecosystems to regional and global climate changes. The moraines are located in the upper forest belt of the North-Chuya Ridge, and a colonization of the moraines began in 60s of the 19th century.

A primary succession on the moraines has been continuing for the last 160 years. They include the formation of pioneer communities on a terrain aged 5–30 years, willow-moss communities on a terrain aged 30–90 years, and a young larch forest with a *Betula rotundifolia*-dominated shrub storey on a terrain aged more than 100 years. The tree storey of this forest is dominated by larch (more than 90%). There is only one adult generation of Siberian larch here. The age of this generation is 60–97 years. The stone pine trees are less numerous (less than 10% of all trees), the age of the trees of an individual adult generation is 50–70 years. A second generation is represented by stone pine trees aged 3–50 years and larch individuals 30–50 years old. Individuals of the second generation may be found even on younger terrains (the terrain age is 30–90 years). Individuals of stone pine may be found on a very young terrain deglaciated less than 30 years ago. These individuals are usually 5 years old and do not survive for a longer time. There are no young larch trees on the moraines: we found only stone pine trees.

On the new moraines, three growth forms (single-stemmed tree, several-stemmed tree, and skirted tree) were identified in Siberian stone pine and Siberian larch. Single-stemmed (Figure 2k) and several-stemmed trees prevail (over 90% of total), their age is less than 100 years. A skirted tree (Figure 2l) is found on the new moraines rather than in the old forest. The reason for its appearance is a combined impact of the winter wind and snow.

The monitoring of the species composition of the primary communities forming on the moraines allowed us to find significant α-diversity. The species density varies from 17 species per 100 $m^2$ for the terrain aged 5–30 years to 29 species per 100 $m^2$ for the moraines more than 100 years old. The general species richness is 146 species of the vessel plants. These species are represented by 25% of the periglacial vessel plant flora of the Altai–Sayan mountain area investigated by N.V. Revyakina [24].

## 5. Conclusions

The monitoring of the modern state of the high-altitude terrestrial ecosystem of the Altai Mountains and supporting retrospective investigations have-been performed at the organism, population, and ecosystem levels of the organization of wildlife for the last 16 years. The primary monitoring areas were old forests (2100–2300 m a.s.l.), forest-tundra ecotones (2200–2500 m a.s.l.), and new moraines with a recently deglaciated terrain (2200–2300 m a.s.l.). The most sensitive indicators of climate changes are Siberian stone pine and Siberian larch available in all monitoring areas.

A retrospective estimation of the radial growth in the old forest allowed finding the intra-century periods of its significant increases and decreases. It also allowed us to reconstruct the fluctuations of spring and summer temperatures for 500 years. The formation of generations of stone pine and larch in these forests was cyclical, and the formation periods were interrupted by 40–50 years of long cooling.

These forest ecosystems demonstrate high stability. They were formed about 1000 years ago in the periglacial area and survived during 3 prolonged strong cooling periods, including the strongest early 19th century cooling in the last millennium. The cause of this stability is the long lifespan of stone pine and larch trees (up to 800 years old), the long generative stage of their life cycle (almost 500 years), the high diversity of the growth forms, and high biodiversity of the lower (shrub and herb) storeys of the ecosystems.

An invasion of stone pine and larch into the ecotone and the formation of the first generation began at the start of a warming in the 60s of the 19th century following a very strong cooling in the first half of the 19th century. The invasion of stone pine into the upper and lower sections of the ecotone was simultaneous. An invasion of larch began 10 years later in the lower section and 20 years later in the upper section of the ecotone. The later stage of stone pine invasion into the ecotone was considerable and fast. It began in the 1920s and has been continuing up to the present day. The invasion of larch was insignificant during the entire 20th century and stopped in the 1980s. Only stone pine continues its invasion now.

The investigation of the early response of the terrestrial ecosystems to the climate warming and monitoring of the ecosystem changes on the Maly Aktru Glacier moraines allowed us to describe the modern primary succession with the successive formation of pioneer communities, willow-moss communities, and young forests. A colonization of the moraines by Siberian larch began about 100 years ago, and a colonization by Siberian stone pine, 30 years later. An effective colonization of larch occurred only during the first 50 years after the beginning of the glacier recession. The colonization of stone pine was weak during the entire period after the beginning of the glacier recession. Only stone pine is colonizing the moraines now, during the modern fast climate warming, which has been detected in the Altai region since the mid-1970s. Stone pine is successfully colonizing the terrain moraines whose age is more than 30 years, and the young individuals found on the younger terrain quickly die. No larch colonization has been observed on the moraines during the 16–year monitoring.

The monitoring of the high-mountain terrestrial ecosystems and the populations of indicator species was supplemented by dendroclimatic investigations and allowed us to understand their dynamics during the previous climate changes and the climatic response in the past during the Little Ice Age period and at present during the modern warming period.

The Aktru mountain glacier basin is a key area of high value, which can serve as a model for investigations of the reaction of highly stable high-altitude ecosystems to climatic changes. The collected data of monitoring are valuable for preserving the biodiversity of the Altai Mountains.

**Acknowledgments**
The study was supported by the Russian Foundation of Basic Research (grants no. 02–05–65178, 03–05–06037, 04–05–79057, 13–05–00762) and the Federal Special Purpose Program "Integration" (project no. E0053).

**References**
[1] Greenwood G 2014 *The Global Fair and Workshop on Long Term Observatories on Mountain Socio-Ecological Systems – Summary of Summaries* (Bern: MRI Publishing)
[2] *Mountains for Europe's Future: A Strategic Research Agenda* 2016 (Bern: MRI Publishing)
[3] Grabherr G *et al* 2005 Long-term environmental observations in mountain biosphere reserves:

recommendations from the EU GLOCHAMORE project, long-term environmental observations in mountain biosphere *Mountain Research and Development* **25** 376–85
[4] Pauli H *et al.* 2015 *Gloria Field Manual: Standard Multi-Summit Approach, Supplementary Methods and Extra Approaches. 5th Edition* (Vienna:GLORIA-Coordination)
[5] Tronov M V 1966 *Ledniki i Klimat* (Leningrad: Hydrometizdat)
[6] Timoshok E E and Skorokhodov S N 2005 Species, cenotic and ecosystem diversity in headwater basin Aktru (Severo-Chuisky Range, Central Altai) *Global Changes in Mountain Regions* ed. M F Price (Duncow: Sapiens Publ.) pp 144–45
[7] Dushkin M A 1967 Geomorfologicheskiy ocherk lednikovogo basseyna Aktru *Glatsiologiya Altaya* vol. 5 (Tomsk: Tomsk State University Publishing) pp 42–65
[8] Galakhov V P *et al.* 1987 *Ledniki Aktru (Altai)* (Leningrad: Gidrometeoizdat)
[9] Sevast'yanov V V 1998 *Klimat Vysokogornykh Raionov Altaya i Sayan* (Tomsk: Tomsk State University Publishing)
[10] Rinn F 1996 *TSAP v3.5. Computer Program for Tree-Ring Analysis and Presentation* (Heidelberg: Frank Rinn Distribution)
[11] Holmes R L 1983 Computer-assisted quality control in tree-ring data and measurement *Tree-Ring Bull.* **43** 69–78
[12] Douglass A E 1919 *Climatic Cycles and Tree Growth: A Study of the Annual Rings of Trees in Relation to Climate and Solar Activity* vol. 1 (Washington: Carnegie Inst.)
[13] Serebryakov I G 1962 *Ekologicheskaya Morfologiya Rasteniy* (Moscow: Vysshaya shkola)
[14] Chistyakova A A 1988 Zhisnennye formy i ih spektry kak pokazately sostoyaniya vida v tsenose (na primere shirokolistvennyh derev'ev) *Bulletin of Moscow Society of Naturalists. Biological Series* (Moscow: Moscow State University Press) **93(6)** 93–105
[15] Korchagin A A 1976 Stroenie rastitelnyh soobschestv *Polevaya Geobotanica* vol. 5 (Leningrad: Nauka) pp 7–130
[16] Okishev P A, Adamenko M F and Narozhniy Yu K 2000 Dinamika lednikov i klimata v gorah Yuzhnoy Sibiri *Regionalny monitoring atmosphery. Chast 4. Prirodno-klimaticheskie izmeneniya* (Tomsk) pp 16–99
[17] Timoshok E E, Nilolaeva S A, Skorokhodov S N, Savchuk D A and Bocharov A Yu 2009 Osobennosty ontogeneticheskih sostoyaniy generativnogo perioda kedra sibirskogo v lesah tsentralnogo Altaya *Rastitelnye Resursy* **45(3)** 204–11
[18] Nikolaeva S N and Savchuk D A 2013 Morphological forms of Siberian Stone Pine trees (*Pinus sibirica* Du Tour) in high altitudinal forests of Severo-Chuisky Range 1. Morphological aspect *Tomsk State University Journal of Biology* **22** 101–14
[19] Kuminova A V 1960 *The Vegetation of the Altai* (Novosibirsk: Nauka)
[20] Timoshok E E, Timoshok E N and Skorokhodov S N 2014 Ecology of Siberian stone pine (*Pinus sibirica* Du Tour) and Siberian larch (*Larix sibirica* Ledeb.) in the Altai mountain glacial basins *Russian Journal of Ecology* **45** 194–200
[21] Tronov M V 1939 Ledniki gornogo uzla Bish-Iirdu *Proc. Tomsk. St. Univ.* **95** 10–27
[22] Gorchakovskiy P L and Shiyatov S G 1985 Phitoindikatsiya Usloviy Sredy i Prirodnyh Protsessov v Vysokogoryah (Moscow: Nauka)
[23] Schweingruber F 1996 *Tree Rings and Environment. Dendroecology* (Berne–Stuttgart–Vienna: Paul Haupt Publ.)
[24] Revyakina N V 1996 *Sovremennaya Prilednikovaya Flora Altae-Sayanskoi Gornoy Oblasti (Proishozhdenie, Stanovlenie, Adaptatsii)* (Barnaul: NIIGP Publ.)

# Response of middle-taiga permafrost landscapes of Central Siberia to global warming in the late 20th and early 21st centuries

**Alexey A Medvedkov**

Department of World Physical Geography and Geoecology, Faculty of Geography, Lomonosov Moscow State University, Leninskiye Gory, 1, Moscow 119991, Russia,

Email: a-medvedkov@bk.ru

**Abstract**. In this paper, regional features of a climatogenic response of the middle-taiga permafrost landscapes of Central Siberia, as well as corresponding transformations of the exodynamic processes, are considered. Lithological-geomorphologic and landscape-geocryological data are analyzed with large amounts of actual data and results of monitoring surveys. Specific features of an ecotone localization of middle-taiga permafrost landscapes and their typical physiognomic characteristics are described. A comprehensive investigation of representative key sites makes it possible to discover the response of different types of permafrost landscapes to regional climate warming. A rapid increase in the active layer depth, slower creep, transformations of the moving kurums, intensive solifluction, and a local replacement of solifluction by landslides-earthflows are revealed within ecotone landscapes of the cryolithozone.

## 1. Introduction

The periphery of the permafrost zone in the Eurasian North is a planetary ecotone with patchy landscapes, especially in areas with complex geological and geomorphologic structure. The study focuses on the Central Siberian segment of the area embracing the left- and right-bank areas of Yenisei, as well as the lower reaches of the Podkamennaya Tunguska River.

A vivid example of diversity of the ecotones landscape is the western part of the Middle-Siberian High Plateau, in the lower reaches of the Podkamennaya Tunguska River. It is important to know how permafrost conditions and landscapes will change in the given part of the ecotone in connection with the global warming. To answer this question, it is necessary to 1) identify permafrost and non-permafrost landscapes and 2) understand whether the stability of permafrost varies with the landscape features. The outcome of this research will provide a basis for developing a scenario for the evolution of morphogenetic systems and landscapes in the ecotone area under study in relation to global warming.

In the monograph of Gorshkov et al. [1] significant amount of geohistorical, paleo-permafrost, geomorphological, and landscape data on this region and research topic was analyzed. According to the results the authors note that solifluction is a very important attribute of the whole system of cryogenic exodynamic processes that worked in unstable conditions during interglacial Pleistocene periods. It is confirmed by the data of stratigraphic location of fossil permafrost traces in the section of Late Pleistocene deposits of the old periglacial zone of the Yenisei Siberia. Sudden climate warming at the beginning of each interglacial period pressed the permafrost deeper, as it also occurs nowadays, and activated the solifluction process. After analyzing numerous monitoring data of the cryolitozone, Konischev [2] notes that immediate response of permafrost to warming climate occurs in autonomous ecosystems (or those close to them). In the study area we observed similar regularities, but with a much more complicated response [3].

The international program of circumpolar monitoring of active layer (CALM) and the international project on the thermal state of permafrost (TSP) are running for more than 15 years. However, Central Siberia is still not covered by geocryological research within the framework of these programs. Therefore, our ongoing landscape-geocryological investigations at representative sites are of special importance. We conducted more than 1000 field observations and measurements. As a result of these observations, we recorded the analyzed changes and their indicators shown in this paper.

## 2. Materials and methods

The field studies (2007–2016) included a landscape–geoenvironmental inventory of natural components at transects through the main types of forest, swamp, and burnt catenas in the left- and right-bank parts of the Yenisei Siberia, boreal taiga subzone (Figure 1). The route profiling was carried out along transects across and along the geomorphological catena, taking into account conjugated relief surfaces, most diverse in morpholithogenic, soil–geographic, and landscape respects. Detailed landscape descriptions along the route were accompanied by the determination of the upper boundary of permafrost with the use of a probe; diagnostics of permafrost and nonpermafrost processes was carried out.

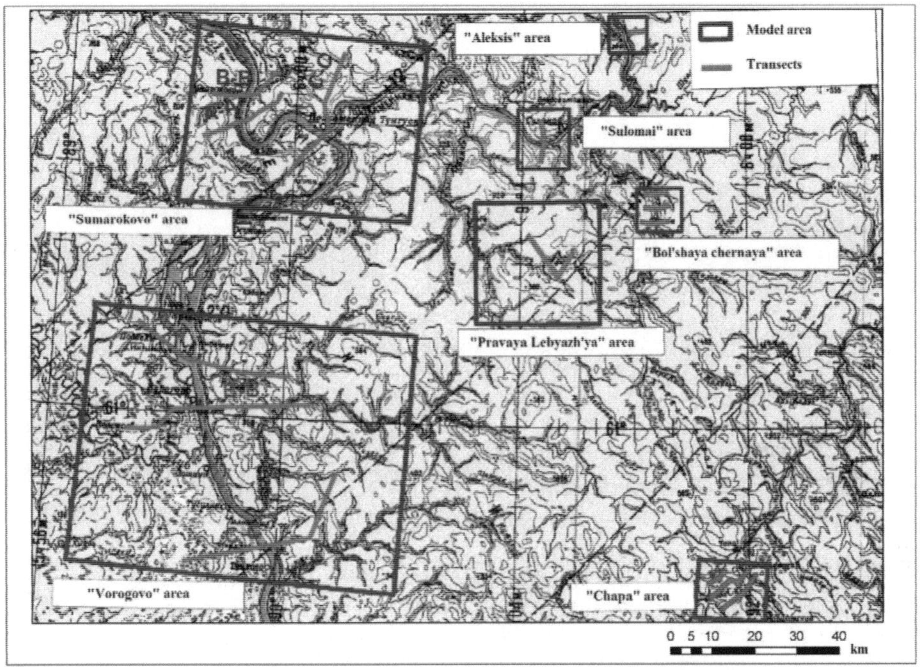


**Figure 1.** Map of key study areas in Yenisei Siberia.

In the course of route surveys special attention was given to field studies of the landscape structure of key areas, taking into account the specifics of surface deposits and the character of their geomorphological differentiation; landscape indication of permafrost natural–territorial complexes, and ranking landscape complexes into permafrost and nonpermafrost. A comprehensive analysis of this data made it possible to identify and outline (at the level of complex stows) landscape complexes most vulnerable to various external impacts (including climatic ones). In addition, the field studies included monitoring of different types of permafrost and nonpermafrost stows in terms of the specifics of their response to climate warming, which were compared with the data obtained in the course of field studies in the 1970s, 1980s, 1990s, and 2000s.

Special attention was given to observations of specific informative natural objects and phenomena.

— *Stone streams.* In the zones of occurrence of active stone streams, the specifics of local overgrowing by moss–lichen cover were studied, and a number of unstable boulders was determined. For low-activity, closed stone streams, the degree and character of forestation and the specifics of stand timber, in particular, the share of inclined trees, were determined. Succession changes in the taiga vegetation within the relatively stable covers and stone-stream–defluction (block structures with loam colmatage) were examined;

— *Soliflual deposits.* Signs of weakening or degeneration of solifluction were assessed, including the presence of inclined trees with vertical tops, as well as the overgrowth of solifluction hollows–disruptions in the above-soil cover. Cases of replacement of solifluction slopes and glacises* on river banks by local landslides–creeps because of a recession of permafrost roof and an increase in the instability of bank massifs were detected;

— *Relic permafrost relief forms.* The cryogenic morphosculpture of the Upper Pleistocene forms of solifluction origin, the signs of thermokarst, and the features of their landscape occurrence were analyzed. Analogous modern formations were sought for and studied;

— *Undergrowth of parvifoliate species (birch, aspen).* The causes of explosive rise of the undergrowth of parvifoliate species in dark and light coniferous forests were studied and analyzed, in particular, in the habitats that had not been typical of such before.

## 3. Results and discussions

### *3.1. Climate changes*

Climate warming in the Central Siberian region has been recorded for more than 30 years since the early 1980s (Figure 2). Here, the mean annual temperature increased by 1–2°C and more as compared with the previous, colder period from the late 1940s to the late 1970s. Since the early 1980s, the winter became warmer, and the spring and autumn, longer.

Years with shorter summer also occurred. In the cold 1974, the minimal mean monthly air temperature in January at Bor Settlement was found to be –35.1°C, while that in the warm 1995 was –17.8°C. At the same time, the difference between the mean July temperatures in the same years did not exceed 1.7°C. The increase in the mean annual temperature in warm years is due to both the higher temperatures in the cold season and the longer warm season.

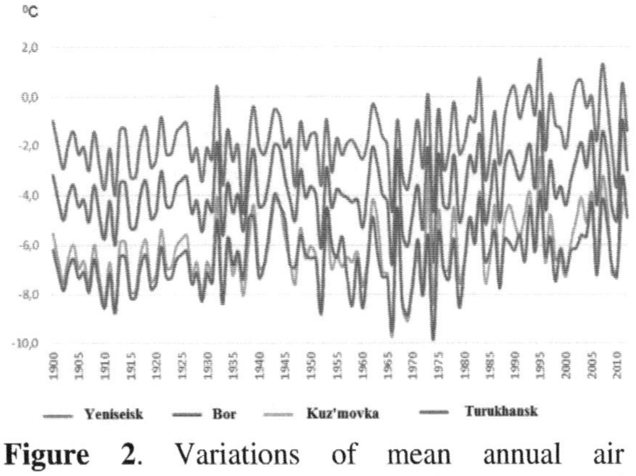


**Figure 2.** Variations of mean annual air temperature for 1900–2012 according to data of hydrometeorological stations of Krasnoyarsk krai in Turukhansk V., Bor Settl., Kuz'movka V., and Yeniseisk T.

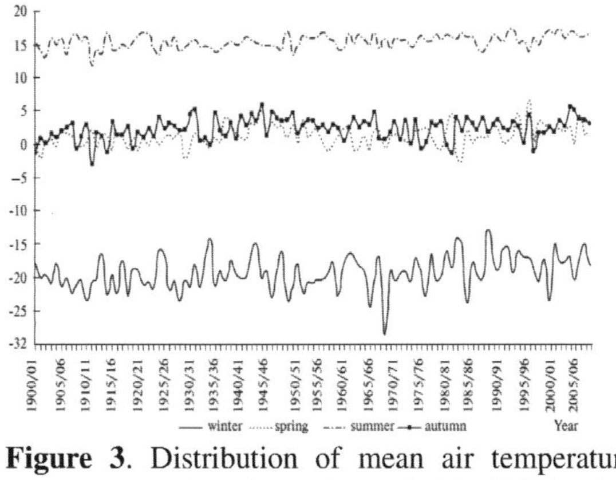


**Figure 3.** Distribution of mean air temperature over seasons based on data of ZGMOS in Bor Settl. (Turukhansk district, Krasnoyarsk krai) for 1900–2009.

Analysis of Figure 3 suggests that the amplitude of variations of winter temperatures is much larger than those in other seasons, a feature that determines their leading role in the annual temperature. It can be clearly seen that the last 20 years of the 20th century can be distinguished by their winter maximums. Note that a concentration of high winter temperatures falls into the period from the late 1980s to the mid-1990s.

### *3.2. Characteristic features of permafrost landscapes and their tolerance to climate warming*

The permafrost landscapes are very sensitive to climate warming, though to a different degree. Of importance in this situation are their indication and the identification of occurrence specifics. Permafrost processes in boreal landscapes often manifest themselves under favorable substrate conditions, irrespective of the heat supply to relief elements. Thus, the lithological–geomorphological and landscape–geographic analyses showed that permafrost stows in the lower reaches of the Podkamennaya Tunguska occur in the surface deposits of aleurite–pelite composition [4]. This demonstrates the priority role of the lithological factor. Therefore, the southern boundary of Upper Pleistocene glacial and lacustrine–glacial deposits (with predominant aleurites and clays containing inclusions of sand, gravel, and sporadic boulders) which cross the area under study is an important landscape boundary too. Previously, landscapes in the lower reaches of the Podkamennaya Tunguska were studied to verify and adjust geological mapping results [5]. The first attempt to differentiate between permafrost and non-permafrost landscapes was made by Gorshkov, Karrasch, Paramonov [4], and Fedorov [6].

In genetically different surface deposits there are significant varieties in seasonal thaw depth. Estimated seasonal thawing of peat soils occurs to a depth of 0.3–0.5 m, silts and clays thaw to depths between 0.8–1.0 m in shadowy and cold conditions, and to 2.0 and even 2.5 m where the ground gets more heat in summer. Coarse grained sand and pebble rock thaw to a depth of 3–4 m. Permafrost varies in thickness from several meters to 25–30 m [7].

Our studies show that permafrost rocks manifest themselves in different stows, which feature dystrophy, suppression, and species poverty, appreciable disturbances of the daily surface, specific soil profile with gley signs, higher watering of soils and surface deposits, and manifestation of cryogenic processes and phenomena.

Thus, the key characteristics of *permafrost* landscapes are:

- *the character of vegetation cover:* suppressed thin taiga with appreciable tilt of trees, sparse stands and dwarf birch thickets, moss–brush vegetation with the predominance of sphagnum mosses (*Sphagnopsida*);
- *specific soils:* cryogenic peat–gley, alluvial–swamp with signs of gley and cryogenic skeletal (stone-stream soils);
- *relief microforms:* solifluction windows–breaks (holes–breaks) (Figure 4), solifluction ledges, frost mounds, thermokarst subsidence;
- *relief mesoforms and their outlines:* solifluction swells (foot plumes); watered and low-mobility stone-streams; hanging bogs; ditch-type channels of creeks and rivers with land-slide signs;
- *the state of surface deposits:* viscous–flow consistency of disperse soils; active and watered stone streams (individual boulders are unstable);
- *the composition of surface deposits:* moraine clays, loams, lacustrine–glacial and alluvial clays, aleurites, frozen peat, solifluction disperse deposits, highly watered boulders of stone-stream fields;
- *higher watering of a stow due to groundwater outcrops.*

**Figure 4.** Solifluction hole–break in ground vegetation filled with cold water on the glacis–floodplain (1–2 m in diameter and 0.5–0.6 m deep in the peat-and-plant layer, usually found every 80–100 m). Water is a sign of the presence of a permafrost aquiclude near the daily surface. (Photo by the author).

*Nonpermafrost* landscapes show
- the presence of full-scale erect tree vegetation;
- high occurrence of hard rocks under a thin mantle of surface deposits;
- relatively good drainage.

*3.3. Sustainability of landscapes to the climate warming*

As to the stability of permafrost landscapes, some differentiation of occurrence toward the south was found. If some type of permafrost stows occurs furthest to the south, then, other conditions being the same, it can be regarded as the most tolerant to climate warming [8]. In this context, three types of stows can be identified in the region under study by their tolerance to climate warming (Table 1):

- *low-stability stows of glacises with good water supply and slopes with open stone flows and summit plains (*at the end of the summers of 2006-2007 the active layer thickness was no more than 1.2–1.5 m in the mentioned landscapes. However, during the same period of 2016 it was about 1.6–1.9 m and sometimes 1.9 m and deep);
- *stable stows of slopes, glacises, and glacis–floodplains*** (Under the forest vegetation within these stows we registered a permafrost table at a depth of 0.5–0.8 m. We suppose that in these stows the active layer thickness increased not more than by 0.2–0.3 m during the last 9 years. More pronounced changes occurred in areas located near thermokarst ponds and where forests or dense bush groves are absent);
- *highly stable-polyfactor permafrost stows on slopes with poor heat supply: "hanging" bogs**** *of heat-deficient slopes (*the location of a permafrost table is approximately at the former level: about 0.4–0.5 m under holes and 0.7–0.8 m under hummocks) and *forested floodplains of large rivers* (in these stows the permafrost table depth is about 0.2–0.3 m under holes and reaches 0.4–0.5 m under hummocks).

Thus, the most stable permafrost stows are hanging bogs of cold slopes and forested floodplains of major rivers (Table 1).

**Table 1.** Dynamics of the top of permafrost in different types of landscapes.

| Types of permafrost landscapes | Depth of the top of permafrost in 2007 | Depth of the top of permafrost in 2016 |
|---|---|---|
| well heat provided summit plains | 1.2–1.5 m | 1.6–1.9 m |
| heat-deficient slopes, glacises | 0.5–0.8 m | 0.7–1.0 m |
| "hanging" bogs | 0.4–0.7 m | 0.4–0.7 m |
| forested floodplains of large rivers | 0.3–0.4 m | 0.3–0.4 m |

* Glacises are accumulative surfaces with slopes from a few degrees to a few tens of minutes, constituted by decerption, solifluction and deluvium, forming foot plumes or occupying valley beds.

** Glacis-floodplain includes areas of glacises near the river bed, which are inundated during spring flood, under meadow-bog or draft birch vegetation on alluvial-gley and peaty-permafrost soils

*** "Hanging" bogs ("hanging" peat) are mostly located on steep near-river slopes with low heat supply. A stone flow underlies a thin peat layer in each of such bogs. The permafrost peatery shows high segregation ice content. The most impressive are ice inclusions with a size of a walnut. The permafrost layer is overlain by ground vegetation, consisting of mosses, lichens, and dwarf shrubs with the abundance of ledum and dwarf arctic birch. However, under the conditions of global warming, the permafrost in the area can aggradate because of the longer vegetation period, resulting in a thicker peat–vegetation layer, which serves as a heat insulator.

*3.4. Climate-driven response of the most vulnerable permafrost landscapes*

In the lower reaches of the Podkamennaya Tunguska, the least stable permafrost is confined to the landscapes of low (200–250 m) summit plains and gentle slopes, composed of clays, loams with inclusions of individual boulders, and aleurolites-fine-sand deposits of glacial complex. Common in such places are bog complexes with low thin (crown density of 40–50%) cedar–fir taiga with birches and larches on a peaty–gley cryogenic soil with a most–brush and, sometimes, lichen ground cover.

At the depth of zero annual temperature variations, the cryogenic soil is cooled to 1°C and even to fractions of degree below zero. The active layer in loams and clays is the layer of seasonal freezing and thawing, which is 0.8–1.0 m in thickness [7], [9]. Since the mid-1990s until now, the top of high-temperature (-0.1 – -1.0°C) permafrost shifted 1.5–2 m and more down, demonstrating the process of permafrost degradation because of the warming of its strata.

The beginning of permafrost degradation had an immediate effect on the appearance of permafrost landscapes, i.e., the disappearance of water in solifluction holes-breaks; fallen trees can be seen with the entire spreading-root assemblage torn out of the earth, because they fall more easily under wind impact, and their root base lost the support of the solid frozen substrate. The result was an increase in the occurrence of relief forms of biogenic origin (the so-called iskors) in the landscapes of the insular-permafrost subzone. Local replacement of solifluction by landslide processes was observed in the zones of river erosion intensification. Because of draining, some plants, primarily horsetail, lose their green color, making the ground vegetation cover yellowish–golden. Small thaw lakes with collapsed and dead forest stand appeared.

In the basements of stone streams and detritus, called kurums, the permafrost retreats downward faster than in the areas of permafrost thin forest. Studies in the key areas show that bald-peak ice melted, small depressions formed, and cold subsurface creeks disappeared in the kurums, primarily on slopes with southern and western aspects. The kurums overgrow with lichens, dwarf shrubs, and individual trees (Figure 5). In the low reaches of the Podkamennaya Tunguska River, the kurums not covered by forest, even on slopes with poor heat supply, on valley beds, on slopes and summit plains with elevations not exceeding 400 m have lost their bald-peak ice. Warm kurums are common in the northern part of the Yenisei Range and the western part of the Central Siberian Plateau up to the Nizhnyaya Tunguska River in the zone of Severnyi kamen' trappean massif, which is as close as 70 km south of the polar circle. The permafrost on the beds of deep valleys and on steep slopes of northern and eastern aspects with low heat supply is more tolerant to warming. The permafrost roof is still stable on the upper plateau of the western Central Siberia with absolute elevations of 550–700 m.

An analysis of the collected material allows us to suggest that a protection response in the form of negative feedbacks forms in the kurums of the boreal taiga subzone. At the first stage, bald-peak ice melts and kurums overgrow with mosses, lichens, and tree species. At gradual overgrowth of a kurum, fine soil accumulates, filling gaps between boulders and forming a more impressive soil profile A1–AB–C of brown taiga skeletal soil [3], [8]. The accumulation of fine earth and the rate of kurum overgrowth were found to increase in the zones of concentration of black crustose lichens. At further accumulation of fine earth, the brown taiga skeletal soil transforms into a peaty brown taiga soil, the share of disperse deposits and the thickness of peat–vegetation layer increase, and the watering of kurum and its isolation from the lower air layer grow. The formation of hanging peat bogs at the second stage is accompanied by aggradation of permafrost because of a growth of icy rocks.

**Figure 5.** A kurum overgrown by lichen and young birches on the right bank in the middle reaches of Bol'shaya Chernaya river. The abundant growth of moss and underbrush vegetation on the kurums leads to the formation of peaty soil layers on slopes which are frozen at depths of 0.3–0.5 m. (The left tributary of Podkamennaya Tunguska river). (Photo by the author).

Our studies show that in 1998–2016 in Central Siberia only high-temperature permafrost experienced degradation: kurums in all subzones of the permafrost zone, though not including the upper peneplain, and the high-temperature permafrost in pelite rocks, only in insular-permafrost zone and further southward [3].

## 4. Conclusions

A large body of data was obtained suggesting the beginning of permafrost degradation in the middle boreal subzone in the Yenisei basin (not less than 70% of the area it occupies within the ecotone).

**Figure 6.** Solifluction landslide in the middle reaches of the Coligny river (right tributary of the Podkamennaya Tunguska river). Landslides - slides are typical for slopes of about 10 to 30 degrees. Such sites have a significant increase in the active layer capacity. (Photo by the author)

The most significant processes of response to climate warming in the boreal landscapes of the Middle Yenisei area include an increase in the thickness of the active permafrost layer and intensification of solifluction (Figure 6);

cases of local replacement of solifluction by land-slide motion of soils in the zones of active river erosion;

anomalously frequent falls of trees that have spreading-type root systems in areas where clay soils are waterlogged, have viscoplastic consistency, and are 1.5 or more meters in thickness;

better drainage on summit plains and adjacent gentle slopes;

higher mobility of large boulders on kurums because of bald-peak ice melting, as well as the number and area of overgrowth spots of mosses and lichens;

depletion of subsurface streams under the kurum boulder cover;

intensification of thermokarst processes within swampy areas.

The ecotone under consideration is located in the western part of the Central Siberian Plateau and the eastern margin of the West Siberian Plain. The response to climate warming causes changes in the permafrost-landscape conditions, exodynamic processes, and the production of natural systems, and affects the life support of the local population [10]. The understanding of the processes taking place in a permafrost ecotone is of importance for assessing the changes in modern boreal landscapes in the Northern Eurasia and the state of its natural-environmental resources in the future.

**Acknowledgements**

The study was supported by the Russian Foundation for Basic Research (project 16-35-00327- mol_a) and the Council for Grants of the RF President (project MK - 7614.2015.5)

**References**

[1] Gorshkov S P, Vadenberghe J, Alexeev B A *et al* 2003 *Climate, Permafrost and Landscapes in the Middle-Yenisei Region* (Moscow: Moscow State University Press) p 96

[2] Konishchev V N 2009 Response of permafrost to the climate warming *Vestn. Mosc. St. Univer. Geogr.* 4 10-20

[3] Medvedkov A A 2016 *Mid-Taiga Geosystems of the Yenisei Siberia under climate change* (Moscow: MAKS Press) p 144

[4] Gorshkov S P, Karrasch H, Paramonov A V 1998 Geomorphologic Indication of Permafrost and Non-Permafrost Landscapes in the middle taiga of Central Siberia *Geomorph.* 4 55-60

[5] Astakhov V I, Gerasimov L M, Eromenko V Y *et al* 1978 *Complex of Remote-Sensing Methods Used in Geological Mapping of Taiga Areas* (Leningrad: Nedra) p 247

[6] Fedorov A N 1991 *Permafrost landscape of Yakutia. Methods of Indication and Problems of Mapping* (Yakutsk: MPI SB RAS) p 140

[7] Leshchikov F N, Shats M M 1983 *Permafrost in the Southern Central Siberia* (Novosibirsk: Nauka Press) p 169

[8] Medvedkov A A 2015 Geoenvironmental response of the Yenisei Siberia mid taiga landscapes to global warming during late XX–early XXI centuries *Water Res.* **42** (7) 922-931

[9] Gorshkov S P 2008 Environmental shock in the Central Siberia: causes and consequences *Geogr. Perv. Sent.* 4 10-16

[10] Medvedkov A A 2013 The Kets ethnos and its "feeding landscape": ecological-geographical and socio-ecological problems under globalization and changing climate *Geogr., Envir., Sust.* **6** (3) 108-118

# Detailed reconstruction of the functional state of the Central Yamal khasyrey as a response to local conditions and regional climate changes in the late Holocene

**Yu I Preis[1], G V Simonova[1] and E A Slagoda[2]**

[1]Institute of Monitoring of Climatic and Ecological Systems SB RAS, 634055, Tomsk, 10/3, Academichesky ave. Phone: +7(3822) 492265, Fax: +7(3822)491950, e-mail: preisyui@rambler.ru

[2] Earth Cryosphere Institute SB RAS, 625026, Tyumen, 86, Malygina str. Phone: +7 (345-2) 688-730, e-mail: eslagoda@ikz.ru

**Abstract.** In this article, data of reconstruction of the dynamics of plant communities, water regimes, geocryological conditions, and peat accumulation of the khasyrey of Central Yamal Peninsula located in the tundra of Western Siberia by traditional paleoecological methods on the basis of detailed studies of the botanical composition, degree of decomposition, density, ash content and $\delta^{13}C$ in the cellulose of peat cut pulp age about 1300 cal. BP are considered. It has been shown that the khasyrey has a sensitive response on century and decadal scales to climate change and the lake regime level. Moreover, the stages of khasyrey functional state are in a good agreement with similar data on mires of the forest zone of Western Siberia and the solar activity periods.

## 1. Introduction

Predictions of the natural landscape changes made on the basis of high-resolution paleoclimatic and paleoecological reconstructions are topical in light of current climate changes. The forecasts for the most sensitive natural landscapes of arctic and subarctic regions are of special interest. At the present time the processes of permafrost degradation have become active. It creates difficulties in the development of these regions and the threat of mass ejection of $CO_2$ and $CH_4$ contained under the permafrost. Long-term sets of data of high-resolution paleoecological reconstructions based on highly sensitive complex of indicators are needed to develop the forecasts. The information about detailed paleoecological studies of mire functional state of Western Siberia tundra is absent.

Studies of permafrost including stratigraphy, dynamics and genesis of palsa are actively carried out around the world [1]. Determination of the optimal environmental conditions for the development of cryogenic processes in soils remains relevant to this day. The palsa in the tundra and forest tundra of Western Siberia are confined mainly to the drained lake basins – khasyreis. The khasyreis are one of the main landscapes of these areas. Periodic filling and draining of lakes cause frequent changes of water regimes and geocryological conditions of khasyreis. The development of khasyreis as well as the permafrost mires occurs cyclically [1, 2]. However the extent of the climate impact on the change of khasyreis functional state in subzones of the permafrost zone is different and isn't known.

The characteristics of tundra zone Holocene paleoclimate are not clarified until the end. The characteristics of mean annual temperature and precipitation amount are practically absent. The high-resolution data of summer temperature have been reconstructed by dendrochronological method for the south of the Yamal Peninsula [3].

The goal of current work is to carry out a reconstruction of paleophytocoenocis, water regimes and cryogenic conditions of khasyrey of Central Yamal as a response to changes of the level of the lake waters and regional climate of the late Holocene according to the detailed data of complex studies of peat deposits.

## 2. Objects and methods of investigation

The study area is located in the central Yamal Peninsula in the continuous permafrost zone with temperatures from -5 to -7 °C [4] on the border of the typical and southern tundra. In winter, temperatures drop to -59 °C. Summer is short and moderately cool. In July, the temperature may rise up to +30 °C. Indicators of the mean annual temperature (from -1.0 to -9.5 °C (1924-2009)) and mean annual precipitation (315-903 mm (1966-2010)) considerably vary [5].

The khasyrey is located on the III-rd terrace of the Central Yamal near Sokhonto Lake (69°08'57,17" N, 70°15'57,67" E). It has a diameter of 300-350 m. Low center of khasyrey is busy by paludification depression and lake with depth of 2-2.5 m and with erosional stream. The rest of the khasyrey is above a decrease on 2.5 m and busy by polygonal-flat-palsa mire with flat cannels and convex polygons of 3-5 m. *Betula nana, Ledum decumbens, Vaccinium vitis-ideae, Rubus chamaemorus, Eriophorum scheichzeri, Carex rariflora* grow in polygons. *Sphagnum majus, S. balticum, S. squarrosum* and *Polytrichum juniperinum, Dicranum angustum, Oncophorus wahlenbergii, Aulacomnium palustre* form a rarefied moss cover on the polygons.

Peat monolith with size of 17.5x20x80 cm from the edge of the khasyrey has been selected and divided into samples of 3.5-0.5cm increments. Research of botanical composition of peat macrofossil by microscopic method, research of decomposition degree (*R*) by centrifugation method, of ash content (*A*) and dry bulk density of peat (P) and its organic matter (OM) have been carried out. The isotopic composition of carbon $\delta^{13}C$ in peat cellulose was determined on the equipment, consisting of a DELTA V Advantage isotope ratio mass spectrometer and Flash EA 2000 elemental analyzer (TomTsKP SB RAS) at the Institute of Monitoring of Climatic and Ecological Systems, SB RAS (Tomsk, Russia). Radiocarbon dating has been made by liquid scintillation method using a radiometer-spectrometer Quantulus 1220 (TomTsKP). The calculation of calibrated age was performed on the program CALIB 7.0.4. [6] using the IntCal13 and MARINE13 data set [7] (Table 1). The comparison of $\delta^{13}C$ in peat cellulose with the reconstruction of summer temperatures in Yamal has been made to revision of obtained dates [3].

**Table 1**. Radiocarbon dates and calibrated ages of peat samples from Khasyrey section.

| Depth (cm) | Age ($^{14}$C year B.P.) | Material | Laboratory code | Calibrated age (year B.P.) | |
|---|---|---|---|---|---|
| | | | | 2δ range | median value |
| 13.5–16 | 113 ± 0.8% | Peat | IMCES–14C805 | 118–65 | 91.5 |
| 26.5–28 | 286 ± 67 | Peat | IMCES–14C789 | 501–267 | 384 |
| 35.5–39 | 639 ± 86 | Peat | IMCES–14C847 | 709–514 | 611.5 |
| 43.5–45.5 | 761 ± 51 | Peat | IMCES–14C785 | 786–651 | 718.5 |
| 70–72 | 1156 ± 122 | Peat | IMCES–14C861 | 1299–899 | 1099 |
| 79–80 | 1377 ± 75 | Peat | IMCES–14C782 | 1415–1172 | 1293.5 |

The calculations of the vertical growth rate (VGR) of peat (mm per year), of the dry peat accumulation rate (PAR), its organic matter (OMAR) and mineral matter (MMAR) ($g \cdot m^{-2}$ per year) are given. The reconstruction of paleophytocenoesis by species composition of plant remains of peat has been performed. The calculations of steps of humidification (*SHd*) [8] and index of humidity (*IHd*) of paleoecotops [9] have been made by standard methods of ecological scales by peat botanical composition for quantitative estimation of the hydrologic regime change of the paleoecotopes. The values of *SHd* and *IHd* have been assigned to some subarctic plants, which is absent in [8, 9]. The belonging of these species to certain ecological groups in relation to the humidification conditions [10, 11] was taken into account. Characteristics of peat properties as additional indicators of hydrological regime were applied. Secondary changes of peat properties have been determined by the absence of an agreed change of *R, P, SHd, SHd* and VGR along depth of peat deposit. The reconstruction of permafrost conditions was performed by the paleosuccessions [1] and peat properties [12]. Paleoprocesses of frost-heave in peat deposit were diagnosed by presence of remains of plant species, which are characteristic for a transition of mire surface from the subaqueous to subaerial condition. The beginning of permafrost degradation was investigated by the appearance of remains of more hydrophilic plants in the peat. The abundance of remains of diatoms seaweed is accepted as an indicator of the presence of thermokarst pool in the past. During wet and/or warm climate periods symptoms of drying of khasyrey surface are evidences about lake drainage.

Climatic conditions for khasyrey development stage have been determined by climate reconstruction data of the forest zone of Western Siberia [13–15], by data of average summer temperatures of the South of the Yamal Peninsula [3] and paleoecotops (by use $\delta^{13}C$). The paleoclimatic curves [13–15] have been digitized and calibrated.

Revealed the zonal features of the dynamics and clarify paleoclimate characteristics of khasyreya development stages by using data on the dynamics of the marshes of the forest zone of Western Siberia [16–19] and the solar activity [20].

## 3. Results and discussion

The khasyrey peat deposit has a depth of 80 cm (Figure 1) and consists of numerous thin layers of transitional peats, that are differed sharply from each other by botanic composition и properties (*R* – 5–35 %, *A* – 6.2–43.1 %, *P* OM – 25–240 g dm$^{-3}$).

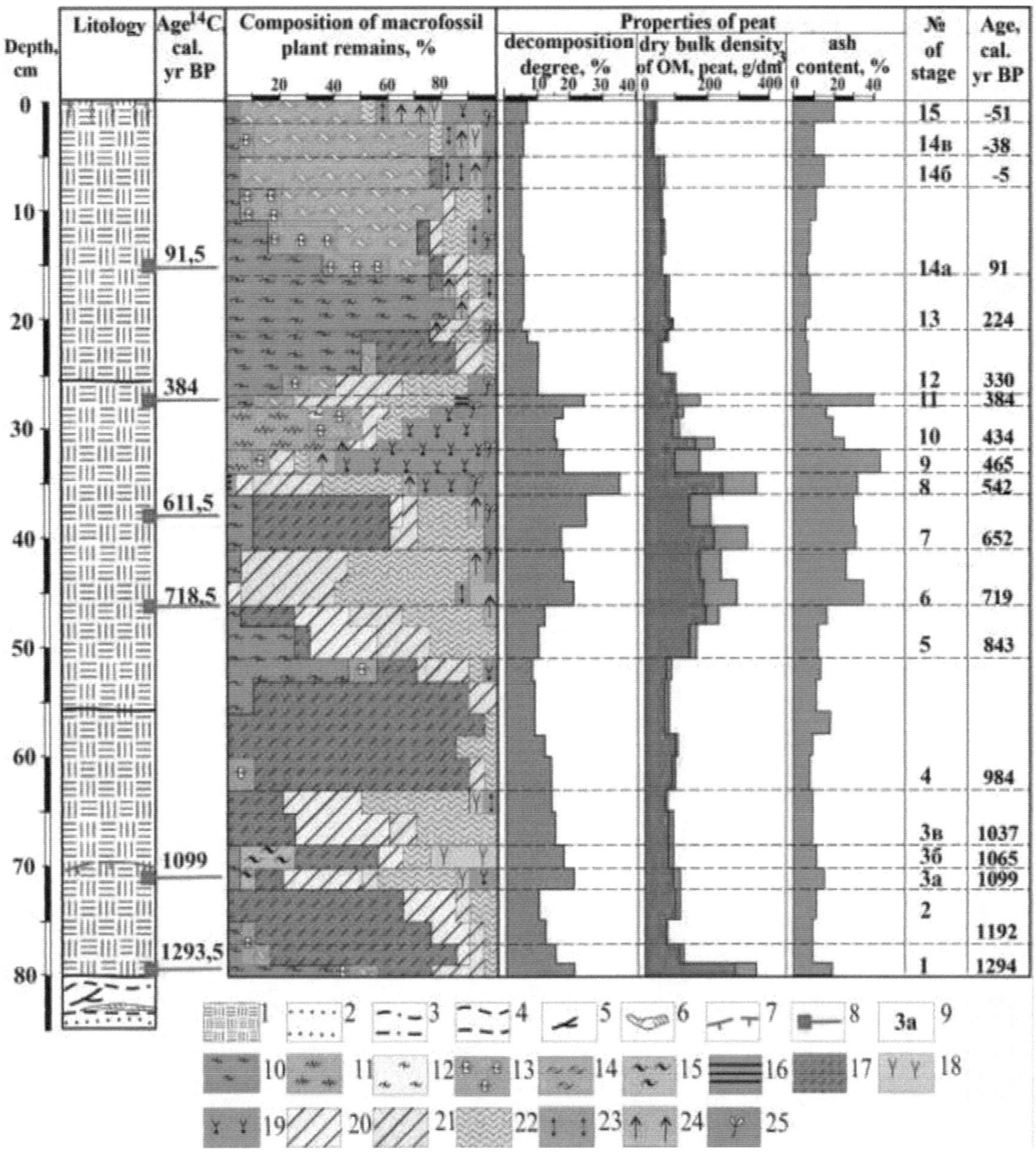


**Figure 1.** Lithology, botanical composition and properties of peat of Khasyrey cutting. Symbols: 1) peat; 2) sands; 3) laminated sandy loam; 4) laminated loam; 5) peat lenses of moss; 6)introduced by water plant remains; 7) the upper limit of the frozen peat (28.08.2014); 8) sampling locations on radiocarbon dating; 9) borders of khasyrey formation stages; plant remains in the peat: 10) *Sphagnum squarrosum*, 11) *S. lenense*, 12) *S. angustifolium*, 13) *S. balticum*, 14) *S. majus*, 15) *S. platyphyllum*, 16) *S. riparium*, 17) *Warnstorfia fluitans*, *Drepanocladus sp.*, *Calliergon sp.*, 18) *Aulacomnium palustre*, *Meesia triquetra*, 19) *Polytrichum juniperinum*, *Dicranum angustum*, *Oncophorus walhenbergii*, 20) *Carex rariflora*, 21) *C. rotundata*, 22) *Eriophorum sp.*, 23) *Betula nana*, 24) *Ericaceae*, 25) *Rubus chamaemorus*.

Three layers differ markedly in their properties. The upper layer (0–26.5 см) has been formed by Spagnum peat with low indexes of properties ($R_{av}$ – 6 %, $A_{av}$ – 10.2 %, $P_{av}$ OM – 65 g dm$^{-3}$), the middle layer (26.5–50.5 см) has been formed by grass, grass-moss and shrub-moss peats with higher indexes of properties ($R_{av}$ – 16 %, $A_{av}$ – 27.6 %, $P_{cp}$ – 215 г/дм$^3$), bottom layer (50.5-80 см) has been formed by Hypnum and grass-Hypnum with lower indexes of properties ($R_{av}$ – 13 %, $A_{av}$ – 12.1 %, $P_{av}$ – 111 g dm$^{-3}$). The most frequent abrupt changes of botanical composition and properties of peat are characteristic of the middle layer. The peat extracted from the depth 0.7 m was frozen (Figure 1). Sandy loams, loamy sands with plant remains and silty sands lie below the peat.

Chronology of peat accumulation is based on 6 calibrated radiocarbon dates. Using satellite imagery Landsat [Google Earth http://www.google.com/earth/explore] from different years it has been revealed that the last wet stage of the khasyrey ended between 2003 and 2008. Age–depth and age–cumulative mass models have a convex shape (Figure 2a), which is typical for mires of continental regions in North America [21] and climatogene type mires of forest zone of Western Siberia [16-19].

a)

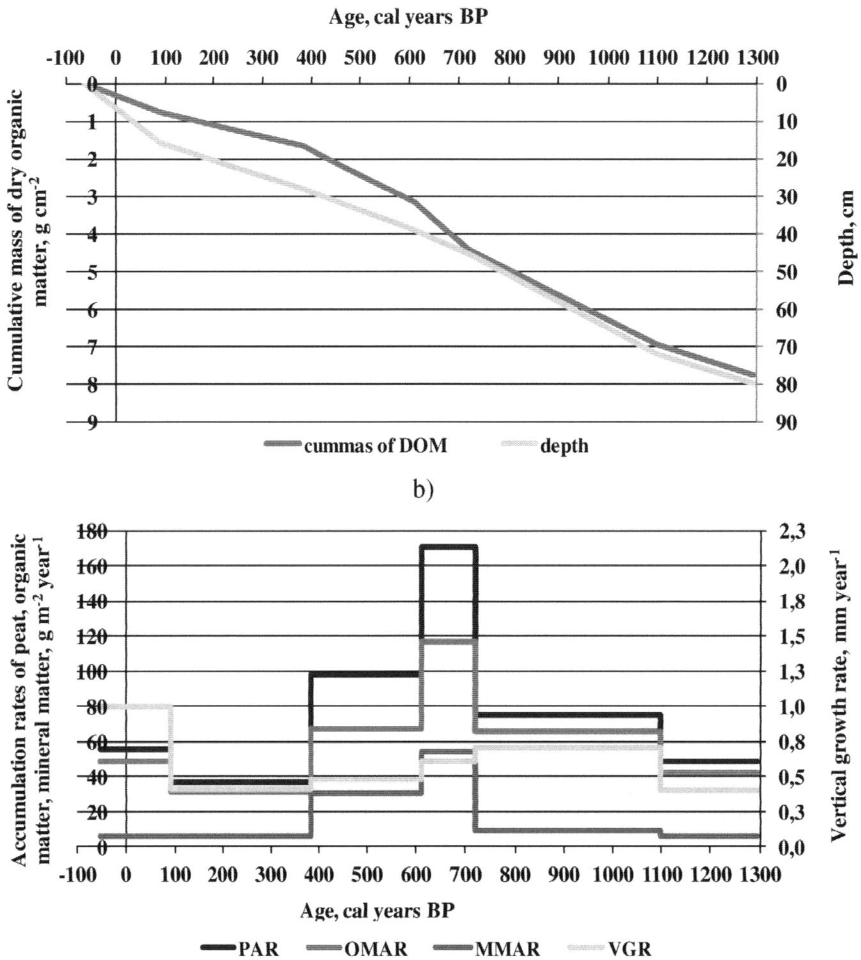


**Figure 2**. Relationship between: a) age and peat depth, cumulative mass of peat organic matter, b) age and vertical growth rate, peat, organic matter, mineral matter accumulation rate for the Khasyrey peat section.

This form (Figure 2a) is caused by differences of primary productivity of paleophytocoenosis, water paleoregimes and paleoconditions of water-mineral nutrition and, consequently, by differences of vertical growth rate and accumulation of peat OM in different periods of formation of khasyrey

(Figure 2b). However these models do not have a plateau due to the absence or short duration of peat accumulation interruptions.

The vertical growth rate (VGR) is 0.59 mm year$^{-1}$. In different layers it varies from 0.4 to 1.0 mm per year. The long-term accumulation rate of OM peat (OMAR) is 57.2 g·m$^{-2}$ per year and varies mainly from 31.1 to 116.6 g·m$^{-2}$ per year in different layers. High values of the dry peat accumulation rate (PAR) in layers are caused by high ash content of the peat. High ash content of the peat and, respectively, rate of mineral matter accumulation (OMAR) (54.2 and 30.8 g·m$^{-2}$ per year) are characteristic for middle layer (26.5-45.5 cm, 719-330 cal. BP).

$\delta^{13}C$ varies in layers from -28.7 to -23.9 ‰. One can clearly see from the Figure 3 that extremes and trends of δ13C curve are consistent with extremes and trends of summer temperature curve (T summer) [3] during the period from 5 to 1135 cal. BP. The difference of ages of their Extreme does not exceed 60 years. Different ages of $\delta^{13}C$ and Tsummer extremes (1153 and 1240 cal. BP) in the bottom layer of peat deposits (Figure 3) are probably due to some rejuvenation of $^{14}C$ radiocarbon date.

Negative signal of $\delta^{13}C$ during the modern warming (since 1988) evidences about significant anthropogenic influence on $\delta^{13}C$ in the atmosphere. Extremes and trends of δ13C curve and solar activity (SA) are consistent during period from 5 to 719 cal. BP (Figure 3). Negative signals of $\delta^{13}C$ are characteristic for the minima of Wolf, Sporer, Maunder and Dalton of the LIA [20], positive ones - for warming between these minima. The maximum positive signal of $\delta^{13}C$ (952-791 cal. BP) is due to the influence of high summer temperatures during a period of low solar activity.

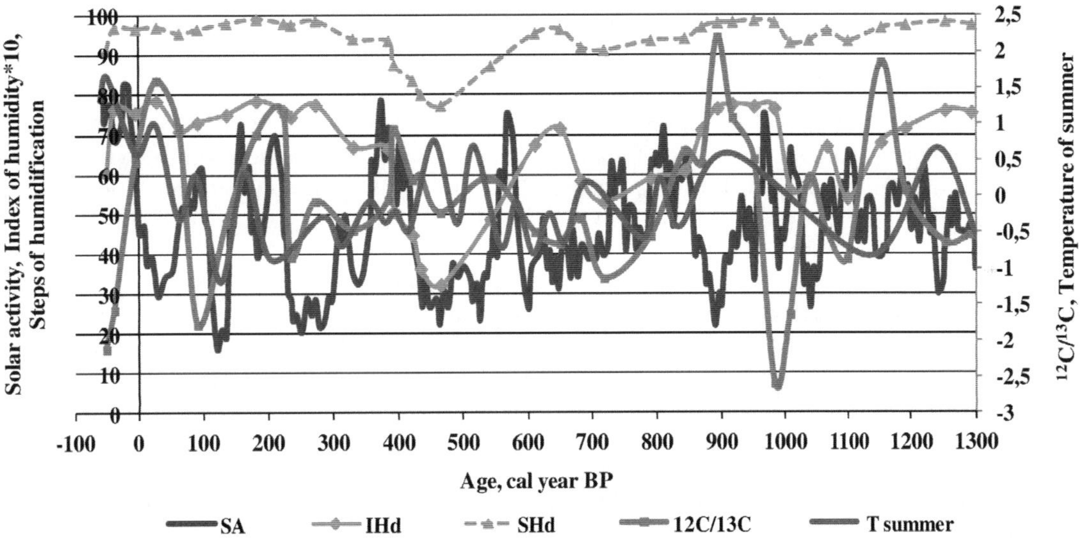


**Figure 3**. Paleogidrological regime dynamics and $\delta^{13}C$ in the cellulose of peat of khasyrey cutting, paleoclimatic curves of summer temperatures on the Yamal Peninsula [3] and curves of the solar activity [20] (summer temperature data are presented as the mean deviation from the windows of 100 years after smoothing, solar activity data are smoothed windows of 22 years).

Paleoclimatic curves of average annual temperature and annual precipitation for the forest zone of Western Siberia [13-15] have insufficiently high resolution and contradictory data for some time periods. This is due to the small number of radiocarbon dates and errors of peat layers ages calculation due to a lack of data on their bulk density. Therefore, we had to accept the extremes of different curves that coincided with *SHd* and *IHd* extremes to analyze the impact of climate on the khasyrey dynamics.

Main peat founders were subhydrophytic and hydrophytic grass: *Carex rotundata*, *C. rariflora*, *Eriophorum scheichzeri*, aerohydrophytic and subhydrophytic mosses: *Warnstorfia fluitans*, *Sphagnum squarrosum* and *S. majus*. Change of the water regime have impacted mainly the change of the ratio of the abundance of these grasses and mosses, as well as on the presence of plant-mesophytes. Seven large phases of khasyrey development and fifteen stage-change with six sub-stages have been identified as a result of reconstruction of phytocenoses, water regimes and permafrost conditions of khasyrey (Figure 1). Stages and sub-stages have intraspecific (13-77 years) and rarely a century (93-140 years) scale.

I phase (1294-1099 cal. BP) - corresponds to an overgrowing of lake by floating in a seasonal freezing conditions. Lying lake sediments below likely were not in the frozen state. The phase has ended by overflow and drainage of the lake due to thermal erosion of its coast in a sharp century warming. 1 stage (from 1294 cal. BP) - greatly drowned moss community from *Sphagnum squarrosum* and *Calliergon stramineum* in conditions of the end of dry cooling. Stage 2 (from1192 cal. BP) - the most drowned Hypnum moss community from *Warnstorfia fluitans*, which has postponed the peat with minimal *R*. The climate humidity has increased. At the end of the stage the abundance of *Carex rariflora* has increased.

II phase, 3 stage (1099-984 cal. BP) corresponds to a mean flooded sedge-cotton grass-Hypnum moss plant community. The decrease of mire water and lake level took place under conditions of dry cooling [15] and the most cold summers (according to the $\delta^{13}C$ analysis). A severe desiccation occurred at 3a and 3b substages: about 1099 (a sharp increase in *R*, *A* and presence of *Oncophorus walhenbergii*), and 1008 cal. BP (traces of fire and the presence of dwarf birch). It is likely that a well-drained conditions of khasyreya is due to the formation of low palsa. High palsa were formed in the forest zone of 1123-1010 cal. BP [17, 18]. More flooded moss community with *Warnstorfia fluitans, Sphagnum platyphillum* и *Meesia triquetra* have been formed on 3b sub-stage in the conditions of increased summer temperatures (according to the $\delta^{13}C$ analysis).

III phase, 4 stage (984-843 cal. BP) corresponds to a strongly flooded *Hypnum* moss floating with *Warnstorfia fluitans.* A significant increase of water table is due to the wet warming [13-15]. At the end of stage (since 869 cal BP) - more drained sedge-moss plant community due to the lake drainage. The beginning of wet period of 997-860 cal. BP have been dated on the mires of the forest zone [16, 17, 19]. This stage corresponds to the Medieval Maximum, which have manifested differently in different regions, including ~ 1050-650 cal. BP in the Polar Urals [22] and 1100-800 cal. BP on the eastern Taimyr and Putoranah [23].

IV phase (843-434 cal. BP) corresponds to progressive draining of the lake drainage, and cryogenic processes in peat and underlying mineral deposits in a predominantly dry cooling. Succession of plants and significant increase of *R*, *P* and *A* of peat evidence it. A high ash content of peat indicates on the arrival of aeolian sand from watershed_areas without vegetation in the summer, and on the diluvium with the spring melt waters. The phase is divided into 5 stages.

Fifth stage (since 843 cal. BP) corresponds to a mean flooded sedge-cotton grass-moss plant community with *Warnstorfia fluitans u Sphagnum squarrosum*. The decrease of water table have been strengthened by lake drainage under conditions of climate change from wet warm to the dry cold one.

Sixth stage (since 719 cal. BP) corresponds to a low flooded sedge-cotton grass plant community with *Carex rotundata.* Mosses disappear. *Betula nana, Ericaceae* and *Rubus chamaemorus* appear. *R*, *P* and *A* of peat increase. The climate was dry cold [13-15] with low summer temperatures (according to the $\delta^{13}C$ analysis). The freezing and the short-term formation of palsa have probably happened. The end of this phase coincides with the start of the Wolf minimum (670-610 cal. BP).

Seventh stage (since 652 cal. BP) corresponds to more flooded sedge-cotton moss community with *Drepanocladus aduncus, Warnstorfia fluitans u Sphagnum squarrosum.* An increase of water content under conditions of a short wet warming [13] and increase of summer temperature (according to the $\delta^{13}C$ analysis) took place. An increase of water content (559 cal. BP) has been also identified in middle taiga mire [16]. Perhaps it was warming between the Wolf and Sporer Minimum.

Eighth stage (since 542 cal. BP) correspond to well-drained sedge-cotton grass community with *Betula nana, Rubus chamaemorus* and *Polytrichum yuniperinum*. Significant surface drying occurred in the conditions of cooling and climate aridity [13-15]. The formation of permafrost and frost-heave of the whole area have begun. A high flooding of preceding stage contributed to this.

Ninth stage (since 465 cal. BP) corresponds to maximum drained community with *Betula nana, Rubus chamaemorus*, *Polytrichum yuniperinum* и *Dicranum angustum.* Icy veins and polygonal mires were formed under conditions of cooling extremum. In the Arctic (from the Urals to the Taimyr Peninsula) 470 cal. BP period was adopted as the year of the beginning of the LIA [24]. The formation of permafrost and poligonal mire coincides with the Sporer Minimum (530-410 cal. BP), which is one of the most continuous and cold sub-periods of the LIA [25]. The growth of continental climate and high summer temperatures [3] favored the cryogenic process. However, the summer temperatures of khasyrey ecotope remained low (according to the $\delta^{13}C$ analysis) due to the presence of permafrost. 615-465 cal. BP was a period of active formation of high palsa in the mires of the forest zone [16-19].

V phase (434-330 cal. BP) correspond to a permafrost degradation, formation and overgrowth of thermokarst pool, firstly under decrease of climate continentality (summer temperatures decrease according to [3] and the δ13C isotopic analysis), then - under decrease of humid warming [13-15].

Tenth stage (since 434 cal. BP) corresponds to more flooded moss community with *Rubus chamaemorus*, cottongrass, *Carex rotundata* and the soil cover of the *Sphagnum lenense* with impurity of mesophitic and hydrophytic mosses. The depth of seasonal thawing of peat deposits have increased. First the height of the polygons decreased and then thermokarst lake was formed. The sudden disappearance of *S. lenense* and high abundance of diatoms peat algae at the depth of 28.5 cm confirm this. Most likely, the filling of the khasyrey lake by water by degrading of permafrost of surrounding upland have begun at the end of this stage.

Eleventh stage (since 384 cal. BP) corresponds to less watered sedge-cotton grass community that was formed at overgrown lakes. The beginning of increase of watering and overgrown of thermokarst lakes (362-304 cal. BP) has been dated in the mires of the forest zone [18, 19].

VI phase (since 330 cal. BP) correspond to a flooding of the basin, a functioning of the lake (depth of slightly more than 2 m), a formation of strongly watered moss and Sphagnum moss floating mostly under conditions of humid warming and cooling [26]. At this stage, the upper limit of permafrost (to a depth of 0.5 m) has decreased. A top of the ice veins have thawed partly and flat-palsa mounds have formed on the place of landfills as a result of thermokarst and erosion. A peat accumulation in thawed state under conditions of relatively high levels of mire water and seasonal freezing have happened.

Tewlfth stage (since 330 cal. BP) corresponds to a strong flooded community of sedge-cotton grass-sphagnum moss and moss floating with *S. squarrosum* and *Warnstorfia fluitans* under conditions of a humid periods of climate: warm and then cold (Maunder Minimum, 305-235 cal. BP).

Thirteenth stage (since 224 cal. BP) corresponds to a strong flooded mesotrophic sphagnum community with *S. Squarrosum* under conditions of warm and cold humid periods of climate (Dalton Minimum, 160-115 cal. BP).

Fourteenth stage (since 91 cal. BP) corresponds to a strong flooded cotton grass-sphagnum moss plant community with *Betula nana* and oligotrophic *Sphagnum majus* mainly under conditions of directed warming later than the LIA.Probably the reduction of the seasonal thawing depth at the end of Dalton Minimum helped to the transition to a more oligotrophic stage of development.

VII phase, 15 stage (since –51 cal. BP to 2014) corresponds to a poorly flooded dwarf birch-shrub-moss plant community with *S. Majus* and impurity of mesophytic and hydromesophytic hypnum moss, under conditions of a sharp decrease of khasyrey surface humidity as result of the lake drainage.

## 4. Conclusions

Thus, a detailed investigation and radiocarbon dating of khasyrey peat deposits have made it possible to obtain data of the dynamics of intra century and century scale of its functional state. A sensitive response of investigated khasyrey on the climate changes of the upper Holocene has been identified. Seven large stages and fifteen stages with six sub-steps of intra century and century scale of changes of the functional state of khasyrey were revealed for 1350- year period of khasyrey formation. Paleoecotopes were most humid in 952-843 and 273- –51 cal. BP, during humid periods of warming and cooling. Paleoecotopes were maximum drained in 1099–1065 and 1037-984, 719-652, 542-434 cal. BP, during periods of the dry cooling. Khasyrey surface desiccation was occurring during periods of wet warming due to the drainage of the lake between 1192 and 1099, 869 and 843 cal. BP, 2003 and 2008. The probability of the short-term formation of low palsa during periods of dry cooling (1099-1065, 1008-984 and 719-652 cal. BP) is high. The formation of polygonal mire was happening in the dry sub-period of the LIA in about 542-434 cal. BP. Permafrost degradation began in the humid warm sub-period of the LIA in about 434 cal. BP. Partial melting of the ice veins with formation of flat-palsa mire happened in 330 cal. BP. The optimal conditions for the formation of palsa and poligonal ice veins in khasyrey of tundra zone were created at increase of continentality of climate and summer temperatures during periods of dry cooling.

Good consistency of the dynamics of water regimes and permafrost conditions of hasyreya and mires of the forest zone of Western Siberia with the data of reconstructions of regional climate and solar activity indicates to the climate conditionality of cyclical character of khasyrey develompent of Western Siberia tundra. Climate influence was direct through the hydrothermal regime, and - indirect through a thaw-permafrost state of peat deposit and the level regime of the lake. It was lake drainage in Medieval Maximum, who reduced the stage of high watering khasyrey significantly.

The work was performed under support of the Basic project of IMCES SB RAS VIII.80.2.3. and the Russian Research Foundation (grant № 14-17-0013).

## References

[1] Vasilchuk Yu K Vasilchuk A K Budantseva N A and Chizhova Yu N 2008 *Palsa of frozen peat mires* ed Y K Vasilchuk (Moscow: Moskow State University Press) p 571
[2] van Huissteden J Berrittella C Parmentier F J W Mi Y Maximov T C and Dolman A J 2011 Methane emissions from permafrost thaw lakes limited by lake drainage Nature *Climate Change* **1** 119–123
[3] Hantemirov R M 2009 *The dynamics of woody vegetation and climate changes in the north of Western Siberia* Author diss PhD (Ekaterinburg: Publishing House UB RAS) p 42
[4] *Geocryology USSR Western Siberia* 1989 ed E D Yershov (Moscow: Nedra Press) p 454
[5] *The All_Russian Research Institute of Hydrometeorological Information – World Data Center.* Available at: http://aisori.meteo.ru/ClimateR (accessed 03 December 2015)
[6] Stuiver M, Reimer P J and Reimer R W 2013 *CALIB 7.0.2 Manual*. Available at: http://calib.qub.ac.uk/calib/download / (accessed 21 Octouber 2015)
[7] Reimer P J *et al* 2013 IntCal13 and MARINE13 radiocarbon age calibration curves 0-50000 years cal BP *Radiocarbon* **55** 869–1887
[8] Ramenskiy L G Tsatsenkin I I Chizhikov O N and Antipin N A 1956 *Ecological Evaluation of the Fodder Lands by Vegetation Cover* (Moskow: Selkhozgiz Press) p 472
[9] Elina G A and Yurkovskaya T K 1992 Methods for determining paleogydlogical regime as the basis objectification of the causes of mire vegetation successions *Botanical J.* **77** 120–124
[10] Lapshina E D 2003 *Flora* of *mires* of *south-east* of the *West Siberia* (Tomsk: Tomsk State University Press) p 296
[11] Korolyuk A Yu Troeva E I Cherosov M M Zakharov V I Gogolev P A and Mironov S I 2005 *Ecological Assessment of Flora and Vegetation of Central Yakutia* (Yakutsk: YaNTs SO RAN) p 108
[12] Jones M C et al 2016 Effects of permafrost aggradation on peat properties as determined from a pan-Arctic synthesis of plant macrofossils *J. Biogeosciences* **121** 78
[13] Blyakharchuk T A and Klimanov V A 1989 The Structure and Development of Wetland cosystems and the Reconstruction of the Paleogeographic Conditions *Proc 10th All-Union seminar tours 30.08-03.09. 1989* Estonian SSR (Tallinn: AS of Estonia Press) pp 45–49
[14] Blyakharchuk T A 2009 Western Siberia, a review of Holocene climatic changes *J. of Siberian Federal University. Biology* **2** 4–12
[15] Volkova V S Gnibidenko Z N and Goryacheva A A 2002 Holocene climatic rhythm of the central part of the West Siberian Plain (palynology, magnetism) *Main Regularities of Global*

*and Regional Climate Change and the Environment in the Late Cenozoic of Siberia* (Novosibirsk: Institute of Archaeology and Ethnography SO RAN Press) pp 48–57

[16] Preis Yu I 2015 Detailed reconstruction of bog functional state as a response to continental climate changes in Holocene (the middle taiga of Western Siberia) *Bull Tomsk Polytechnic University* **2** 90–102

[17] Preis Yu I Bobrov V A Budashkina V V and Grishin V M 2010 Estimate of flows of mineral matter by the properties of peat deposits of Bakchar bog (southern taiga of Western Siberia) *Bull of the Tomsk Polytechnic University* **316** 43–47

[18] Preis Yu I and Kurina I V 2012 High-resolution reconstruction of mire paleoekotopes of Southern taiga of Western Siberia as a response to Holocene climate change *Research of Natural and Climatic Processes on the Great Vasyugan Mire* ed M V Kabanov (Novosibirsk: SO RAN Press) pp 14–38

[19] Preis Yu I 2016 Dynamics of ridge-pool complex of Iksinskoye bog (Western Siberia) as a response to climate change in the second half of the Holocene *Geography and Natural Resources* **2** 94–103

[20] Levi K G Zadonina N V Yazev S A Voronina V I Naurzbaev M M and Hantemirov R M 2012 *Heliogeodynamics: Natural Aspects of Global Solar Minima* (Irkutsk: Irkutsk State University Press) p 462

[21] Yu Z C 2006 Modeling ecosystem processes and peat accumulation in boreal peatlands *Boreal Peatland Ecosystems* eds Wieder R K, Vitt D H *Ecological Studies Series* (New York: Springer Press) **188** pp 313–329

[22] Shiyatov S G 2003 Rates of changes in the upper treeline ecotone in the Polar Ural Mountains *PAGES News* **11** 8–10

[23] Naurzbaev M M and Vaganov E A 2000 Variation of early summer and annual temperature in east Taymir and Putoran (Siberia) over the last two millennia inferred from tree rings *J. of Geophysical Research* **105** 7317–26

[24] Bolshiyanov D Yu Makarov A S Morozova E A Pavlov M V and Savatyugin L M 2009 Development of natural environment of the Earth polar regions of last millennium according to the study bottom sediment of lakes *Problems of the Arctic and Antarctic* **1** 108–115

[25] Wanner H Solomina O Grosjean M Ritz S P and Jetel M 2011 Structure and origin of Holocene cold events *Quantery Science Reviews* **30** 3109–23

[26] Khotinsky N A 1977 *The Holocene of Northern Eurasia* p 198

# Isotopic composition of winter precipitation in Altai foothills

**N S Malygina, A N Eirich, and T S Papina**

Institute for Water and Environmental Problems SB RAS, Ul. Molodezhnaya 1, Barnaul, 656038, Russia

E-mail: natmgn@gmail.com

**Abstract.** In recent years, some general circulation models have been improved by using stable water isotopologues, such as HDO (δD) and $H_2{}^{18}O$ ($\delta^{18}O$). In this paper, the results of a study of the isotopic composition of precipitation sampled in the winter seasons of 2014-2015 and 2015-2016 in Altai foothills are presented. The isotopic composition varied greatly: from -17.4 ‰ to -30.63 ‰ for $\delta^{18}O$ and from -132.1 ‰ to -235.0 ‰ for δD in 2014-2015; from -15.9 ‰ to -26.8 ‰ for $\delta^{18}O$ and from -118.5 ‰ to -207.7 ‰ for δD in 2015-2016. The weighted average values of $\delta^{18}O$ and δD for both seasons were close to each other (-21.2 ‰ and -160.1 ‰ for the first season and -20.4 ‰ and -157.5 ‰ for the second season), while the values of $d_{exc}$ differed significantly. A trajectory analysis (by the Hybrid Single Particle Lagrangian Integrated Trajectory – (HYSPLIT) model) has shown that the difference in the isotopic composition of the precipitation is largely associated with the change of the main source regions of atmospheric moisture, namely, the change of the North Atlantic waters (the 2014-2015 winter season) for the inland waters (the 2015-2016 winter season). A comparison of the results of an isotopic analysis has shown good agreement with data of the Global Network of Isotopes in Precipitation (GNIP) interpolated for 1960-2010. Thus, with a proper interpretation the data of the isotopic composition of the winter precipitation in Altai foothills can be used for this territory as an alternative source of the GNIP network data.

## 1. Introduction

Several general circulation models (GCMs) of the ocean and atmosphere have been improved by a possibility to explicitly simulate the hydrological cycle of two stable water isotopologues (hereafter called "water isotopes"), namely, HDO and $H_2{}^{18}O$ [1-3]. Global modeling of the isotopic water composition can, therefore, help in interpreting the observed isotopic changes in various archives [4]. A number of studies have already clearly demonstrated this possibility to improve the interpretation of the detected water isotope variability in terms of climatic and environmental changes by using appropriate GCM simulations [1]. At present, there exist about a dozen of GCMs with explicit isotope diagnostics [5]. An example is ECHAM5-wiso; its simulation results on a global scale are in good agreement with the available observations of the isotopic composition [1-2]. However, when modeling on a regional scale, ECHAM5-wiso slightly underestimates the temperature over Europe (the mean difference: ~ 0.4 °C) and overestimates the modeled precipitation up to 30 mm/month in the central part of Western Europe [6].

With GCMs taking into account the isotopic composition of precipitation, the Global Network of Isotopes in Precipitation (GNIP) data are used first [7]. The GNIP has been monitoring the stable isotopes in precipitation since 1961. In the Russian Federation (USSR), there have been no more than 40 stations sampling precipitation for longer than a year [7]. Since early 2000s, no continuous monitoring of the isotopic composition of precipitation according to the GNIP recommendations has been conducted. The research results mostly cover plain lands of the country [8-9].

The proportion of stable water isotopes in precipitation is determined by an empirical formula called the global meteoric water line (GMWL) [10-11]:

$$\delta D = 8 \times \delta^{18}O + 10\%, (1)$$

Using the deviation of the isotopic ratios between $\delta^{18}O$ and $\delta D$ from the GMWL data, we can estimate the isotopic fractionation in the region under study, which is described by their local meteoric water line (LMWL), i.e. by the regional ratio between $\delta^{18}O$ and $\delta D$ in atmospheric precipitation. The deuterium excess ($d_{exc}$)

$$d_{exc}=\delta D-\delta^{18}O\times 8, (2)$$

proposed by W. Dansgaard [11] on the basis of the GMWL-dependence to characterize the regional peculiarities of atmospheric water vapor or precipitation has been successfully used to identify their sources [12-13]. Thus, the changes in the isotopic composition of atmospheric precipitation in Russia, especially in piedmont and mountainous areas with adverse climatic and orographic conditions, are of utmost importance. Our aim is to study the formation of the isotopic composition of winter precipitation in the Altai foothills.

## 2. Materials and Methods

### *2.1. Study area*

The Altai foothills are located between the Ob plateau, Prealtai plain, Biya-Chumysh upland, and Salair ridge. This area is of scientific interest, since it is in front of the Altai major orographic barriers and serves as a point of convergence of the oceanic and continental air masses responsible for wide variations in the isotopic composition of precipitation.

Atmospheric precipitation (snow) samples were collected in the Altai foothills (Figure 1) immediately after their formation in the periods from December to February of 2014-2015 and 2015-2016.

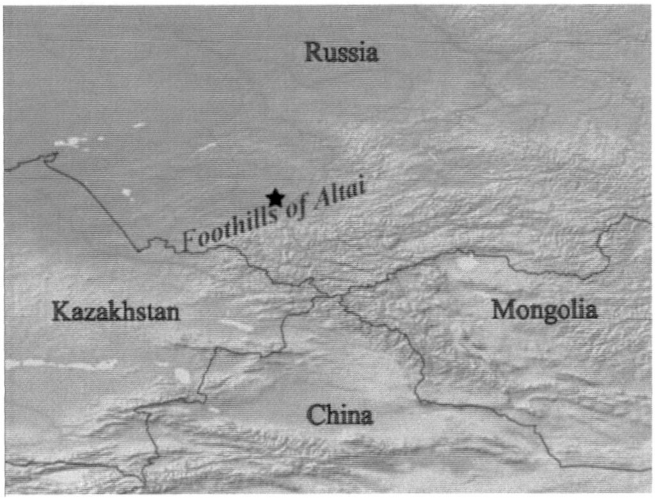


**Figure 1.** Altai foothills (*black asterisk: precipitation sampling point*).

### *2.2. Isotope analysis*

An isotope analysis of water samples was performed at the Chemical-Analytical Centre of the Institute for Water and Environmental Problems of the Siberian Branch of the Russian Academy of Sciences with IR-laser absorption spectroscopy on a device Picarro L2130-i equipped with a WS-CRDS system (Wavelength-Scanned Cavity Ring Down Spectroscopy). The use of the WS-CRDS system made it possible to eliminate the spectral overlay [14] and achieve high accuracy and reproducibility in estimating $\delta^{18}O$ and $\delta D$ in the analysed samples. The detection limit was 0.4 ‰ and 0.1 ‰ for $\delta D$ and

$\delta^{18}O$, respectively. The analysis was performed according to the V-SMOW-2 (Vienna Standard Mean Ocean Water 2) standard.

### *2.3. Synoptic analysis*

The synoptic situation in the precipitation period was analyzed on the basis of meteorological observations at the precipitation sampling point and using data of the Russian Federal Service for Hydrometeorology and Environmental Monitoring (Roshydromet) [15] and the National Oceanic and Atmospheric Administration (NOAA) [16]. Besides, we used NCEP CFSR data [17] on the wind direction and speed at 10 meters.

### *2.4. Trajectory analysis*

The moisture transport history of the precipitation in the observation period was traced by using a HYbrid Single-Particle Lagrangian Integrated Trajectory (HYSPLIT) model developed at the National Oceanic and Atmospheric Administration Air Resources Laboratory (NOAA ARL) [18]. We used Global Data Assimilation System (GDAS) data of 1 degree, global, from 2006 to the present time. The backward trajectories were calculated for each precipitation day. The backward trajectories started at the beginning of precipitation from the sampling point in the Altai foothills. Each backward trajectory covered a 240-hour interval. All the backward trajectories data were combined into a single directory based on the Google Earth. Then, using a program called ArcGIS (ESRI, USA) we completed the conversion of the data in a map data model. Cartographic visualization allowed us to analyse the time and space changes in the parameters under study, namely, to trace the paths of backward trajectories of the air masses.

## 3. Results and Discussion

### *3.1. Synoptic situation*

The total precipitation amounts for the two winter seasons being analyzed differed significantly from each other. In the winter of 2015-2016 precipitation was 30% less than that in the 2014-2015 season (Figures. 2a and 2b). The monthly distributions of precipitation were also different. In the winter of 2014-2015, there was more precipitation in late January and early February, while the 2015-2016 precipitation was mostly in December. The temperature conditions of precipitation were different as well. On average, the precipitation temperature during 2015-2016 was two times higher than the average one for the 2014-2015 winter season. At the same time, in 2015-2016 (in contrast to the 2014-2015 season) snow was observed even at positive temperatures (above 0°). An analysis of maps of average wind speed and direction at 10 m (the data are from NCEP CFSR) showed no significant differences between the data of the winter seasons (Figure 2).

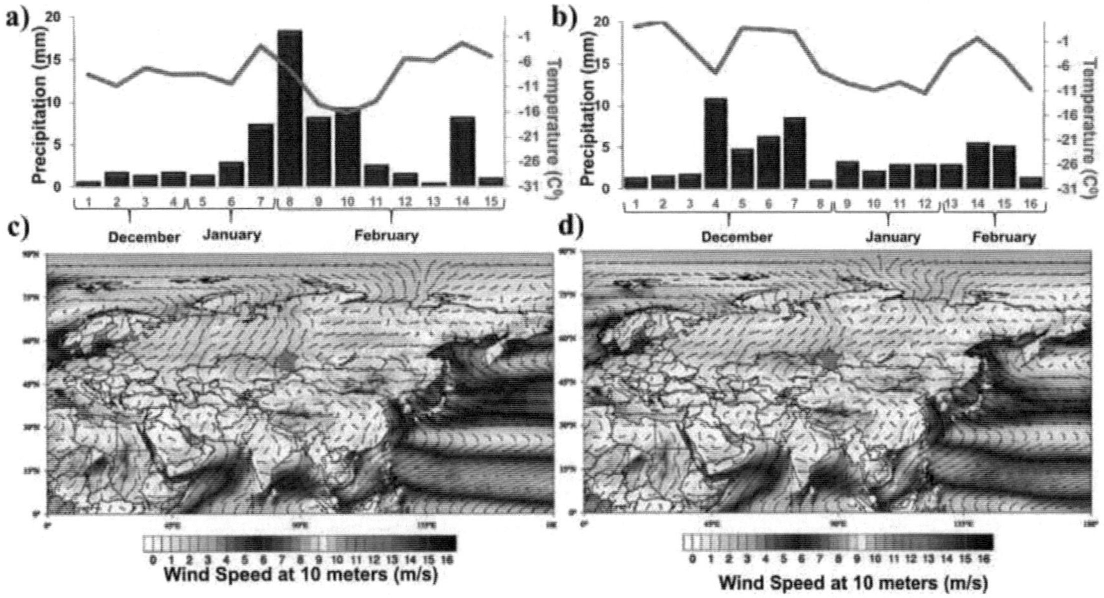


**Figure 2**. Precipitation and temperature in December-February of 2014-2015 (a) and 2015-2016 (b) and average wind speed and direction at 10 m in the first (c) and second (d) seasons being analyzed in the Altai foothills (*red asterisk: precipitation sampling point*).

### *3.2. Isotopic composition of precipitation*

The isotopic composition of precipitation in the winter of 2014-2015 ranged from 17.4 ‰ to -30.63 ‰ for $\delta^{18}O$, and from -132.1 ‰ to -235.0 ‰ for $\delta D$. The deuterium excess for this winter season ranged from 0.1 ‰ to 16.3 ‰. The isotopic composition in the winter of 2015-2016 varied in a smaller range both for $\delta^{18}O$ (from -15.9 ‰ to -26.8 ‰) and $\delta D$ (from -118.5 ‰ to -207.7 ‰); $d_{exc}$ showed both negative and positive values (from -5.3 ‰ to +11.9 ‰).

The weighted average isotopic composition of precipitation in the 2014-2015 winter season reached -21.2 ‰ for $\delta^{18}O$ and -160.1 ‰ for $\delta D$, and $d_{exc}$ was 9.4 ‰. In the winter of 2015-2016, the weighted averages for $\delta^{18}O$ and $\delta D$ were close to those of the previous season with -20.4 ‰ and -157.5 ‰, respectively, while $d_{exc}$, being equal to 5.7 ‰, decreased by almost 4 ‰. This allows us to assume greater isotopic fractionation of precipitation in the winter of 2015-2016 as compared to 2014-2015 and/or a change in the precipitation sources from season to season.

The calculated LMWLs for the winter seasons of 2014-2015 and 2015-2016 are described by the following regression equations (Figure 3):

$$\delta D = 8.2 \times \delta^{18}O + 13.4‰, (R^2 = 0.99), (3)$$

$$\delta D = 7.8 \times \delta^{18}O + 2.6‰, (R^2 = 0.97), (4)$$

The deviation to the right of the LMWL for the 2015-2016 winter season from the GMWL is due to the influence of evaporative fractionation on the formation of the isotopic composition of precipitation.

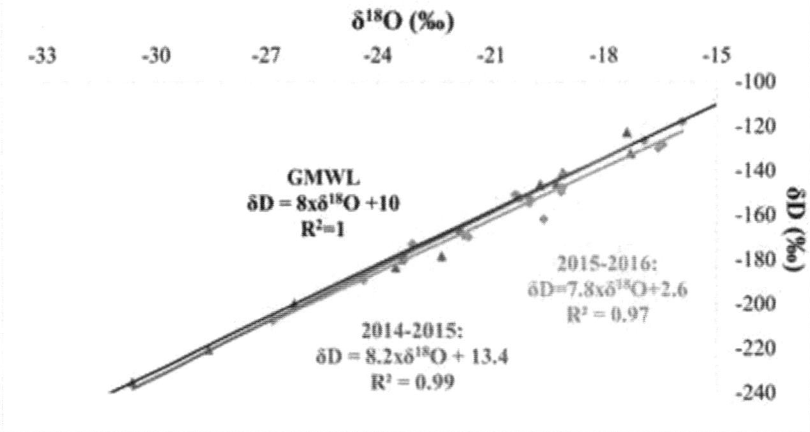


**Figure 3.** GMWL (*black line*) and LMWL of precipitation in December-February of 2014-2015 (*blue line*) and 2015-2016 (*turquoise line*).

The weighted average isotopic composition of precipitation was compared with the GNIP interpolated data for the period of 1960-2010 (Table 1) obtained for the Altai foothills with the help of an online calculator [7].

**Table 1.** Weighted average isotopic composition of precipitation in Altai foothills (December-February) according to GNIP interpolated data (1960-2010) and observational data of 2014-2015 and 2015-2016.

| | GNIP | 2014-2015 | 2015-2016 |
|---|---|---|---|
| $\delta^{18}O$ (‰) | -22.5 | -21.2 | -20.4 |
| $\delta D$ (‰) | -176.3 | -160.1 | -157.5 |
| $d_{exc}$(‰) | 3.9 | 9.4 | 5.7 |

The isotopic compositions of winter precipitation in the Altai foothills are in good agreement with the results of the GNIP. The difference for $\delta^{18}O$ is ~ 2‰, and for $\delta D$, ~19‰. The values of $d_{exc}$ for the winter precipitation in 2014-2015 and 2015-2016 differed significantly from the GNIP interpolated data. The differences may be due to a change in the atmospheric precipitation sources from season to season and the fact that different time series are analyzed (some seasons and averages are for 1960-2010).

*3.3. Precipitation sources*

Using the HYSPLIT model, we calculated backward trajectories of air masses for each of the precipitation times to assess the change in their sources. An analysis of the backward trajectories (the 2014-2015 winter season) showed that the air masses responsible for precipitation were transported mainly from northwest (Figure 4a), in particular, from the North Atlantic Ocean. This is confirmed by the weighted averages of $d_{exc}$ (9.4 ‰) for precipitation, which is not significantly different from the $d_{exc}$ value for the North Atlantic Ocean [19]. In the 2015-2016 winter season the trajectories of air masses responsible for precipitation were mostly of south-western and western directions (Figure 4b). These air masses passed over some inland water bodies (such as the Caspian Sea), whose evaporating moisture could change the isotopic composition of water in the air masses responsible for precipitation in the Altai foothills. This is confirmed by the fact that the weighted average of $d_{exc}$ decreased in the 2015-2016 winter season.

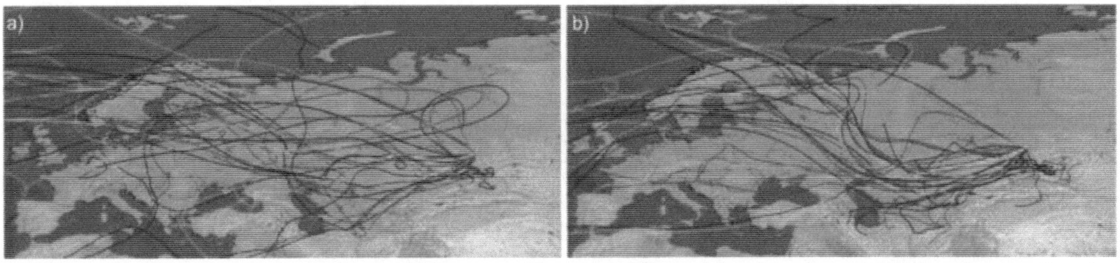


**Figure 4.** Backward trajectories at 500 m (*red lines*), 1500 m (*blue lines*), and 3000 m (*green lines*) for air masses which determined precipitation in December-February 2014-2015 (*a*) and 2015-2016 (*b*) in the Altai foothills.

Thus, the trajectory analysis of air masses for the 2014-2015 and 2015-2016 winter seasons allowed us to identify the main source regions of precipitation whose change led to isotopic fractionation of precipitation, exhibited by variations in $d_{exc}$ (of about 4 ‰) between the two winter seasons.

## 4. Conclusions

The isotopic composition of precipitation sampled in the Altai foothills in the winter seasons of 2014-2015 and 2015-2016 varied over a wide range. The maximum values were ~15 ‰, ~117 ‰, and ~22 ‰ for $\delta^{18}O$, $\delta D$, and $d_{exc}$, respectively. The weighted averages of $\delta^{18}O$ and $\delta D$ for the two seasons were close to each other (-21.2 ‰ and -160.1 ‰ for the first season and -20.4 ‰ and -157.5 ‰ for the second season), while $d_{exc}$ varied significantly. The results of an isotopic analysis showed good agreement with the GNIP interpolated data for 1960-2010, however, $d_{exc}$ only for the second winter season was close to the GNIP interpolated data. A trajectory analysis (with the HYSPLIT model) showed that the differences in the isotopic composition between the seasons were largely due to a change in the main source regions of atmospheric moisture, namely, the replacement of North Atlantic waters (in the winter of 2014-2015) with inland waters (in the winter of 2015-2016). Thus, the data on the isotopic composition of precipitation in the Altai foothills with a correct interpretation of the results can be used as an alternative source of information for the GNIP data.

## Acknowledgements

This work was supported by the Russian Foundation for Basic Research (project no. 16-35-00188 mol_a), government order on project VIII.77.1.5 "Climatic and Ecological Changes in Siberia according to Data of Glacio-Chemical, Diatomic, and Pollen Analysis of Glacier Core Samples" (no. 0383-2014-00052013).

## References

[1] Eichinger R, Jöckel P, Brinkop S, Werner M and Lossow S 2015 *Atmos. Chem. Phys.* **15** pp 5537 – 55
[2] Gribanov K, Jouzel J, Bastrikov V, Bonne J-L, Breon F-M, Butzin M, Cattani O, Masson-Delmotte V, Rokotyan N, Werner M and Zakharov V 2014 *Atmos. Chem. Phys.* 14 pp 5943 – 57
[3] Jouzel J, Russell G L, Suozzo R J, Koster R D, White J W C and Broecker W S 1987 *J. Geophys. Res.* **92(D12)** pp 14739 – 60
[4] Jouzel J, Hoffmann G, Koster R D and Masson V 2000 *Quat. Sci. Rev.* **19** pp 363–379
[5] Sturm C, Zhang Q and Noone D 2010 *Clim. Past.* **6(1)** pp 115 – 129
[6] Langebroek P M, Werner M and Lohmann G 2011 *Earth Planet. Sc. Lett.* **311** pp 144–154
[7] http://www-naweb.iaea.org/napc/ih/IHS_resources_gnip.html
[8] Ferronsky V I and Polyakov V A 2012 Isotopes of the Earth's Hydrosphere (Springer Netherlands)
[9] Vasilchuk Yu K 2014 *Doklady Earth Sciences* **459** pp 109 – 111

[10] Craig H 1961 *Science* **133** pp 1702 – 1703
[11] Dansgaard W 1964 *Tellus* **16** pp 436 – 468
[12] Fricke H and O'Neil J 1999 *Earth Planet. Sci. Lett.* **170** pp 181 – 196
[13] Merlivat L and Jouzel J 1979 *J. Geophys. Res.* **84** pp 5029 – 33
[14] http://www.esrl.noaa.gov/
[15] http://rp5.ru/docs/about/ru
[16] http://www.noaa.gov
[17] http://cfs.ncep.noaa.gov/cfsr/
[18] https://ready.arl.noaa.gov/HYSPLITtraj.php
[19] Pfahl S and Sodemann H 2014 *Clim. Past.* **10** pp 771 – 81

# Development and study of a parallel algorithm of iteratively forming latent functionally-determined structures for classification and analysis of meteorological data

**V A Sorokin[1], Yu V Volkov[2], A I Sherstneva[3], and I A Botygin[2]**

[1] Undergraduate, National Research Tomsk Polytechnic University, Tomsk, Russia
[2] Senior Researcher, Institute of Monitoring of Climatic and Ecological Systems SB RAS, Tomsk, Russia
[3] Assistant professor, National Research Tomsk Polytechnic University, Tomsk, Russia

E-mail: vas1@tpu.ru, yvvolkov@tpu.ru, sherstneva@tpu.ru, bia@tpu.ru

**Abstract.** This paper overviews a method of generating climate regions based on an analytic signal theory. When applied to atmospheric surface layer temperature data sets, the method allows forming climatic structures with the corresponding changes in the temperature to make conclusions on the uniformity of climate in an area and to trace the climate changes in time by analyzing the type group shifts. The algorithm is based on the fact that the frequency spectrum of the thermal oscillation process is narrow-banded and has only one mode for most weather stations. This allows using the analytic signal theory, causality conditions and introducing an oscillation phase. The annual component of the phase, being a linear function, was removed by the least squares method. The remaining phase fluctuations allow consistent studying of their coordinated behavior and timing, using the Pearson correlation coefficient for dependence evaluation. This study includes program experiments to evaluate the calculation efficiency in the phase grouping task. The paper also overviews some single-threaded and multi-threaded computing models. It is shown that the phase grouping algorithm for meteorological data can be parallelized and that a multi-threaded implementation leads to a 25-30% increase in the performance.

## 1. Introduction

The fundamental scientific researches include tasks of forecasting weather, predicting climate and global changes in the atmosphere. Their effective solution requires high-performance computing systems that allow parallelization. Due to global changes in climate, it is very relevant to study the multidimensional geophysical data and detect important dependences (usually, in time) of different meteorological parameter fields or other parameters for regions of different scales. For example, studying various time trends in a given area of distribution, since trend mathematical description allows to make prognoses on studied parameters. Clustering technology is one of the basic approaches for solving such tasks, especially for processing experimental data of geophysical nature. For this problem the parallel computing is necessary due to big volume of processed data. For a single location, the number of geophysical data bins may reach a few thousands, while properly analyzing climatic changes might require processing a huge number of locations. Besides having to store a significant amount of data, it is also necessary to have a non-trivial approach to mathematical processing. To summarize, implementing parallel computing is one of the most important tasks for meteorological data classification and analysis [1-4].

As mentioned above, the cluster analysis is a possible approach to mathematical processing of geophysical data (meteorological data in particular). In general, the method consists of the following steps: selecting objects for cluster analysis, forming the criteria and metrics for object evaluation, determining objects' identity according to metrics and criteria, using identical objects to form clusters (groups). The described steps are executed iteratively. The optimal result might require altering the key parameters at every clustering step.

Depending on the way of data processing, the cluster analysis methods are divided into hierarchical and non-hierarchical. The hierarchical methods consequently merge smaller clusters into the bigger ones (agglomerative) or split big clusters into the smaller ones (divisive) [5].

Depending on the way of data analysis, there are non-overlapping (precise) and overlapping (fuzzy) clustering algorithms, as well as single-step or multi-step methods depending on the number of iterations [6].

BIRCH (Balanced Iterative Reducing and Clustering using Hierarchies) clustering algorithm is based on double-step hierarchical clustering and can effectively be used for very large volume of data. This algorithm, however, cannot work with anything except numerical data.

K-means method decomposes a set of objects into a preset number of clusters. The algorithm is easy to use while sensitive to emissions (which can distort mean values) and requires knowing the exact number of clusters [7].

CLOPE (Clustering with slope) algorithm is intended to work with transaction data, well-scalable and can boast a high performance and clustering quality. The method automatically selects a required number of clusters and is widely used in data mining [8].

PAM (Partitioning around medoids) algorithm is a modification of K-means method. It is very effective for smaller amounts of data [9] and widely used in such fields, as text analysis, bioinformatics and intelligent transport systems [10].

Konohen self-organizing maps is one of the neural algorithms without teacher. It allows to obtain a set of objects with reduced dimensionality compared to the original [11].

Concluding this brief overview, we have to point that there is no universal clustering algorithm. It is proven by the existence of a whole number of heuristic criteria as well as algorithms that do not have a clear criterion, yet are capable of doing a reasonable clustering. Moreover, the clustering result significantly depends on metrics that is usually selected by expert's subjective judgement [12]. As such, finding the optimal approach for a given task is only possible by analyzing the nature of data and task specifics.

## 2. Grouping algorithm

This work studies thermal fields with the goal of finding latent functionally determined structures, which would allow a new way to analyze and classify sets of measurement data. Generally, natural climate changes show a cyclic behavior, so it is possible to do a research using signal spectral analysis methods. The authors of this work suggest an original phase grouping algorithm based on one of the possible ways of representing oscillations (analytical signal of Nobel laureate Zoltán Pál Dienes Gabor). Phase is the main parameter of oscillation process. While amplitude describes variable parameter range, phase contains an information on periodicity of variable change over time. Using cluster analysis methods and algorithms, researching process phase allows to obtain new estimates that describe the behavior of meteorological parameters over a given period of time.

A study of monthly average temperature data over the 56 years at 818 weather stations shows that data changes form an oscillating process with quasi-period of one year. The temperature values deviating from an annual cycle may possibly describe seasonal climate changes. However, finding consistent dependencies proves a complicated task [13]. The researches show that frequency spectrum of temperature oscillating process is narrow-banded and has only one mode for most weather stations, which allows to use causality conditions and introduce oscillation phase [14]. The yearly phase component, being a linear function, was removed from the phase for 56 years period of time. Afterwards, the remaining phase fluctuations allow to consistently research their coordinated behavior and timing using the Pearson correlation coefficient for evaluation.

We've developed an algorithm to compute phase values for each set of temperature values. Its general scheme is shown in figure 1 below.

Step 1. Fast Fourier transform (moving from time domain to frequency domain). After input data is uploaded, every set of temperature values is converted into a sum of sinusoidal functions with Fourier discrete transform (1).

$$X_k = \sum_{n=0}^{N-1} x_n e^{-\frac{2\pi i}{N}kn}, k = 0 \dots N-1 \quad (1)$$

Fourier discrete transform is based on fast Fourier transform algorithm.

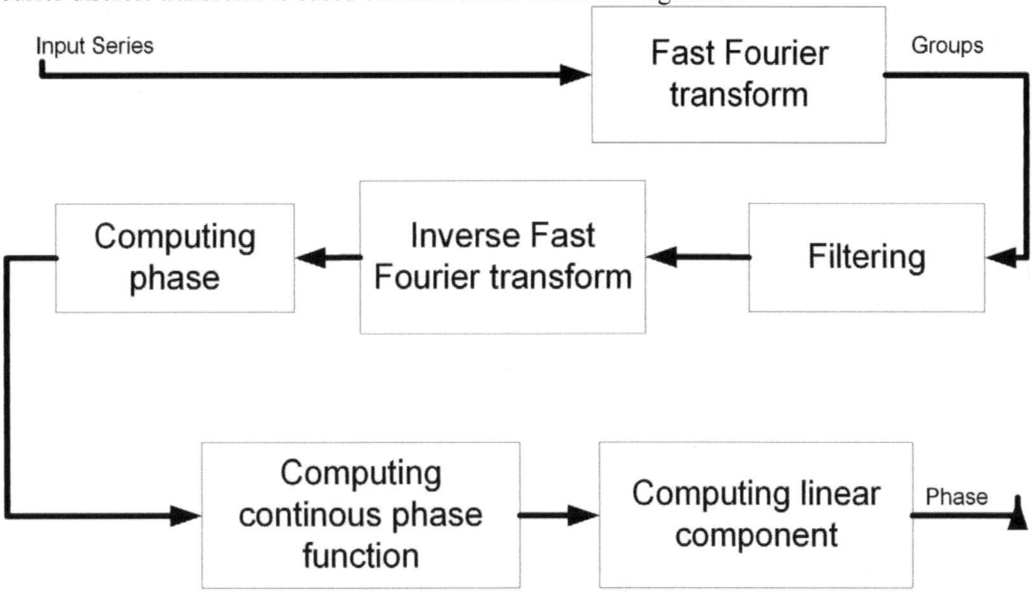


**Figure 1.** General scheme of computing phase values for sets of temperature data.

Step 2. Filtering. Median filtering is used to reduce data noise. One-dimensional median filter, based on heuristic processing method, uses sliding-window method to average segments of temperature data.

Step 3. Inverse Fourier transform. This algorithm is used to return from phase domain into time domain.

$$x_n = \frac{1}{N}\sum_{n=0}^{N-1} X_k e^{\frac{2\pi i}{N}kn}, n = 0 \dots N-1 \quad (2)$$

Step 4. Computing phase. Using the analytic signal theory, the phase value is found as an argument to the ratio of the real part to imaginary part. From the computing point of view, the analytic signal corresponding to the given real function is found through Fourier transform and following inverse Fourier transform by positive phases. The phase value obtained at this step is one of the main values of arc tangent function within yearly period of time (figure 2).

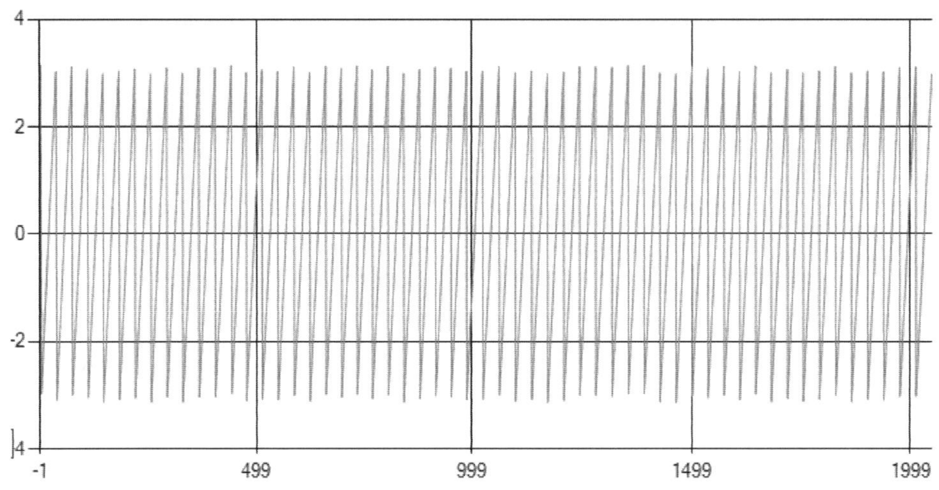


**Figure 2.** Main values of arc tangent function within yearly period of time.

Step 5. Computing continuous phase function over the entire observation interval.

Step 6. Computing the linear component. It is subtracted from the continuous phase function, which allows to form phase fluctuations of the given temperature data set (figure 3).

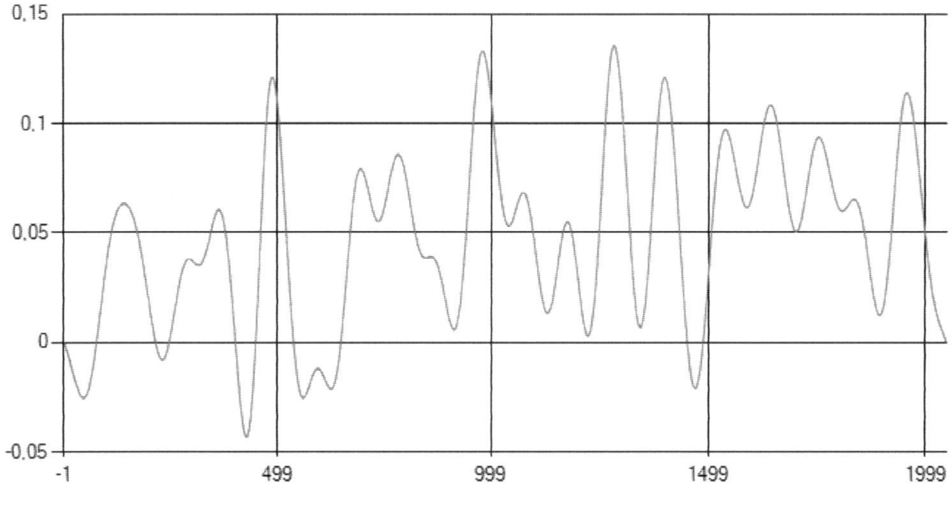


**Figure 3.** Phase fluctuations.

An algorithm to properly group weather stations with similar climate was developed. It compares phases by calculating pairwise correlation coefficient for each one, uses them to obtain standard phases and generates the final station grouping accordingly.

A general scheme of standard phase calculation is shown in figure 4 below.

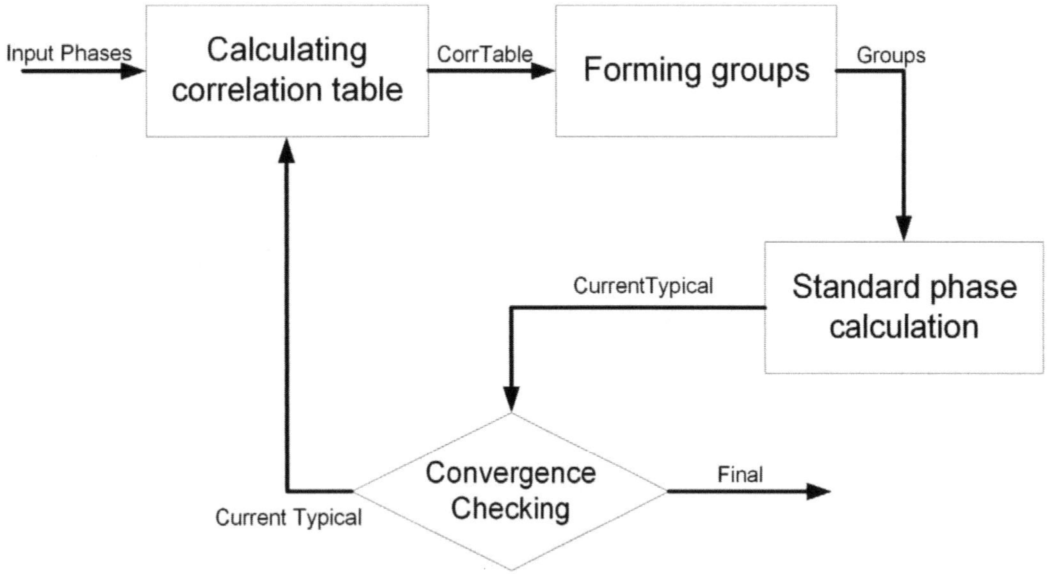


**Figure 4.** General scheme of standard phase calculation.

Stage 1. The phase values computed for each group are sent to the correlation table calculation node input, which outputs pairwise correlation coefficients.

Stage 2. The resulting phase correlation table is sent to the group forming node. The groups are created from the temperature data sets which, alongside group-forming data set, have correlation coefficients above given level.

Stage 3. The standard phase computing node uses the created groups to calculate standard phase values for the current iteration.

The described procedures are done iteratively, with previous level standard phase values as input data and next level standard phase values as output data. The algorithm is executed for each set of temperatures and, consequently, for each weather station. Additionally, the groups are changed according to the given threshold criterion – primarily, through inclusion of new elements which did not belong to the group at the previous stage. The threshold determines group size and coherence between current phase and previous ones during the iteration process.

Every calculation in algorithm is divided into two parts: temperature data phases are computed during the preparation part, grouping is performed during the main part.

## 3. Parallel computing

Let us look into the possibility of parallel computing in the phase clustering algorithm suggested above. During the preparation phase the calculations are linear and each step requires results from the previous step. Therefore, parallel computing is not possible here. However, since it is necessary to process a lot of temperature data sets (above three hundred) it might be expedient to use parallel computing to prepare each set.

All of the parallel computing steps are based on MapReduce model shown in figure 5 below.

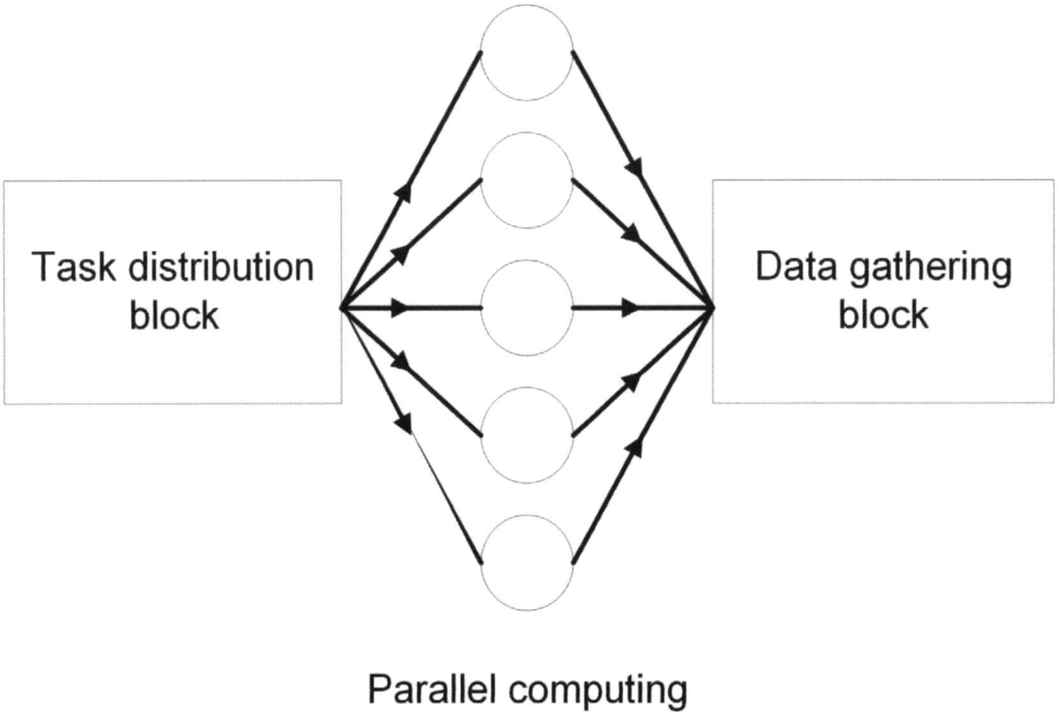


**Figure 5.** General scheme of parallel computing.

The task distribution block decomposes input data by available computing power to calculate phases of temperature data sets (Map procedure). Afterwards, the results obtained by each node are collected in data gathering block to prepare for further analysis (Reduce procedure).

The parallel computing is also not possible for all of the phase grouping algorithm, because this process is iterative as well (each iteration requires result from the previous step). However, it is possible to implement parallel computing for certain algorithm steps. In the phase grouping block diagram (figure 2), the parallel computing is possible for «Calculation of correlation table» and "Calculation of standard phases" steps. The general scheme of parallel computing (figure 5) is being used in these cases.

The process of correlation table calculation can be distributed between available calculating powers. After the data is gathered into a single table (based on final table analysis), it is possible to implement parallel computing for standard phase calculation process.

- Multi-threaded computing in standard mode;
- Multi-agent computing based on distributed computational system.

The efficiency analysis has shown that phase grouping algorithm for mete

Over the course of this work, we've conducted program experiments aimed to evaluate computation efficiency in phase clustering task. The following computing models were reviewed:

- Single-threaded computing in standard mode.

Phase algorithm groups using meteorological data lends itself to parallelization, and a multi-threaded implementation provides increased productivity by 25-30%.

## 4. Results

The steps described above were used as a phase clustering task solution for 818 weather station over the 56 year observation period. A text file with every row containing monthly average temperatures for weather stations was used as an input data.

To form the groups of stations with similar climate, we've developed an algorithm based on selecting standard phases, related to main groups with the help of pairwise correlation table.

The first stage of algorithm inputs computed phase values into correlation table calculation node, resulting in pairwise correlation coefficient for each weather station. It uses Pearson correlation coefficient:

$$r_{xy} = \frac{\sum_{i=1}^{m}(x_i - \bar{x})(y_i - \bar{y})}{\sqrt{\sum_{i=1}^{m}(x_i - \bar{x})^2 \sum_{i=1}^{m}(y_i - \bar{y})^2}} \quad (1)$$

After obtaining the correlation table, algorithm forms weather station groups with coefficients of correlation to other weather stations above given $r$ value. The phase value for every group is average for every phase in the group. The default phase function for grouped stations are replaced with current ones. Then the process is repeated and groups change according to $r$ criterion by including new elements that group didn't contain in the previous iteration. Eventually, current phases will have their standard properties defined and stop changing due to process convergence.

Afterwards, the duplicates are removed from the group list and it allows to calculate the correlation table of default phases together with computed standard phases. The result of the algorithm is shown in figures 6 and 7.

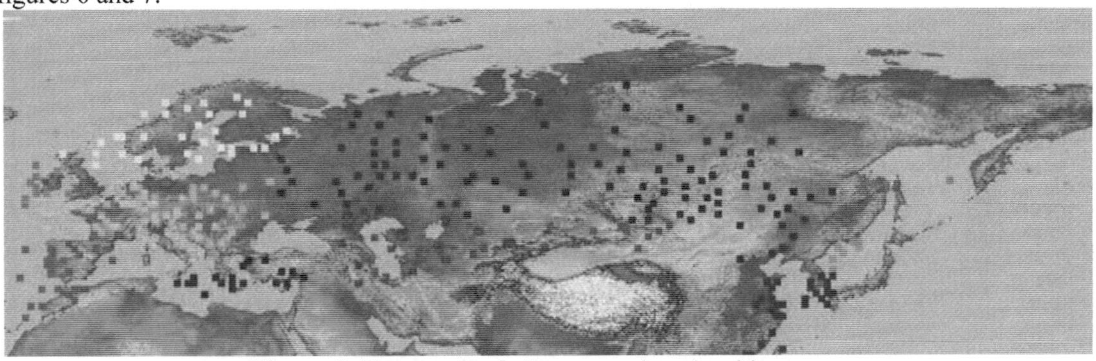

**Figure 6.** First result of phase grouping algorithm.

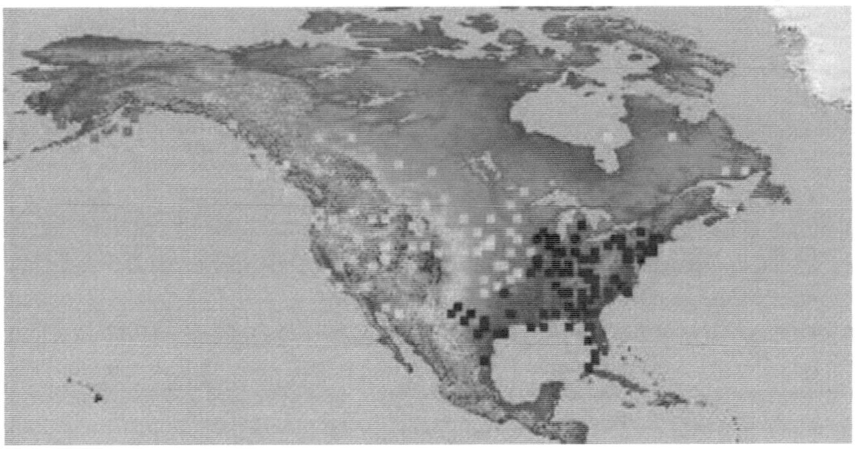

**Figure 7.** Second result of phase grouping algorithm.

## 5. Conclusions

The monthly average changes over the studied period form an oscillation process with quasi-period of one year. The temperature change deviation from yearly cycle is a definitive feature of weather. However, determining consistent dependencies is complicated. The analysis of climate field structure (zoning, classification etc.) is aimed to find regularities of climate type formation in global climate system. On the other hand, defining the territorial boundaries of climate types varying by properties allows to decompose the massive amounts of stored climate data into a smaller number of information structures.

This work suggest an original method for determining climate fields which allows to solve both tasks for any spatial and temporal scales. The method is based on the fact that thermal oscillation process phase spectrum is narrow-banded and has only one mode for most weather stations. This allows to use the causality conditions and introduce the oscillation phase by using analytic signal theory. The yearly phase component is a linear function which was removed from phase by least square method. The remaining phase fluctuations allow to consistently research their coordinated behavior and timing, using the Pearson correlation coefficient for dependence evaluation.

In the case of surface air temperature, the developed algorithm allows to find climate structures with coordinated changes in thermal field, to make conclusions on climate uniformity for given region and to research climate changes over time by analyzing standard group deviations.

The results are obtained within the framework of the project "Development and research of intelligent information-analytical system for analysis and prediction of climatic processes on the basis of high-performance clusters" of RAS 1.33P fundamental research program "Fundamental problems of mathematical modeling. Fundamental problems of factorization methods in different fields. Algorithms and software for ultra-high performance computing systems".

**References**

[1] Sherstnev V S, Sherstneva A I, Botygin I A and Kustov D A 2016 High Technology: Research and Applications (Key Engineering Materials vol 685) (Switzerland: Trans Tech Publications) pp 867–871

[2] World Community Grid *http://www.worldcommunitygrid.org/* Accessed: January 2016

[3] Botygin I A, Popov V N, Tartakovsky V A and Sherstnev V S 2015 *Proc. 21st Int. Symp. Atmospheric and Ocean Optics: Atmospheric Physics* vol 9680 (Tomsk: SPIE) pp 96800J1–96800J4

[4] Popov V N, Botygin I A and Koshelev N V 2016 *High Technology: Research and Applications (Key Engineering Materials)* vol 685 (Switzerland: Trans Tech Publications) pp 925–929

[5] Hierarchical agglomerative methods *http://stu.sernam.ru/book_fan1.php?id=81* Accessed: January 2016

[6] Neysky I M Classification and comparison of clustering methods h*ttp://it-claim.ru/Persons/Neyskiy/Article2_Neiskiy.pdf* Accessed: January 2016

[7] K-Means Clustering means *http://home.deib.polimi.it/matteucc/Clustering/tutorial_html/kmeans.html* Accessed: January 2016

[8] Clustering of categorical data: scalable algorithm CLOPE *https://basegroup.ru/community/articles/clope* Accessed: January 2016

[9] Introduction to Modern Data Mining *http://statistica.ru/local-portals/data-mining/* Accessed: January 2016

[10] Rechkalov T V 2015 *Supercomputer days in Russia 2015* (Moscow: Publishing house of the Moscow State University) pp 632–640

[11] Kohonen self-organizing maps: mathematical apparatus *https://basegroup.ru/community/articles/som* Accessed: January 2016

[12] Clustering *http://www.machinelearning.ru/wiki/index.php?title=%D0%9A%D0%BB%D0%B0%D1%81%D1%82%D0%B5%D1%80%D0%B8%D0%B7%D0%B0%D1%86%D0%B8%D1%8F* Accessed: January 2016

[13] Volkov Yu V and Tartakovsky V A 2013 *Energy: Efficiency, reliability, safety* (Tomsk: SCAN) pp 177–178

[14] Vakman D E and Weinstein L A 1977 *Physics-Uspekhi* **123** 657–682

IOP Conf. Series: Earth and Environmental Science **48** (2016) 012013 doi:10.1088/1755-1315/48/1/012013

# Simulation of Arctic and North Atlantic ocean water and ice seasonal characteristics by the INMIO-CICE coupled model

**K V Ushakov[1,2,3], T B Grankina[1,2], R A Ibrayev[1,3,2,4], and I V Gromov[4]**

[1] Hydrometeorological Research Centre of Russian Federation, 11-13 Bol'shoi Pedtechensky per., Moscow 123242, Russia

[2] P.P. Shirshov Institute of Oceanology RAS, 36, Nakhimovsky pr., Moscow 117218, Russia

[3] Institute of Numerical Mathematics RAS, 8, Gubkina str., Moscow 119333, Russia

[4] Moscow Institute of Physics and Technology, 9, Institutsky per., Dolgoprudny 141701, Russia

Email: ushakov@ocean.ru

**Abstract**. The first results of simulation of the seasonal variability of the Arctic and North Atlantic ocean waters and ice by a coupled model based on a full three-dimensional ocean dynamics model INMIO4.1 and a sea-ice model CICE5.1 are considered. The coupled model can be run on massively parallel computers under control of the Compact Modelling Framework CMF2.0. The numerical experiment is performed according to the CORE-II protocol with 1948 atmospheric forcing data. Possible causes of the deviation of the model solution from the ERA-20C reanalysis and WOA09 climatology are discussed.

## 1. Introduction

The Arctic region is an important territory both scientifically and economically. The Arctic Ocean (AO) is sometimes called the Arctic Mediterranean Sea or the Atlantic Ocean estuary. Due to its relatively small volume and long interface with the Atlantic Ocean, the processes taking place in the North Atlantic have a substantial influence on the Arctic Ocean. On the other hand, it is affected by relatively fresh Pacific Ocean waters coming through the Bering Strait.

The Arctic Ocean is bounded by Russia, Denmark, Iceland, Norway, Canada, and the USA. For the most part of the year it is used for sea cargo transportation and mineral and biological resources extraction. For the maintenance of economic activity it is important to have accurate and reliable operational information on the sea environment, including the ice conditions, which can be obtained by mathematical modelling. Seasonal numerical forecasting of the Arctic system is also very important because of the observed trends of regional warming, decrease in the sea ice area and volume and increase in the frequency of extremal weather condition events.

This paper is aimed at gaining the first experience on a coupled ocean-ice numerical simulation of the Arctic Ocean and model investigations of the mechanisms of intra-annual variability of the sea environment. Our numerical experiment is a calculation of the quasi-stable Arctic climatic system produced by cyclic annual variations of the atmospheric parameters and solar radiation taken from the CORE-II database [1] for the year 1948. The results obtained by the model are validated against the ERA-20C reanalysis [2] and WOA09 climatology [3,4] data.

In Section 2, ocean and ice models and a coupling technology are described. Section 3 shows the setup of the numerical experiment. Section 4 presents the results of the calculations. Section 5 contains conclusions and plans to develop the model.

## 2. Coupled model

The numerical experiment is performed with a coupled ocean-ice model at a nominal spatial resolution of 0.25°. The ocean component is a model called INMIO4.1 developed at the Institute of Numerical Mathematics of the Russian Academy of Sciences (INM RAS) and the Shirshov Institute of Oceanology (IO RAS). The sea ice is described by a model called CICE5.1 of the Los Alamos National Laboratory, USA. The coupled model is implemented for massively parallel computers under control of the Compact Modelling Framework CMF2.0 developed at INM RAS.

### *2.1 INMIO ocean model*

The INMIO numerical model has been developed to investigate the general ocean circulation in a wide range of spatial and temporal scales. The full 3D PEM system of equations in the Boussinesq and hydrostatic approximations is implemented by a box-method for an arbitrary B-type horizontal grid and vertical z-axis. The water-air interface is described by free-surface conditions and explicit definition of heat, momentum, and water fluxes by the CORE atmospheric boundary layer bulk formulas [5]. At the solid lateral boundaries free slip and zero heat and salt flux conditions are imposed. To provide numerical stability, biharmonic filters are included in the momentum, heat, and salt transfer equations. The model has a built-in ice thermodynamics solver [6], which is also used in the calculations.

The momentum transfer term is approximated with a centered difference scheme. For the transport of heat, salt, and other tracers a flux-corrected transport scheme [7] is used. The water and tracer transport schemes are numerically consistent. In the time derivative approximation we use a leap-frog scheme, periodically applying an Euler step scheme for mutual relaxation of the computational modes. To use the computer resources effectively, the fast barotropic ocean processes are modelled separately from the baroclinic processes by solving a system of shallow water equations using a predictor-corrector scheme [8] with small time steps. The vertical turbulent mixing is parameterized by the Munk-Anderson scheme with convective adjustment of the unstable water columns. The background vertical viscosity and diffusion coefficients are $10^{-4}$ and $10^{-6}$ $m^2/s$, and the maximal values in regions with low Richardson numbers are $10^{-2}$ and $10^{-3}$ $m^2/s$, respectively. Except for the vertical mixing, all the processes are described by explicit numerical methods, which made it possible to effectively parallelize the program code [9]. A more detailed description of the model is presented in [10].

### *2.2 CICE ice model*

The CICE sea ice model has been widely used by scientific groups all over the world. For instance, it was used in the AOMIP project [11]. The model defines the states of ice and snow in terms of their distribution functions $g(x,t,h)$ in time, geographical space, and thickness space. The description of ice includes several (by default, 5) thickness categories. The main prognostic variables for each category are the ice compactness (concentration), grid cell-mean ice and snow thickness, ice and snow internal energy, ice salinity, and surface temperature [12].

In this paper, the CICE model was run in a thermodynamic regime which includes modelling of the local processes of production, melting, and mutual conversion of ice and snow cover according to precipitation, diffusive heat conductivity, propagation of several radiation bands, heat exchange with the atmosphere and liquid water, evaporation and the corresponding latent heat flux, water and salt exchange with the ocean, and other processes with the exception of moving in space and internal stresses. The thermodynamics is modelled in a mushy approximation which describes the sea ice as a fresh ice medium containing brine pockets defining the distribution of salinity and enthalpy [13]. For the thermal conductivity, a bubble parameterization [14] is used. In addition to snow cover, the model simulates the development of melt ponds as tracers including their drainage to the ocean. The solar radiation absorption is parameterized by the Delta-Eddington scheme.

*2.3 Coupled model*

The coupled INMIO-CICE ocean and ice model is implemented for massively parallel computer systems by means of the Compact Modelling Framework (CMF), which provides fully parallel interprocessor communications, multilevel data interpolation, and asynchronous I/O operations [15]. The CMF allows running an abstract coupled model consisting of an arbitrary number of components. Each of them, in turn, links in program libraries of a particular model (ocean, ice, atmosphere, land, etc.) and defines its own time cycle for solving the physical equations and periodical exchange of the two-dimensional data fields describing the interfaces with the other model components. In the exchange, these fields are remapped by the CMF from the source to the receiving grids.

In the coupled model, the data sent from the ice to the ocean includes ice concentration, ice-water stress components, fluxes of heat, salt, fresh water, and penetrating solar radiation. These quantities are summed over all ice categories. The ocean component sends to the ice component the fields of sea surface temperature, salinity, and velocity components, the surface tilt, and the freezing-melting potential. The latter is calculated before the data transmission and is equal to the coupling interval time mean heat flux necessary for the surface grid cell temperature to be brought to the freezing point. In the case where the potential is positive (i.e. the temperature is below the freezing point) the model instantly changes the water temperature to the freezing point, and then the ice model gradually calculates the corresponding changes in the ice volume and concentration, salt, and fresh water fluxes.

The atmosphere component is implemented by means of a CMF function which reads the forcing data from the NetCDF files and sends them to the ocean and ice components with remapping to the corresponding grids. The ocean and ice components receive the air temperature, specific humidity, and wind velocity components at a height of 10 m, downwelling short- and long-wave radiation, rain and snow intensity.

## 3. Numerical experiment setting

The purpose of the numerical experiment was to simulate the seasonal variability of the water circulation and ice of the Arctic Ocean under some defined annual variations of the atmospheric circulation. The experiment setting corresponds to the CORE-II protocol, which provides the daily mean downwelling radiation fluxes, monthly mean precipitation, continental runoff, and the diurnal cycle of atmospheric parameters (temperature, humidity, and wind) taken from the NCEP/NCAR reanalysis for the year 1948. The data also contains some modifications [16] for the ocean model to run correctly without the interactive atmospheric component. The monthly mean continental runoff is defined as the precipitation spread over the ocean surface close to the coasts and river estuaries. We have performed 5 model years of calculation with cyclic specified forcing data.

To avoid the model sea level drift, a surface water flux normalization condition is applied, i.e. the global mean value is subtracted from the total water flux (Prec − Evap + Runoff). The initial temperature and salinity conditions are taken from the annual mean WOA09 data, and the initial ocean current velocities are zero. The initial sea ice distribution fills the ocean north of 70° N with uniform 2m-thick ice. The ocean bottom topography is interpolated from the ETOPO5 database [17] without the internal continental water reservoirs and the Black Sea. The model time step is 10 min for the baroclinic and ice processes, and 20 s for the barotropic processes.

The horizontal model grid is three-polar [18] with poles in Siberia, North-Western Canada, and at the true geographic South Pole. The grid resolution in the spherical part (south of 60° N) is 0.25°. The grid cell dimensions vary from 28 km × 28 km at the equator to 6 km × 28 km on the Antarctic coast, 8 km × 17 km at the North Pole, and 11 km × 8 km along the continental coast of the Arctic Ocean. The vertical discretization includes 49 levels with steps from 6 m in the upper layer to 250 m in the ocean depth. The ice model works in the area north of 42° and uses the same horizontal grid. Since the Antarctic is not the subject of the current study, it is handled with a simple built-in ice model [6] to reduce the computational expenses.

The space-variable horizontal heat and salt diffusion coefficient is taken proportional to the square root of the horizontal grid steps product, and the biharmonic diffusion coefficient is proportional to the 3rd power of this value. The simple and biharmonic background viscosity coefficients are scaled due to the grid steps in the same way. The corresponding diffusion values on the equator are $10^3$ $m^2/s$ and $-9\cdot10^9$ $m^4/s$. The horizontal viscosity coefficient is parameterized by the Smagorinsky method [19] with the coefficient $C^2$=4, and the background equatorial value is $10^3$ $m^2/s$. For the biharmonic viscosity the coefficient is the same, and the background equatorial value is $-27\cdot10^9$ $m^4/s$. The shallow water equations use viscosity with a uniform coefficient of 100 $m^2/s$.

## 4. Results of the numerical experiment

Let us consider the last 5th year of the experiment and compare the model ice concentration field with the ERA-20C data for the months of the ice cover maximum and minimum, March and September. For the March ice concentration there is a good agreement of the model results with the reanalysis data (figures 1a and 1b). The only exception is the North Pacific region where the model shows slightly more ice. This confirms a hypothesis of [20] that to simulate the winter ice extent in the Northern Hemisphere it is generally sufficient to use a thermodynamic ice model.

The Arctic and North Atlantic ice extent in September is overestimated by the model, possibly because of the insufficient northward heat transport due to the low resolution, high model ocean viscosity, and the absence of southward ice transport, because the dynamic ice submodel was turned off, for instance, by the Labrador Current. On the other hand, the Pacific sector of the Arctic Ocean lacks ice, which may be explained by the incorrect simulation of the Bering Strait water transport or by peculiar features of the ocean weather in this year which cannot be simulated without taking into account the interannual variability. Since the 1990s there are many measurement data estimating this transport on a level of approximately 0.8 Sv northward, for instance, [21]. For the mid-20th century there are less data and their spread is larger, from 1 to 2-3 Sv [22]. Therefore, the estimate of 1.2 Sv by our model seems realistic. However, this question should be investigated in more detail to assess both the interdecadal variability and the accuracy of measurements and reanalysis data. Note that a close value of the Bering Strait transport (1.3 Sv) was obtained in [23] by the eddy-resolving INMIO model configuration with the ERA40 forcing.

Figure 1c shows the deviations of the modelled monthly mean temperatures from the corresponding fields of the ERA-20C reanalysis data for the chosen year. Inside the Arctic Ocean the deviations are rather small, since the surface temperatures are fixed by the freezing point equation. As for the neighboring basins, the large anomalies are close to the ice boundaries in the Labrador and Greenland Seas. Such seasonal conditions are very hard to simulate without an interactive atmospheric model. The mechanisms of model solution biases in this case are discussed in [16].

The March temperature distribution is greatly affected by the vertical mixing processes, since in this month it is usually most intensive. Consider the March upper mixed layer depth (MLD) based on the model solution and on the WOA09 data (figure 2). According to [24], let us define the MLD boundary as the depth at which the potential density (calculated from the monthly mean temperature and salinity) deviates from its surface value by 0.125 σ-units. The plotted depth range is limited to 1000 m. The model reproduces the MLD field reasonably well, except for the region of east-Atlantic mid-latitudes, possibly due to the weather peculiarities of the current year.

IOP Conf. Series: Earth and Environmental Science **48** (2016) 012013 doi:10.1088/1755-1315/48/1/012013

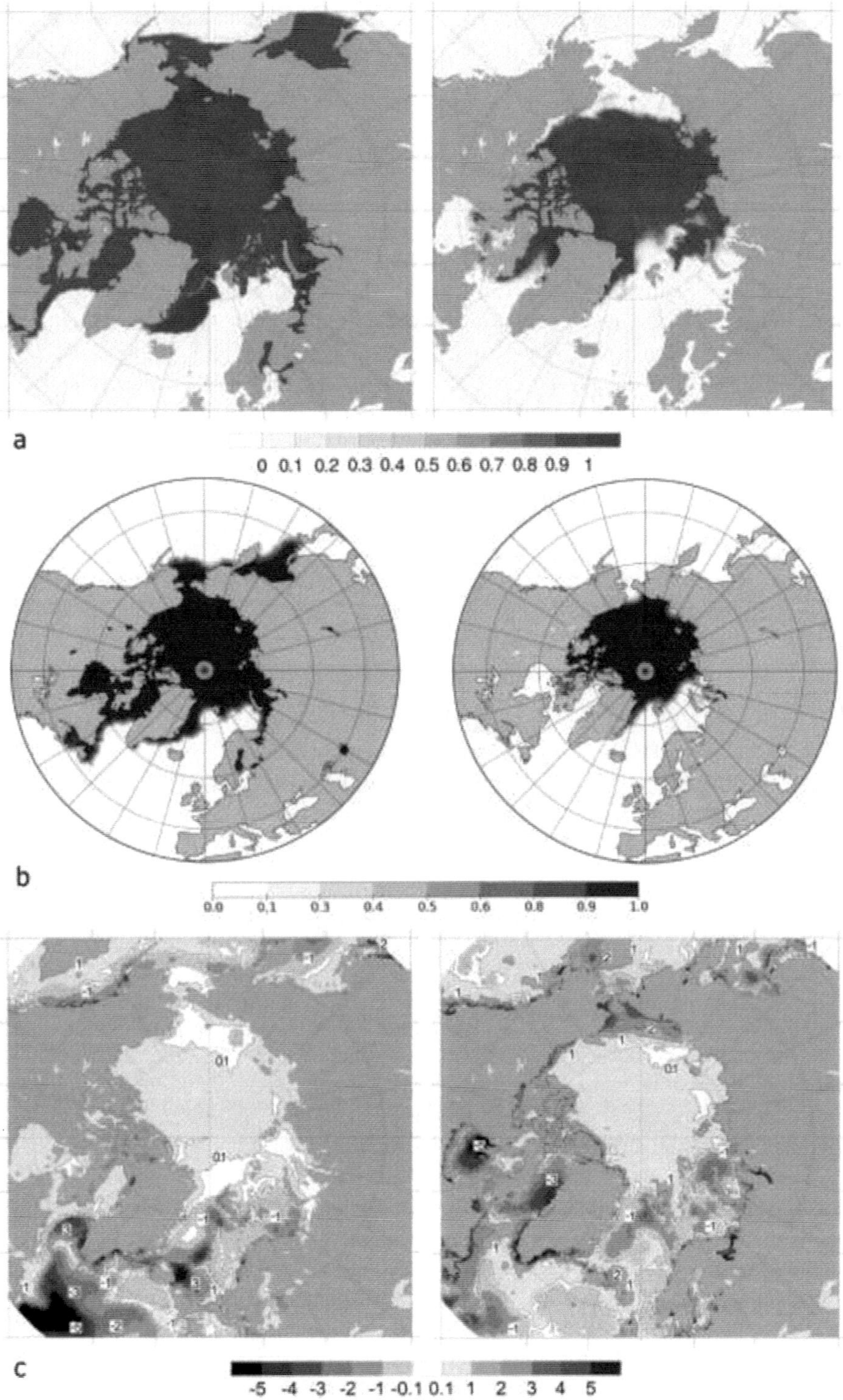


**Figure 1.** Monthly mean fields for March (left) and September (right): a – model ice concentration, b – ERA-20C ice concentration, c – model SST (°C) deviation from ERA-20C.

The horizontal boundaries of the surface warm bias in the Labrador Sea are close to the boundaries of an area with deep mixing in the model solution and no such mixing in the WOA09 data. For example, in the north of the Labrador Sea (62°N, 54°W) the model convective adjustment penetrates up to the 25$^{th}$ model vertical level (550 m), while below this level the model and climatology data are close. The model MLD at this point is 2875 m, which coincides with the full model ocean depth. The climatological MLD is only 100 m, and the difference between the surface and bottom potential densities is 0.5 σ-units. Thus, it can be assumed that the March warm surface bias in the north of the Labrador Sea is caused by incorrect reproduction of the water stratification by the model. To correct such errors, a better vertical mixing parameterization should be used. A similar conclusion can be made from an analysis of the vertical T,S profiles off the south-west Svalbard coast, where the cold season convection leads to the development of a cold model bias persisting in summer.

On the other hand, a similar analysis of the vertical T,S distributions to the east and north-east of Iceland shows that vertical mixing is not directly associated with the development of a "dipole" surface temperature bias existing throughout the year. This effect is most pronounced in winter and is probably caused by the incorrect simulation of the complex structure of ocean currents near the ice edge and polar front [25].

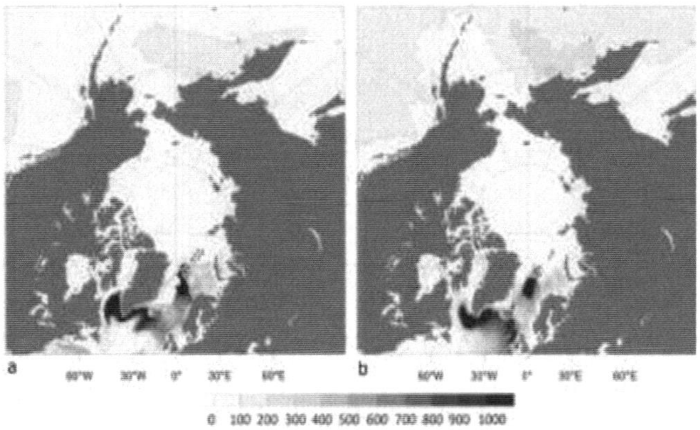


**Figure 2.** Ocean surface mixed layer depth in March (meters, a – based on model solution, b – based on WOA09 data).

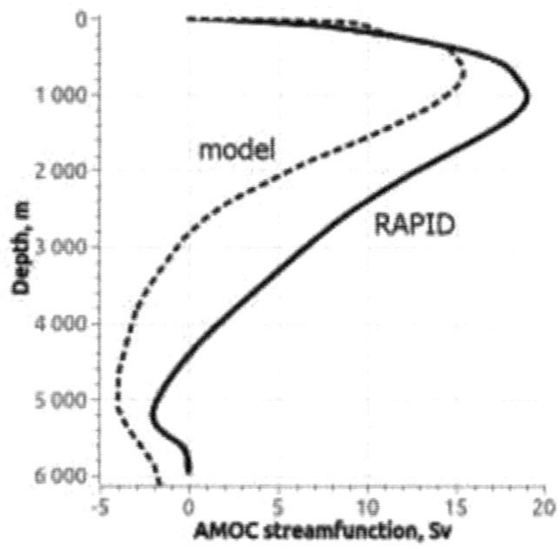


**Figure 3.** Vertical profile of the AMOC stream function at 26.5°N based on model solution and RAPID data.

Figure 3 presents a comparison of vertical profiles of the Atlantic meridional overturning circulation (AMOC) stream function at 26.5°N based on the model solution and RAPID 2004-2008 observation program data taken from [26]. The stream function is defined as the integral Atlantic northward water transport at the chosen latitude between the surface and the chosen depth. The model function differs from zero at the bottom, which corresponds to the contribution to AMOC from the Bering Strait transport, the continental runoff, and the surface water fluxes. The RAPID data include a spatially uniform corrective addition to the velocity field, such that its stream function is zero at the bottom.

Despite the different time periods and uncertainties in the RAPID data (velocity correction, use of the reference level method in open deep ocean), we see that the coupled model underestimates the meridional water transport (the stream function maximum is 15.4 Sv in the model solution and 18.6 Sv in the RAPID data).

The North Atlantic deep water (NADW) penetration depth, determined according to the zero isoline of the AMOC stream function, reaches about 3000 m, which is the value produced by most CORE-II models of [26] but less than that produced by the observational data. Therefore, the underlying Antarctic bottom water (ABW) occupies a larger depth range. The maximal ABW southward transport in the model solution (4 Sv at 5000 m) is correctly located on the vertical axis but overestimated about 2 times; it lies at the upper boundary of the model results range in [26]. Since the main source of NADW are dense and cold Nordic Seas' waters overflowing the Greenland-Scotland Ridge [27], to improve its simulation it may be necessary to use a parameterization of the overflowing process.

Finally, we consider the mechanism of development of the September warm bias in the Chukchi Sea. Figure 4 shows the annual variations of the surface ocean parameters at (69°N, 170°W). According to the CICE conventions, fluxes are defined per unit area of ice, and ice and snow thicknesses are grid cell-averaged.

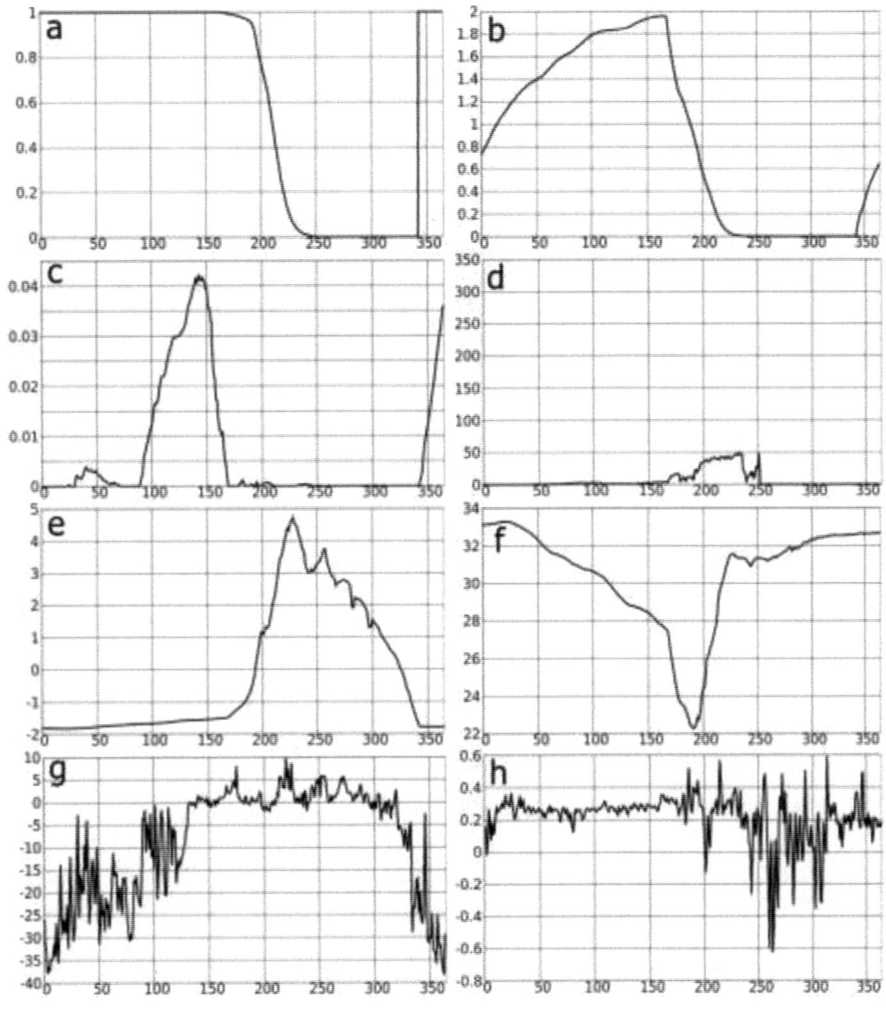


**Figure 4.** Annual variations of surface ocean parameters at (69°N, 170°W): a – ice concentration, b – ice thickness (m), c – snow thickness (m), d – ice-penetrating shortwave radiation (W/m$^2$), e – sea surface temperature (°C), f – sea surface salinity (psu), g – air temperature (°C). In the narrowest section of the Bering Strait: h – section-averaged northward surface velocity (m/s).

It can be seen that the dramatic change in the surface ocean characteristics in the chosen location is on the 170$^{th}$ day of the year, when the sea surface temperature exceeds the freezing point and the ice concentration and thickness start to decrease quickly. Before that time the ocean is covered with ice and snow, and the ocean surface heat and radiation fluxes are close to zero. Thus, we may conclude that the gradual pre-170$^{th}$ day heating of the ocean surface layer may be caused by warm Pacific water whose transport in the Bering Strait is relatively stable in this period. A contribution can also be made by the snow cover melting, which starts on the 140$^{th}$ day, when the air temperature rises above zero and leads to an increase in the solar radiation absorbed by the ice and ocean water. This process is accompanied by gradual ocean freshening, which speeds up when the ice starts melting and stops on the 190$^{th}$ day, when the Bering Strait current slows down and becomes unstable. Thus, the Chukchi Sea ice discharge in September can be qualitatively explained by two successive processes: 1) gradual Pacific water heating and penetrating solar radiation due to melted snow cover; 2) start of fast ice melting and open water heating by solar radiation. In [28] it is pointed out that precipitation plays an important role in the ocean freshening in this period, which deserves further investigation.

## 5. Conclusions

The work has yielded the first results of a calculation of the Arctic Ocean characteristics using a coupled simulation by the INMIO-CICE model with a spatial resolution of 0.25°. The 5-year numerical experiment is the first important step in the development of a modern coupled ocean-ice model. Its results show that on the space and time scales considered the coupled model can reproduce the Arctic Ocean thermohydrodynamic variability qualitatively well. The causes of the main model solution biases from the "real" ERA-20C reanalysis data and WOA09 climatology have been identified. The ice-ocean model can be used as a powerful tool for the investigation of large-scale circulation and regional forecasts. Further work will include tuning of the model in an eddy-permitting regime with ice dynamics, improvement of the parameterizations, and use of a higher resolution.

**Acknowledgements**

This work was supported by the Russian Science Foundation (project no. 14-37-00053) and performed at the Hydrometeorological Research Centre of the Russian Federation.

## References

[1] Large W and Yeager S 2009 The global climatology of an interannually varying air–sea flux data set *Clim. Dyn.* **33** (2-3), 341-364.
[2] Poli P et al 2016 An Atmospheric Reanalysis of the Twentieth Century *J. Climate* **29**, 4083–4097.
[3] Locarnini R A et al 2010 World Ocean Atlas 2009, Volume 1: Temperature. S Levitus Ed. *NOAA Atlas*, 184 pp.
[4] Antonov J I et al 2010 World Ocean Atlas 2009, Volume 2: Salinity. S Levitus Ed. *NOAA Atlas,* 184 pp.
[5] Large W and Yeager S 2004 Diurnal to decadal global forcing for ocean and sea-ice models: the data sets and flux climatologies *NCAR Technical Note: NCAR/TN-460+STR* (CGD Division of the National Center for Atmospheric Research) 105 pp.
[6] Schrum C and Backhaus J 1999 Sensitivity of atmosphere-ocean heat exchange and heat content in North Sea and Baltic Sea. A comparative Assessment *Tellus* **51**A 526-549.
[7] Zalezak S T 1979 Fully multidimensional flux-corrected transport algorithm for fluids *J. Comput. Phys.* **31** 335-362.
[8] Griffies S M 2015 Elements of the Modular Ocean Model (MOM) (2012 release with updates) *GFDL Ocean Group Technical Report* **7** 645 pp.
[9] Kalmykov V V 2013 Program complex for numerical modelling of the coupled system of ocean-atmosphere on massively parallel computers, PhD thesis (Moscow, INM RAS, in Russian).
[10] Ibrayev R A, Khabeev R N and Ushakov K V 2012 Eddy-resolving 1/10° model of the World Ocean *Izv. Atmos. Ocean. Phys.* **48(**1) 37-46.
[11] Johnson M et al 2012 Evaluation of Arctic sea ice thickness simulated by Arctic Ocean Model Intercomparison Project models *J. Geophys. Res.* **117** C00D13.
[12] Hunke E C et al 2015 CICE: the Los Alamos Sea Ice Model Documentation and Software

User's Manual Version 5.1 (Los Alamos National Laboratory).
[13] Feltham D L, Untersteiner N, Wettlaufer J S and Worster M G 2006 Sea ice is a mushy layer *Geophys. Res. Lett.* **33** L14501.
[14] Pringle D J, Eicken H, Trodahl H J and Backstrom L G E 2007 Thermal conductivity of landfast Antarctic and Arctic sea ice *J. Geophys. Res.* **112** C04017.
[15] Kalmykov V V Ibrayev R A 2013 A framework for the ocean-ice-atmosphere-land coupled modeling on massively-parallel architectures *Vychislitel'nye Metody i Programmirovanie* **14** 88-95.
[16] Griffies S M et al 2009 Coordinated ocean-ice reference experiments (COREs) *Ocean modelling* **26(**1-2) 1-46.
[17] ETOPO5 1988 Data Announcement 88-MGG-02. Digital relief of the Surface of the Earth (NOAA, National Geophysical Data Center. Colorado).
[18] Murray R J 1996 Explicit Generation of Orthogonal Grids for Ocean Models *J. Comp. Phys.* **126**(2) 251-273.
[19] Griffies S M and Hallberg R W 2000 Biharmonic Friction with a Smagorinsky-Like Viscosity for Use in Large-Scale Eddy-Permitting Ocean Models *Mon. Wea. Rev.* **128** (8) 2935–2946.
[20] Ushakov K V, Ibrayev R A, Kalmykov V V 2015 Simulation of the World Ocean climate with a massively parallel numerical model *Izv. Atmos.Ocean. Phys.* **51**(4) 362-380.
[21] Roach A T et al 1995 Direct measurements of transport and water properties through the Bering Strait *J. Geophys. Res. Oceans* **100(**18) 443–458.
[22] Bloom G L 1964 Water transport and temperature measurements in the eastern Bering strait, 1953–1958 *J. Geophys. Res.* **69** (16) 3335–3354.
[23] Khabeev R N 2013 Features of the North Atlantic water circulation in the 3-dimensional eddy-resolving World Ocean model PhD thesis (Moscow, MSU, in Russian).
[24] Monterey G and Levitus S 1997 Climatological cycle of mixed layer depth in the world ocean. (U.S. government printing office, NOAA NESDIS, Washington, DC) 5 pp.
[25] Wanamaker A D et al 2012 Surface changes in the North Atlantic meridional overturning circulation during the last millennium *Nat. Commun.* **3** 899.
[26] Danabasoglu G et al 2014 North Atlantic Simulations in Coordinated Ocean-ice Reference Experiments phase II (CORE-II). Part I: Mean States *Ocean Modelling* **73** 76-107.
[27] Dickson R and Brown J 1994 The production of North-Atlantic deep waters – sources, rates, and pathways *J. Geophys. Res. Oceans.* **99** (C6) 12319–12341.
[28] He Y et al 2015 Observed features of temperature, salinity and current in central Chukchi Sea during the summer of 2012 *Acta Oceanol. Sin.* **34** (5) 51–59.

# Peculiarities of Lake Baikal water level regime

**Darima B Dabaeva, Bair Z Tsydypov, Alexander A Ayurzhanaev, Sergey G Andreev, and Yendon Zh Garmaev**

Baikal Institute of Nature Management SB RAS, Sakyanovoi st., 8, Ulan-Ude, 670047, Russia

dabaevadarima@gmail.com

**Abstract.** A 15-year-old low-water period in the basin of Lake Baikal established an endurance record in the entire history of observations. It began in the mid 90s of the last century. With some probability it may continue in the following years. An analysis of meteorological series of air temperature and precipitation in the region is conducted. A statistically significant trend of increasing temperature and decreasing rainfall is revealed. Atmospheric precipitation affects the long-term fluctuations in the river run-off to a greater extent than the other elements of the water balance. An analysis of the inflow of water into Lake Baikal is performed. It is found that the water level of the lake almost directly depends on the water content of the Selenga River. The minimal run-off in dry periods, as well as the annual run-off, tends to decrease. It is a continuous series of low run-off, which provided the negative trend in the minimal run-off. A dendrochronological reconstruction of the Selenga River run-off is made. A statistically significant trend of decreasing Selenga River run-off is revealed in the recent decades, and an analysis of temperature and precipitation for the basin on the Russian side is made.

## 1. Introduction

Lake Baikal is the largest freshwater lake by volume in the world, containing roughly 20 % of the world's unfrozen surface fresh water.

With a volume of 23.6 thousand km$^3$ of fresh water, it contains more water than all the North American Great Lakes together. With a maximum depth of 1642 m, the Baikal is the world's deepest lake. It is considered to be one of the clearest lakes and it is the world's oldest lake – 25 million years old. It is the seventh-largest lake in the world by surface area.

Known as the 'Galapagos of Russia', its age and isolation have produced one of the world's richest and most unusual freshwater fauna, which is of exceptional value to evolutionary science.

Lake Baikal together with the adjacent territories in 1996 was included in the UNESCO World Heritage List. In connection with this, the problem of the protection and conservation of lake water resources is acquired exclusively acute [1].

The Baikal Lake basin is located almost in the center of the vast Asian continent and covers an area of 545 km$^2$ (excluding water area of Lake Baikal), 45 % of which lies within the Russian Federation, the rest is on the territory of Mongolia. About 73 % of river waters are formed on the territory of Russia, 27 % – Mongolia [1].

Lake Baikal and its basin is a peculiar and very fragile natural ecosystem, which provides a natural process of water formation, famous for its transparency and purity.

Comprehensive study of the lake has a particular importance due to the problem of acute shortage of fresh water. And according to experts, the world may face it in the coming decades. The most significant man-made interference with the natural state of the lake system was the construction of the Irkutsk hydroelectric stations and Baikal cardboard mill [2].

Because of the global warming in Transbaikalia the temperature of air limit layer increases, precipitation decreases [3]. All this has led to the low water content of the rivers flowing into the Baikal. There has been steady declining trend of the main flow of the Baikal tributary – the Selenga River due to the lack of rainfall [4, 5].

## 2. Materials and methods

An analysis of the meteorological series of air temperature and precipitation was held along the whole basin of Lake Baikal based on the initial data of weather stations and a global database of geospatial meteorological parameters Climate Research Unit – CRU (spatial resolution of 0.5×0.5 of angular degree) [6].

The Analysis of the hydrometeorological information was held on the basis of historical data and weather gauging stations of the Republic of Buryatia, Transbaikalia region and Mongolia.

## 3. Climate change in the Baikal region

On the basis of the annual CRU data a statistically significant trend of increasing temperature and decreasing rainfall in the basin of Lake Baikal was found.

Two humidification periods were allocated: 1980-1998 years – wet period (figure 1, a); 1999-2015 years – dry period (figure 1, b). The wet period is characterized by positive trends of precipitation totals for almost the entire basin of Lake Baikal, with the highest growth rate observed for Khamar-Daban ridge – 14 mm, and for the area of the Middle Baikal – 10 mm for the period under review (figure 1).

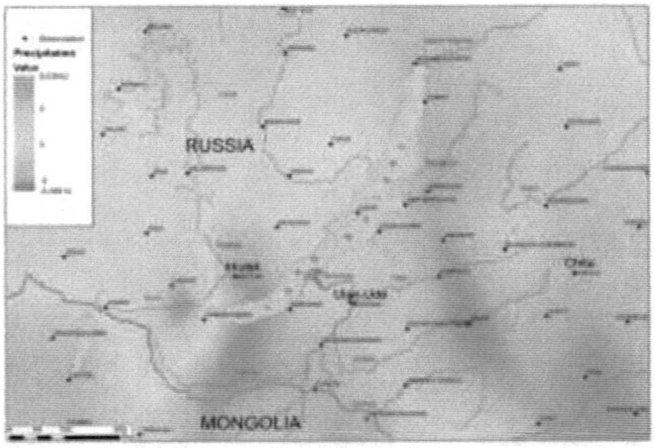


)

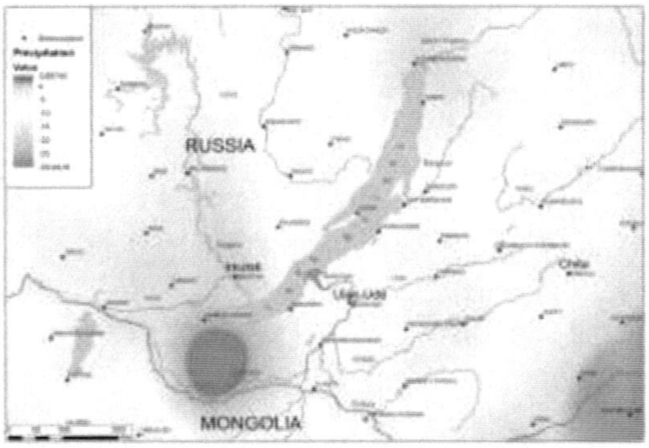


b)

**Figure 1.** Map of linear trends of precipitation, mm for the period under review : a) 1980-1998 years; b) 1999-2015 years.

During the dry season in 1998-2015 years there is a negative trend in the amount of precipitation for the basin of Lake Baikal. Extreme reduction in moisture for the dry period under review is marked for Khamar-Daban ridge and amounts to -30 mm. In the Upper Angara and the Barguzin river basins the rate of decreasing reaches -18 mm. Atmospheric precipitation effects the long-term fluctuations in river flow to a greater extent than other elements of the water balance.

From 1885 to 2012 years the average annual air temperature in the Trans-Baikal region increased by 2.0 °C [5] (note that at the same time, the annual average temperature around the globe increased by 0.85 °C) [7]. The temperature increase is observed for the entire catchment area of Lake Baikal, while the areas with low growth rates of surface temperature are interspersed with areas with high growth rates, both in latitudinal and longitudinal directions (figure 2).

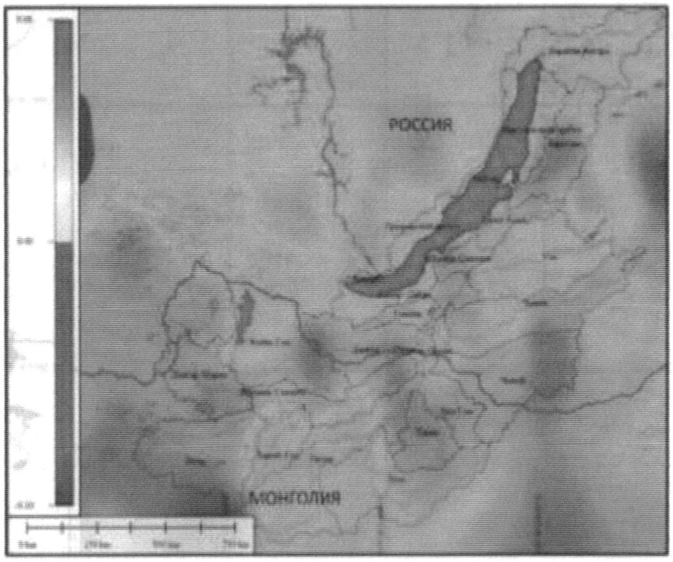


**Figure 2.** Map of linear trends of surface air temperature 2000-2014 years. (Winter months), °C/year.

### 4. Peculiarities of Lake Baikal water level regime

After the construction of the Irkutsk hydroelectric station (1956) Lake Baikal became an artificial water reservoir, as its level is being determined, to a greater extent, not by natural factors but the interests of the hydropower industry [2].

It should be noted that since 1996 Lake Baikal is UNESCO World Natural Heritage site. Due to the fact that the Baikal is an artificial reservoir, it can lose this honorary status, which may lead to the risk of inclusion of Lake Baikal to the list of World Heritage in Danger.

The actual water levels of Lake Baikal in vivo (1900 – 1956 yrs.) ranged from 454.93 m (historical minimum was recorded in 1904) to 457.10 m (maximum was recorded in 1869). In regulated conditions (1960 – 2015 yrs.) the minimum mark was registered in 1982 – 455.27 m, the maximum in 1988 – 457.42 m (hereinafter used Pacific Heights system).

Average annual useful inflow into the lake during the observation period (1900-2015) is 1872 $m^3/s$, in vivo – 1916 $m^3/s$, minimum average annual inflow into the lake was observed in 1903 – 1106 $m^3/s$, the maximum was in 1932 – 3251 $m^3/s$; over-regulated in the period – 1824 $m^3/s$, the minimum inflow was recorded in 1979 – 1244 $m^3/s$, the maximum – in 1973 – 2848 $m^3/s$ [8].

If the outflow of Lake Baikal will be controlled, it may gradually lose its uniqueness and over the time this can lead to the loss of biodiversity up to the complete disappearance of many species. When the level exceeded 457.0 m, during the high-water years in the middle of the 90-ies of the last century, low-lying coastlines of the east coast were destroyed (coastal forests, recreation zones, beaches and coastal constructions), it caused environmental damage throughout the natural and biological complex of lake system [9]. It should be noted that the real threat of complete destruction of Yarki island ridge that separates the open Baikal from Verkhneangarsky shallows exists. When approaching the lake level to the mark 457.0 m and a corresponding wave of activity for 3-5 days Yarki island group would be flooded, as a result Lake Baikal would increase to the north up to 40-50 km and Verkhneangarsky sor and the delta of the Kichera River, the Upper Angara River will completely disappear. Cold Baikal water will destroy all unique ecosystems of shallow waters in the delta of the Upper Angara and the Kichera including Verkhneangarsky omul race. At the same time the parameters of the lake water bowl and consequently the level regime will change [10].

If the level of Lake Baikal is extremely low, we can observe changes in the groundwater regime and groundwater lowering; violation of the existing mechanisms of Baikal water purification; reduction in water exchange of the sor system with open Baikal, the increase in average temperatures and heavy overgrowth of shallow water; the death of aquatic organisms on the shore of the lake and the coastal system as a result of drying and freezing of the habitats, which are responsible for the processing of organic matter; peat fires in the Selenga River delta [11].

The spring-and-summer period in 2015 was characterized by great losses of forests from fires in the Republic of Buryatia. The dynamics of forest fires on the lands of the State Forest Fund of the Republic for the period from 1936 to 2015 was analyzed. According to the Federal Forestry Agency of the Republic of Buryatia from April to October of 2015 Forest fires caught the area of 750,500 hectares, that exceeds the long-term annual average by 12 times. The total damage caused by the forest fires was 20.46 billion rubles (according to the minimum rates of payment for standing timber) and 203.36 billion rubles (according to the auction prices of standing timber). It should be noted that the calculations do not include the basic costs of firefighting.

Before the construction of the Irkutsk hydroelectric complex (in natural conditions) the level of the lake varied in the range up to 2.17 m. During the period of operation of the Irkutsk hydroelectric station until 2001 the level of the Baikal was 17 times higher than the mark of 457.0 m and 18 times it fell below 456.0 m [12]. After acceptance in March 26, 2001 of the Russian Federation Government Decree № 234 "On the limit values of the water level in Lake Baikal in the implementation of economic and other operations" the level fluctuated in the meter range (456.0–457.0 m), but on 25.02.2015, for the first time since 2001, it fell below the mark of 456.0 m. This is primarily the result of the low water period, which was established in the middle of the 90-ies of the last century and was the highest rate ever recorded. The situation with the low-water level of Lake Baikal basin escalated in June of 2014, continued throughout the summer and autumn of the last year and continues in 2016. With a certain degree of probability it can continue in the following years.

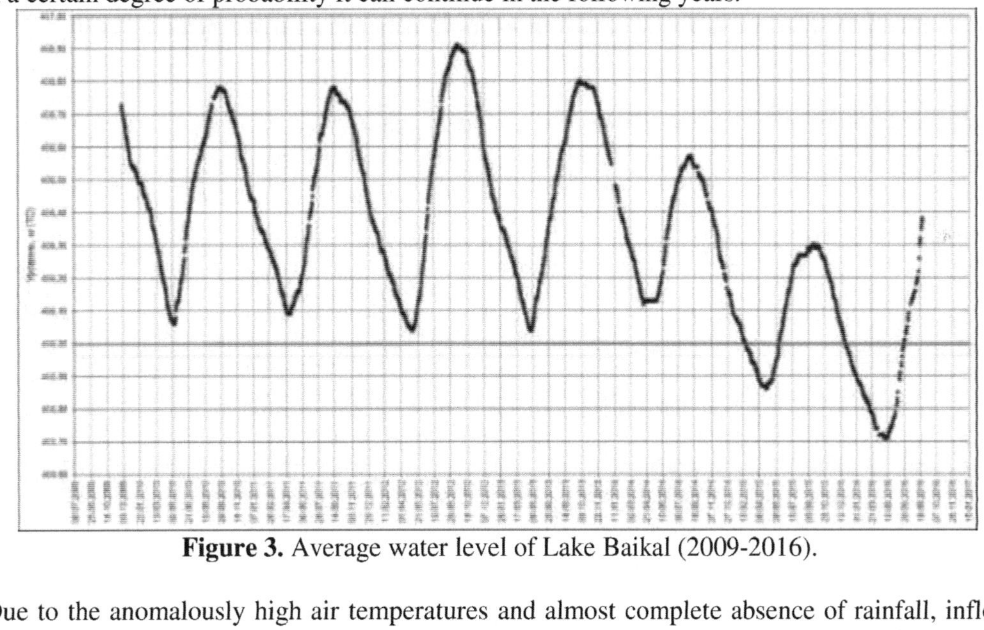

**Figure 3.** Average water level of Lake Baikal (2009-2016).

Due to the anomalously high air temperatures and almost complete absence of rainfall, inflow of water into the lake in the last 2015 was extremely low. The volume of inflows in 2015-2016 aquicultural years expected to be close to a minimum, which observed in the 1903-1904 years and amounted 34.7 $km^3$. In the current aquicultural year, inflow is expected to reach 35.3 $km^3$ at long time average annual value of the normal equal to 61.9 $km^3$. Thus, the useful inflow deficit could reach more than 26 $km^3$.

It should be noted that in 2015 the level below 456.0 m was observed (with a recession) from February 19 to (at rise) June 5, 108 days duration. The Minimum level mark 455.86 m was detected on April 26-28. The maximum level of Lake Baikal water was 456.30 m, which is the lowest value since 2001. In 2016 the level below 456.0 m was observed (with a recession) December 26 to (at rise) June, 29, 184 days duration. The Minimum level mark 455.71 m was detected on April 28 to May 06.

An analysis of the inflow of water into Lake Baikal was carried out. The minimum and maximum water flow, volume of the flow of the major rivers of Lake Baikal basin were studied. It was found that the run-off of the Upper Angara and the Barguzin rivers in the last 20 years has remained within the long-term average rate, the Selenga run-off was reduced and now stands at 65 % of the normal value. Together, these three rivers provide 70 % of the annual inflow of water into Lake Baikal. The level of the lake almost depends on the water content of the Selenga. A good alignment between the inflow fluctuations in Lake Baikal and the Selenga river run-off was determined, which was confirmed by the high values of correlation of the coefficients between these variables: for the observation period (1934-2014) – 0.85, during the dry periods (1954-1958, 1976-1982, 1996-2014) – 0.68 [8].

The formation of water resources in the catchment area of Mongolia which accounts 67 % of the watershed district of the Selenga was studied. The dynamics of the Selenga and the Orkhon rivers flow, its main tributary, was studied. High-water period of the Selenga River was observed from 1979 to 1995. In recent years shallow period in the Selenga River basin is observed. Thus for the last 20 years about 700 rivers and 450 lakes dried up in Mongolia according to the Institute of Meteorology and Hydrology of Mongolia (Ulaanbaatar). The Government of Mongolia intends to start the construction of hydroelectric power station "Shuren", designed in the middle reaches of the main riverbed of the Selenga. Besides "Shuren" HPS it is planned to build hydroelectric power plants on the tributaries of the Selenga – Orkhon, and Delgermuren and Egiin-Gol. Let us note that such actions of Mongolia and the World Bank violate international convention on transboundary waters and a number of other international agreements.

The minimum flow during the dry periods, as well as the annual flow, tends to decrease. It is a continuous period of low flow which provided a negative trend of the minimum flow. Thus, the inflow of water into Lake Baikal in recent years is the record-breaking minimum for the whole period of instrumental observations.

## 5. Reconstruction of the water flow of the Selenga river basin

Unfortunately, for a full analysis of the history of the Baikal fluctuations, we have a too short period of instrumental observations – only 110 years. Considering that our object under study is Lake Baikal, which is the oldest lake in the world (20-35 Million years), this short period of observation data does not allow us to get a more or less objective picture of the changes in its level for a longer time. In this regard, only long-term series of observations will allow to define changes in moisture regime correctly. An analysis of trees annual rings will allow to extend essentially climatic characteristics data [14-18]. For example, the reconstruction of the Baikal level dendrochronological data is presented in the work of Andreev SG [19].

According to gauging stations and dendroclimatic stations, a spatio-temporal reconstruction of river flow parameters in the Selenga river basin was conducted [20]. As a result, water flow model reconstruction of the Selenga River and its tributaries – rivers: Uda, Khilok, Chikoy, Dzida, Orkhon and Kharaa was obtained (figure 4).

For individual watercourses the time-series until 1666 have been restored. This allowed us to reveal regularities of moisture fluctuations in Baikal Asia in retrospective.

In general, figure 4 shows that there is a spatial and temporal coherence of growth of trees with water flow dynamics of the Selenga River and its tributaries. This is particularly evident in the dry and humid periods. The correlation values for individual tributary basins are analyzed. The maximum value of 0.66 is a generalized chronology obtained through 8 local chronologies and "responsible" for the valley of the Jida River. Jida River Valley is an isolated area, surrounded by high mountain ranges. Here, the formation of run-off is within the valley. This explains the close relationship with the growth of water consumption of trees, depending in turn on the incoming water from the atmosphere. High correlation can be traced to the small rivers in the steppe and forest steppe zones.

For the Chikoy River connection with wood rings is smaller ($R = 0.47$). This value can be explained by the fact that the formation of the Chikoy river flow is under the influence of the Pacific air masses coming in the second half of the summer on the east and affecting the upper catchment area, thereby weakening the bond dendrochronological series of stations located in middle and lower parts of the valley.

It should be noted that the reconstruction of the Kharaa River water flow, built on the same stations as for the Chikoy River, shows a similar picture while taking into account that Kharaa basin is located 300 km to the south. This coupling is determined by the general laws of the Chikoy and Haran rivers run-off formation.

Annual fluctuations in water of the Orkhon River flow showed a close relationship (R = 0.53) with stations located in the Jida River Basin. The values obtained for the correlation between the indices of growth of annual rings and instrumental measurements of river flows on Mongolian part of the Selenga River Basin (for the river Kharaa - 0.47 for the Orkhon River - 0.53) allow us to expand our network of stations in the basins of these rivers and get more reliable model reconstruction.

XX century in comparison with the previous was provided with water. The general trend of the XX century has the following character: until mid-century, water availability has increased, reaching a peak in the mid-1940s, and then began to slowly drop. The modern period is characterized by an increase in water content.

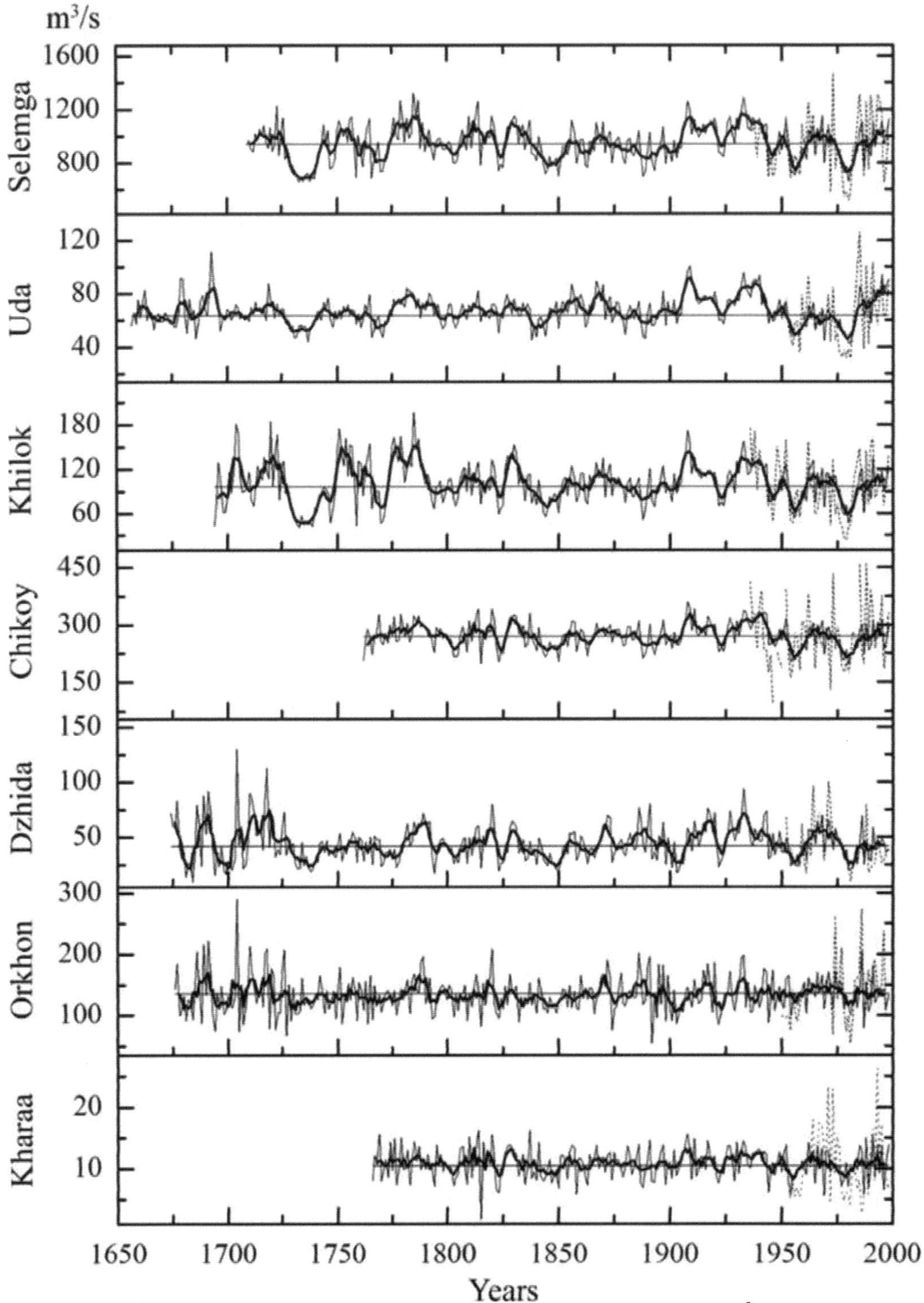


**Figure 4.** The reconstruction of the Selenga River basin rivers water discharge ($m^3/s$). Legend: dashed line – instrumental measurements of water flow for gauging stations, thin curve – year-by-year reconstruction, bold curve – smoothed reconstruction over 5 years, the horizontal line – water flow rate during the instrumental period.

It is interesting to compare the data of historical chronicles with the reconstructed series of water flow.

## 6. A comparison of the historical chronicles with a time series of water discharge

Detailed chronological comparison of historical records and moisture regime on the basis of the reconstruction of the water regime was carried out [21]. Analysis of historical records is an indirect verification of the received time-series for the reconstruction of water flow in the Selenga River Basin.

There was a relatively high-water period from 1700 to 1725. For example, 1715 - "is a very rainy year in Mongolia", falls on the peaks of the reconstructed water consumption [21]. From another source [22] we know that Petropavlovskaya Fortress was built at the confluence of the Chikoy and Selenga rivers and it was first mentioned in 1713. Due to regular annual flooding in 1726 S.L. Vladislavich-Raguzinsky ordered the transfer settlement on 2 mile upstream of the Chikoi River. In 1727, I.D. Buchholz built Petropavlovskaya Fortress in a new place. Petropavlovskaya Fortress increasingly served not as a military target, but as trading posts, where caravans were formed for trade with China.

The period 1726-1745 was characterized as arid and is marked on our reconstructions. In the historical sources for the West Transbaikalia and Mongolia, these years are marked as dry and barren ("great drought") for decades [21]. A similar period, but for a shorter development time was observed in the modern period - in the late 70's - early 80's of the XX century.

Period 1746 - the end of the 50s of the XVIII century was marked by floods in the Selenga River. They peak in 1751, when the great flood was recorded.

Time intervals 1758-1761, 1778-1810 - periods of drought in Mongolia.

The following period was characterized by registered floods. The peak recorded in 1830 - "... great flood on the Selenga River. In Selenginsk most of the right-bank structures were washed away. As a result, in 1841 the city had to be moved to the left bank ... "[21].

1860 year is a year of the summer drought in Central Mongolia. 1862 year in the Trans-Baikal region was marked by the drought: "... unbearable heat, terrible drought, so the crops are bad, a little grass ...". It is noteworthy that in the same year on the Gusinoe lake islands had disappeared, that is, the lake was filled. This fact is explained by the intensification of tectonic processes in the Baikal region during this period (the formation of the Gulf of failure, etc.), and is not related to moisture regime. There was a "great flood on the territory of Buryatia" in the spring and summer of 1866. In 1869: "... the unusual flooding in Irkutsk Province and West Transbaikalia, the level of the lake rose nearly seven feet under the usual ..., ... a great flood on the Selenga ...,. ... level in Verkhneudinsk increased up to1.81 fathoms, part of the city and many villages were flooded ..." [23]. The end of the XIX century as a whole is characterized by the historical chronicles as a dry period. The beginning of XX century was characterized by a sharp increase in water consumption values, which is consistent with our obtained reconstructed series of streamflow.

Thus, considering the high correlation of the Selenga run-off and the lake level, we can indirectly have an idea of the fluctuation of the water level (inflow of water) in the lake for the last 300 years. For example, we already have the opportunity to respond specifically, whether high or low water levels in the lake have been observed in the middle of the XVIII and XIX centuries. This good correspondence of water level in the lake with a volume of annual runoff of the Selenga opens up new opportunities for data on the level of the lake regime over the past three centuries.

## 5. Conclusions

Thus, a significant transformation of the natural environment is currently observed in the Baikal Lake basin.

In particular, a statistically significant trend of increasing temperature and decreasing precipitation is established with the identification of wet and dry periods. In recent years, due to abnormally high air temperatures and almost complete absence of atmospheric precipitation, water flow into the Lake Baikal has been extremely low during the whole period of instrumental observations. As a result, there were problems with the level of the Baikal Lake regime.

There is a high year and seasonal growth in response to the current draft in the dendrochronological series for specific areas. Spatio-temporal coherence growth trees with water flow dynamics of the Selenga River and its tributaries was revealed. So statistical models of reconstruction of water flow in the article show good correlation (r = 0.47 ... 0.66) between the increase in the annual rings of Scots pine and annual consumption of water.

For example, the Mongolian part of the Selenga River Basin shows that the spatial relationship between the water drain and dendroclimatic stations located on the territory of Russia, can be traced for hundreds of kilometers. The data flow rates for a large time interval can serve as a basis for the planning of water resources management. This aspect is particularly relevant for the Selenga River, because of its cross-border location.

It was found that for forest-steppe regions of the Baikal Asia as limiting factor for tree growth are the atmospheric precipitations which form the water flow of the Selenga River Basin rivers.

These historical records without pretending to accuracy of the tool are an illustration of indirect, fragmentary verification of the received time series for the reconstruction of water flow in the Selenga River Basin. However, a significant correlation in the reconstruction of models allows you to use the data for retrospective analysis of periods of droughts and floods.

## 6. Acknowledgments

These investigations were partially supported by the Russian Foundation for Basic Research (grant no. 13-05-41378).

## References

[1] Afanas'ev A N 1976 *Water Resources and Water Balance of the Baikal Lake Basin* (Novosibirsk: Nauka) p 239 *(In Russian)*

[2] Atutov A A, Pronin N M and Tulokhonov A K 1999 *Hydropower and the Lake Baikal ecosystem condition* (Novosibirsk: Publishing House of SB RAS) p 281 *(In Russian)*

[3] Budyko M I and Izrael Yu A 1987 *Antropogenic Variations in Climate* (Leningrad: Gidrometeoizdat) p 406 *(In Russian)*

[4] Garmaev Yen Zh 2010 *River flow of Baikal Lake Basin* (Ulan-Ude: Buryat State University Press) p 272 *(In Russian)*

[5] Obyazov V A 2013 *Russia. J. Proceedings of the RAS* **450** (6) 713 *(In Russian)*

[6] Harris I, Jones, Philip, Osborn, Timothy and Lister, David 2014 *Int. J. Climatol.* **34** 623-642

[7] Kokorin A O 2014 *Climate change: a review of the Fifth IPCC Assessment Report* (World Wildlife Fund) p 80

[8] Bychkov I V and Nikitin V M 2015 *Russia. J. Geography and Natural Resources* **3** 5-16 (*In Russian*)

[9] Bezrukov L A and Nikol'sky A F 1995 *Russia. J. Geography and Natural Resources* **1** 125-134 *(In Russian)*

[10] Tulokhonov A K, Garmaev Yen Zh and Tsydypov B Z 2013 *Russia. J. Geography, Environment, Sustainability* **2** (6) 20-27

[11] Sinyukovich V N 2016 *Russia. J. Water Sector of Russia: problems, technologies, management* **1** 42-51 *(In Russian)*

[12] Nikitin V M, Saveliev V A, Berezhnykh T V and Abasov N V 2015 *Russia. J. Region: economics and sociology* **3** (87) 273-295 *(In Russian)*

[13] Garmaev Yen Zh and Khristophorov A V 2010 *Water resource of Baikal Lake Basin,* ed A K Tulokhonov (Novosibirsk: Academic Publishing House "Geo") p 231 *(In Russian)*

[14] Fritts H C 1991 *Reconstructing Large-Scale Climatic Patterns From Tree-Ring Data* (Tucson: Univ. of Arizona Press) p 286

[15] Cook E R and Kairiukstis L A 1990 *Methods of Dendrochronology. Applications in the Environmental Sciences* (Dordrecht: Kluwer Academic) p 393

[16] Schweingruber F H 1996 *Tree Rings and Environment: Dendroecology* (Berne: Haupt Publ) p 609

[17] Shashkin A V and Vaganov E A 1993 *Russia. J. of Ecology* **5** 3-10 (*In Russian)*

[18] Vaganov E A and Shiyatov S G 1999 *Russia. Siberian Journal of Ecology* **2** 111-116 (*In Russian)*

[19] Andreev S G 1999 *Russia. J. Proceedings of the RAS* **368**(3) 400-403 *(In Russian)*

[20] Garmaev E Zh and Andreev S G 2013 *Runoff reconstruction of major rivers of the Selenga river basin* Certificate of state registration database № 2013621267, Russian Federation. Proposal № 2013620567 /.; Rightholder: Baikal Institute of Nature Management SB RAS; Receipt date: 06.06.2013; Registered in the Registry Database: 26.09.2013 (*In Russian*)

[21] Levi K G, Zadonina N V, Berdnikova N E, Voronin V I, Glizin A V, Jazev S A, Baasandzav B, Ninzbadgar S, Balzinnyam B and Buddo V Yu 2003 *Modern geodynamics and geliogeodinamika. 500-year history of anomalous phenomena in nature and society of Siberia and Mongolia* (Irkutsk: IrGTU) p 383 (*In Russian*)

[22] Artemyev A R 1999 *Cities and forts Trans-Baikal and the Amur Region in the 2nd half of the XVII-XVIII centuries* (Vladivostok: Primorye. polygraph. Plant) p 336 (*In Russian*)

[23] Zadonina N V and Levi K G 2008 *Chronology of natural and social phenomena in Siberia and Mongolia* (Irkutsk: Irkut. state. University Press) p 759 (*In Russian*)

# Vegetation cover dynamics of the Mongolian semiarid zone according to multi-temporal LANDSAT imagery (the case of Darkhan test range)

**M A Zharnikova, ZH B Alymbaeva, A A Ayurzhanaev and E ZH Garmaev**

Baikal Institute of Nature Management SB RAS, Sakyanovoi st., 8, Ulan-Ude, 670047, Russia
Tel .: +7 924 352-72-78, fax: (3012) 433742, e-mail: rita_zharnikova@mail.ru

**Abstract.** At present much attention is given to the spatio-temporal dynamics of plant communities of steppes to assess their response to the current climate changes. In this study, a mapping of a selected modeling polygon was carried out on the basis of data decoding and field surveys of vegetation cover in the semi-arid zone. The resulting large-scale map of actual vegetation reflects the current state of the vegetation cover and its horizontal structure. It is a valuable material for monitoring of changes in the chosen area. With multi-temporal satellite Landsat imagery we consider the vegetation cover dynamics of the test range. To analyze the transformation of the environment by the climatic factors, we compared series of NDVI versus the precipitation and of NDVI versus the temperatures. Then we calculated the degree of correlation between them.
Keywords: desertisation, semiarid zone, vegetation cover transformation, succession, Landsat, mapping of vegetation.

## 1 Introduction

During the last decades the problems of desertification and transformation of the environment are still relevant. Climate changes and irrational nature management lead to the transformation of the vegetation cover. During the last decades an increased grazing pressure on ecosystems has changed nature balance because of higher vulnerability of semiarid and arid ecosystems, entailing their degradation and desertification [1].

Mongolia has a unique complex of natural and climatic conditions and inimitable landscapes. Which have been causing interest for the scientists of different disciplines for ages. The steppes are sensitive to the effects of anthropogenic factors and climate change. They serve as an indicator in the study of spatio-temporal dynamics of landscapes and their response to the current climate changes. Spatially-structural studies of vegetation are particularly important, especially for the monitoring and are presented as an essential condition of an ability to predict the dynamic processes in the landscape sphere.

The aim of this work is to study the structure and dynamics of vegetation of steppe ecosystems of semi-arid zone of Northern Mongolia as in the case of Darkhan testing range. During the study a map current vegetation model site was compiled, the spatial dynamics of vegetation was revealed using the multi-temporal satellite imagery and the correlation values of NDVI with precipitation and temperature was analyzed.

## 2 Materials and methods

*2.1 Characteristics of the study area*

To study the dynamics of degradation and desertification processes of arid lands we have founded a number of testing ranges in different latitudinal zones (51-44° N) on the meridional Baikal Gobi transect (105-107° E). The site, which is located 3 km far to south-west of the city of Darkhan, 19.16 $km^2$ in area, was selected as a testing range (Figure 1). The bigger part of the site is situated in the portioned slopes of small-altitude Orkhon-Kharaa River interfluve (absolute altitude marks are 810-870 m.) and partially at the eastern part of Bayan-Daba mountain foot and on the second Kharaa River terrace above the flood plain. According to botanical-geographical zoning, this area refers to the Mongolian steppe province of the Srednehalhaskaya Subprovince of the Central Asian (Daurian-Mongolian) subregion of the steppe region of Eurasia [2]. The climate of dry Mongolian steppes is characterized by strongly pronounced continentality which is caused due to the location of the country in the centre of Eurasian mainland. This type of climate is formed under the influence of a whole number of factors. The location of dry steppes, 50° N further southward, is conductive to a great amount of solar radiation during the vegetation period. But isolation from the main moisture sources leads to climate aridisation. With increase of climate aridity precipitation deficit increases [3], which affects the composition and quality of phytocenosis.

**Figure 1.** The Darkhan testing range.

*2.2 Satellite data*

To obtain basic information on the spatial organization of ecosystems satellite imagery from Landsat (sensor TM, ETM +, OLI) with a spatial resolution of 30 m were used (available on the website http://glovis.usgs.gov).

To assess the spatial-temporal dynamics of vegetation we used the Landsat imagery, obtained at approximately the same period of the phases of vegetation development: September 20, 2000 September 10, 2002 and August 31, 2013 and July 14, 2014. Processing was carried out using a standard method of classifying images without training ISODATA. The combination of multi-temporal vector layers for each category of vegetation and detection of changes were implemented through the ArcGIS software package.

To assess the current status of vegetation cover we used as a basis the Landsat 8satellite image from 14 July, 2014. On the initial stage the image was classified into 20 classes. Then we combined them into corresponding vegetation types based on the field studies. For each class we made sampling of pixels brightness values of each image channel belonging to some or other groups of plant communities. At the final stage of creation of a geobotanical map we conducted generalization of the image, i.e. the combination of the pixels in uniform contours.

We used 16-day composites of NDVI (product MOD13Q1 [4]) to assess the correlation between rainfall and temperature [5].

To highlight the different ecotypes (sloping surfaces, gullies and ravines systems, etc.) we used materials of radar topographic mapping (SRTM srtm.usgs.gov).

*2.3 Geobotanical data*

Geobotanical study was conducted on the basis of original materials collected during the field work.

In total, according to standard methods, 83 full geobotanical descriptions were performed, 3 geobotanical routing profiles from 2 to 4 km length were conducted, geobotanical monitoring areas were laid (10x10 m), biological productivity of characteristic phytocenoses was defined. Lines of route profiles were laid with the seizure of the most different clusters of the satellite image.

For primary sorting of the descriptions of phytocenoses and their processing using ecological scales the IBIS program was used. 7% of the descriptions were rejected. Summary tables of the species composition, geographical and ecological spectrum were obtained. The types of plants were ranked on scales of moisture and richness, soil salinity.

While identifying the species composition of the vegetation we used abstract of flora [7] and key to identification of plants of Mongolia [8] and Buryatia [9].

## 3 Results

In the analysis of vegetation the array of geobotanical descriptions was differentiated by the principle of uniformity of floristic composition, the prevailing types of horizontal addition of communities, the uniformity of habitat into several groups. Obtained a set of communities which are similar according to the composition and structure are defined as coenoflore, which are comparable in volume to the units of the average rank of vegetation classification, in particular to formations. To highlight the associations, the descriptions were subjected to table processing using the dominant-determinant principle.

Mapping of the selected testing range was carried out on the basis of the data decoding and field surveys. The resulting large-scale map of actual vegetation (M 1:20000) reflects the current state of vegetation cover and its horizontal structure. It is a necessary material for the monitoring of the changes in the chosen area (Figure 2).

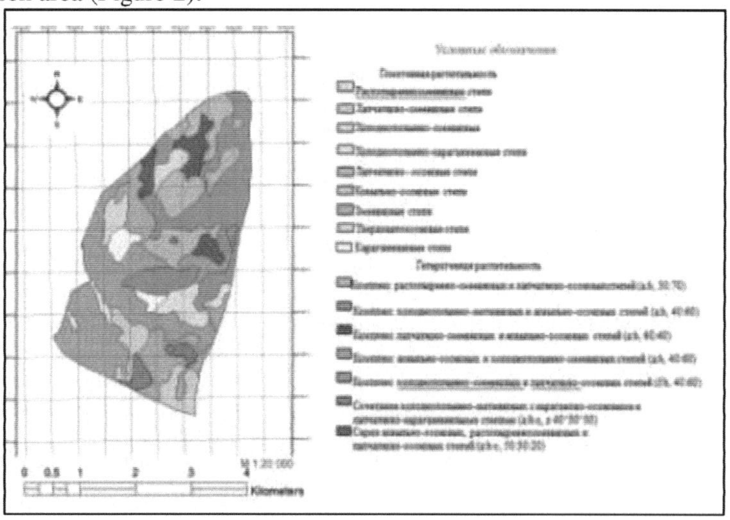

**Figure 2.** Map of vegetation Darkhan testing range.

IOP Conf. Series: Earth and Environmental Science **48** (2016) 012015 doi:10.1088/1755-1315/48/1/012015

Almost everywhere the current vegetation cover is presented by modified communities, many of them are the stages of degradational successions, formed as a result of constant (seasonal and year-round) cattle grazing. Because of the overgrazing the vegetation of these landscapes is on the stage of middle digression. Vegetable species are in bad vital condition. General projective herbage doesn't grow under 40%, while the average value is 15-20%. Assessment of the species coenotic value was made basing on activity of the species. Comparative analysis of middle projective herbage showed, that 4 species have maximal coenotic value: Carex duriuscula (C.A. Meyer), Caragana microphylla (Pall.) Lam., Artemisia frigida (Willd), Potentilla acaulis. Their middle projective herbage is quite big (6-10%). It is established that coenotic value of these species considerably change in the communities of different types of relief.

In the process of mapping in this model range 9 homogeneous (76.8% the occupied territory) and 7 heterogeneous cenoses were identified. The homogeneous type of vegetation is characteristic of sloping areas in the foothills and leveled tops advanced ridges and also on their feet.

The NDVI vegetation index characterizes the content of photosynthetically active biomass. To analyze the interannual dynamics of vegetation depending on climatic factors, we compared the values of NDVI of Darkhan testing range, averaged for the vegetation period (May-September), with rows of precipitation, summed over August and September for each year (Figure 3).

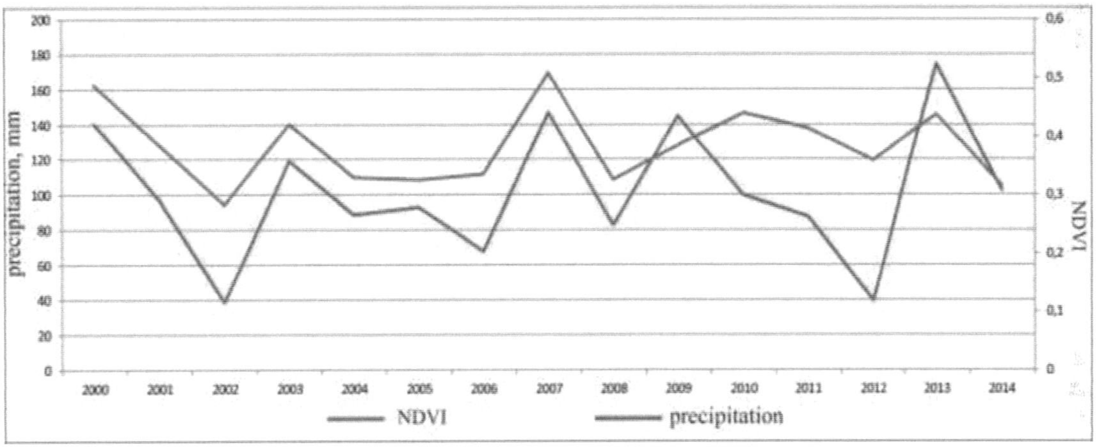


**Figure 3**. The time series of NDVI, precipitation and temperature data.

Correlation analysis revealed that the highest correlation coefficient was observed for NDVI and precipitation total for the August - September and equals to 0.71. The correlation coefficient of NDVI and precipitation amounts during the vegetation period is equal to 0.36, the correlation of NDVI and annual precipitation is equal to 0,41. However, the correlation between NDVI and average temperatures during the summer and autumn periods is very low and does not exceed 0.2. The greatest amount of precipitation falls on was during 2000, 2007, 2009 and 2013. The least amount of precipitation occurs in 2002 and 2012.

To assess the spatial dynamics of vegetation the images for moist years - 2000, 2013, and the dry years - 2002 and 2014 (Figure 4) were used. Unfortunately, due to the high percentage of cloud cover was not available pictures for 2007, 2009 and 2012. They could give the most complete picture of the dynamics of vegetation distribution.

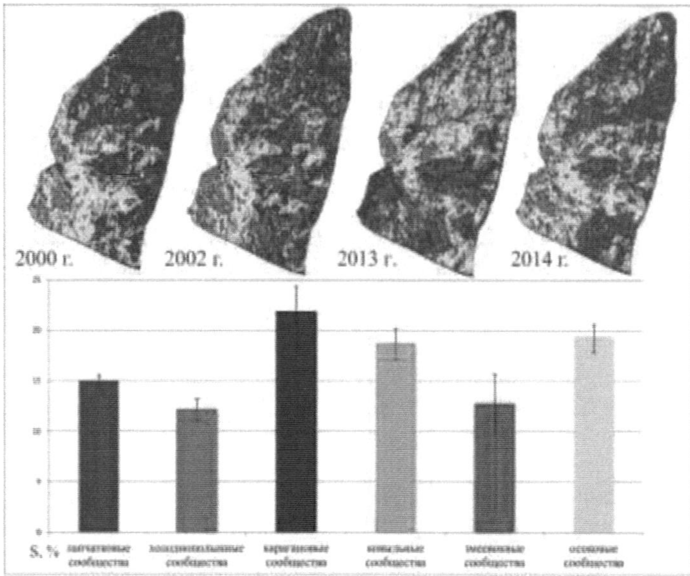


**Figure 4**. Models vegetation testing range on the results of the interpretation of satellite images and the histogram distribution of classes by pixels.

The results of analysis of satellite imagery have identified that the greatest changes in the vegetation areas are observed in Cleistogenes and Caragana communities. Thus, in dry years the area occupied by Caragana communities significantly increased in comparison with the humid years - from 18.6% in 2000 to 24.3% in 2002 and 20.3 percent in 2013 to 24.3% in 2014, while Cleistogenes community reduced the area from 15.7% in in 2000 to 11.6% in 2002 and 14.2% in 2013. to 9.8% in 2014 (Table 1). The intensity of growth Cleistogenes communities is closely linked to the presence or absence of precipitation. In humid years their condition is improved and the amount of biomass increases. During droughts growth communities is severely depressed [10].

**Table 1.** The ratio of the area of plant communities of the model range in percent.

| | 2000 | 2002 | 2013 | 2014 |
|---|---|---|---|---|
| Potentilla communities | 15.6 | 14.9 | 15.5 | 14.3 |
| Artemisia communities | 12.2 | 13.3 | 12.2 | 11.0 |
| Caragana communities | 18.6 | 24.3 | 20.3 | 24.3 |
| Stipa communities | 20.1 | 17.1 | 18.0 | 19.9 |
| Cleistogenes communities | 15.7 | 11.6 | 14.2 | 9.8 |
| Carex communities | 17.8 | 18.9 | 20.4 | 20.6 |

Caragana has a high stress tolerance to drought, however, it is an invasive species, able to occupy the vacated niche. In addition, it is considered that the invasion of adventitious species is one of the leading factors in the transformation of natural ecosystems [11]. Potentilla and Artemisia frigida community remain relatively stable.

Another factor affecting the degradation of plant cover is a high anthropogenic load. According to the Mongolian State National Statistical Committee of the territories of Selenge aimag and Darkhan-Uul in the first decade of this century, there is a steady trend of increasing of the number of farm animals [12]. Express surveys of herdsmen who keep their farms within the territory of Darkhan testing range modeling were conducted. It appears that about half of them migrated here (closer to the main markets for livestock products) with their herds in the 2000s from other peripheral aimags of

Mongolia: western (mostly) and southern - Gobi [12]. Increasing overgrazing pressure leads to the vegetation cover degradation and its change. Pasture desertification acquires the features of an irreversible and progressive character in terms of the dry multi-year period.

## 4 Conclusions

Based on the use of remote sensing data for the first time was examined dynamics of various areas of steppe plant communities on the testing range of Northern Mongolia. Large-scale map of actual vegetation based on remote sensing methods, reflects the current state of cenotic diversity and condition of steppe ecosystems. The results demonstrate a close correlation relationship between the state of the vegetation cover and the amount of precipitation. In drought conditions, the most valuable in fodder grasses partially drop out grass. Their place is taken by digress-active invasive species. Climate-caused vegetation changes occur on the background of increasing anthropogenic load – increase in the number of livestock.

It was possible to assess the current state of vegetation with high precision using information methods. Based on the data of field descriptions of ecotope distribution of steppe ecosystems was identified and the classification of the steppes was carried out. On the basis of processing of satellite images retrospective assessment of the dynamics of vegetation was obtained. The combination of remote sensing and field research methods allowed more efficiently and accurately assess the condition and character of the use of steppe ecosystems.

## References

[1] Bazha S N, Gunin P D, Danzhalova E V, Kazantseva T I, Bayasgalan D and Drobyshev Yu I 2010 Contemporary processes of degradation of pastoral steppe ecosystems in Mongolia *Proc. Intern. Confer.* Vol 1 pp 59–64

[2] Lavrenko E M, Karamyisheva Z V and Nikylina R I 1991 *Eurasian steppes* (Leningrad: Science press) p 146

[3] Beresneva I A 2006 *Climates of arid zone of Asia* (Moscow: Science press) p 288

[4] Huete A, Justice C and Van Leeuwen W 1999 *MODIS Vegetation Index: Algorithm Theoretical Basis Document (ATBD).* Version 3.0. 25 (URL:http://modis.gsfc.nasa.gov)

[5] Hijmans R J, Cameron S E, Parra J L, Jones P G and Jarvis A 2005 Very high resolution interpolated climate surfaces for global land areas *International Journal of Climatology* **25** 1965–78

[6] Zverev A A 2007 *Information technology in vegetation studies* (Tomsk: TML-press) p 311

[7] Gybanov I A 1996 *Abstract flora Outer Mongolia (vascular plants)* (Moscow: Valang press) p 136

[8] Grybov V I 1982 *Key to Identification of Vascular Plants of Mongolia* (Leningrad: Science press) p 442

[9] Anenkhonov O A 2001 *Key to Identification of the Plants in Buryatia* (Ulan-Ude: Republic press) p 672

[10] Kazantseva T I 2009 *The productivity of plant communities of zonal steppes and deserts of Gobi region of Mongolia* (Moscow: Science press) p 336

[11] Alimov A F and Bogutskaya N G 2004 *Biological invasions in aquatic and terrestrial ecosystems* (Moscow: KMK press) p 436

[12] Tsydypov B Z, Garmaev E ZH, Alymbaeva ZH B, Batotsyrenov E A, Ayurzhanaev A A, Sayapina D O, Zharnikova M A, Sodnomov B V and Tylokhonov A K 2016 Spatial-temporal assessment of the change in the plant cover of arid climate zones along the Baikal-Gobi transect *Journal of Science Review* **5** 8-16

# Landscape dynamics assessment of dry climatic zones on the Baikal-Gobi transect from NDVI time series and field investigations data

**D O Sayapina, M A Zharnikova, B Z Tsydypov B V Sodnomov and E Zh Garmaev**

Baikal Institute of Nature Management SB RAS, Sakyanovoi st., 8, Ulan-Ude, 670047, Russia

bz61@mail.ru

**Abstract.** Starting in the eighties of the 20th century, the scientists of the Baikal Institute of Nature Management (BINM SB RAS) have been conducting field observations of the Transbaikalia geosystems transformation due to the change of climate and nature management. An utmost importance is placed on the study of a negative response of the land geosystems. This is shown through their deterioration, degradation, and desertification in particular. Through the years of research (1985-2015) in dry areas of the north of Central Asia, the scientists of the BINM SB RAS established a network of key sites for contact monitoring of the status and dynamics of the geosystems and the negative natural-anthropogenic processes along the Baikal-Gobi meridional transect (51-44° N, 105-107° E). The monitoring of the status and dynamics of the vegetation cover of some key sites is conducted by processing and analysis of multitemporal and multispectral Landsat and MODIS Terra imagery. An automatic analysis of the time variation of NDVI and a comparison with the progress of the index in the previous seasons are performed. The landscape indication of the key sites is made on the basis of satellite imagery and complete geobotanical descriptions. Landscape profiles and facies maps with natural boundaries are created.

## 1. Introduction

According to the United Nations Convention to Combat Desertification (UNCCD), land degradation in arid, semi-arid, and dry sub-humid areas – also referred to as drylands – may result from various factors, including climatic variations and human activities [1, 2].

It includes diverse processes, ranging from changes in plant species composition to soil erosion, and reduces the land's productive potential [3-5]. Land degradation may diminish the land's resilience, making it more vulnerable and reducing its capacity to recover from disturbances. Furthermore, land degradation can have negative effects on other resources, such as water, soil, flora, and fauna [6, 7].

Causes of land degradation can be natural or man-made. Natural causes include periodic stress from extreme and persistent climatic events, aridity, and droughts. Man-made causes include unsustainable human land use – for example, overgrazing, deforestation, and over-cultivation – as well as indirect socio-economic drivers, such as unstable food market prices and political instability or changes [8-10].

Land degradation and desertification processes are widespread in Russia and Mongolia. Over 85% of the Mongolian territory is located in dry climate zones: dry sub-humid, semiarid, arid, and extra-arid.

The vegetation is the first experience the impact of desertification [11], and that's why investigation of its changes have a particular importance.

Our research deals with vegetation dynamics of dry climatic zones of Mongolia from field investigations and NDVI (Normalized Difference Vegetation Index - a standardized index, showing the presence and condition of vegetation) time series analysis. The tasks of this study included: 1) field geobotanical studies on testing ranges and classification of vegetation; 2) analysis of the dynamics of the NDVI vegetation index according to satellite data; 3) analysis of the spatial structure of phytocenosis. A number of model monitoring polygons and key sites of contact monitoring has been established in different latitudinal zones according to the longitudinal transect (105-107° E, 51-44° N). These polygons include a wide range of territories with dry climate conditions. This work revealed the main factors, agents, and trends of development for desertification processes in different climate zones.

Two model polygons were considered in Central Mongolia: 1) the Kharaa River downstream basin and the Orkhon River right feeder; 2) the central part of Dundgovi aimag (Mid-Gobi). The first polygon is situated in a semiarid climate zone with grassland and bunchgrass steppes. The second is located in an arid climate zone; from botanical and geographical points of view, its main feature is the prevalence of desertified steppes.

## 2. Materials and methods

Remote sensing methods, especially satellites, have provided opportunities to organize immediate vegetation monitoring.

Of particular importance in the establishment of a system of remote sensing monitoring is the possibility of organizing completely automated satellite data processing. Over the last years, such technologies were actively being established and developed in the Russian Space Research Institute, Russian Academy of Sciences (IKI RAS). They allowed the creation and actualization of the archive of constant satellite observations on the territory of Russia and adjoining states for the period from 2000 until the present. The basis of automatized technologies established by the IKI RAS is the analysis of NDVI values and comparison with the index variation of the previous years [12].

The temporal dynamics of the vegetation was estimated using archive geoportal data of the IKI RAS for the period of 2001-2015. (http://pro-vega.ru).

NDVI areas were mapped with satellite images for the first half of September for a 25-year period from 1990 to 2015. Necessary multispectral images made by the spectroradiometer TM (Landsat-5 satellite) and the spectroradiometer OLI (Landsat-8) were downloaded from the geoportal of the US Geological Survey, using GloVis search (http://glovis.usgs.gov) for the selected polygons.

Height radar data of the Digital Elevation Model SRTM (Shuttle Radar Topography Mission) were downloaded from the FTP-server of the U.S. Geological Survey. To conduct relief morphometric analysis, a number of the corresponding morphometric maps were established and analyzed (along with the data of field observations): hypsometry, slopes, and aspects. Also, topographic modeling of three-dimensional images was conducted. To identify patterns of spatial distribution of plant communities the method geobotanical profiling was used. As a result of field studies we conducted analysis of phytocenotic diversity on the basis of geo-botanical and landscape descriptions.

While identifying vegetation species, Mongolian [13] and Buryatia [14] keys to identification of plants were used.

## 3. Semiarid climate zone

Among the desertification types discovered by FAO-UNEP [15], on the greater part of the Selenge River basin, especially in its Mongolian part, the vegetation cover degradation is the most widely spread.

It is revealed in structural changes of the steppe, forest steppe, and pasture phytocenosis, the successions of their species by synanthropic ones, and decreased projective cover and grass height. In places with the highest development of degradation and land desertification processes caused by a high density of grazing animals on easily deflated sand and sabulous soils, digression processes of steppes and forest steppe geosystems are increased even more. As a result, soil cover is being destroyed, affected by trampling and subsequent deflation, while the existing types of microrelief are being reformed. We noticed an especially strong manifestation of these processes in the poorly populated pine *Pinus sylvestris L.* and drought resistant elms *Ulmus pumila L.* in the Kharaagol sandy area (absolute height values of 735-815 m).

According to the scheme of the landscape and the ecological district division located in the central part of the Selenge River basin, the Kharaagol model polygon is situated in an area of high ecological intensity, caused by both natural and anthropogenic factors. Let us consider the key site of 19.16 km$^2$ situated in the northern part of the Kharaagol polygon, 3 km southeast of the Darkhan city. The biggest part of this site, located on the separated slopes of low hill terrain of the Orkhon-Kharaagol interfluve (maximal height marks are 810-870 m), has steppe caragana-cereal-fringed sagebrush (*Artemisia frigida – Leymus chinesis – Caragana microphylla*) vegetation. As geobotanical studies have shown, the small-leaved caragana *Caragana microphylla (Pall.) Lam.* dominates in the projective vegetation cover (17-20 %). Fringed sagebrush, *Artemisia frigida Willd.* is also widely spread, as well as Chinese wild rye *Leymus chinensis (Trin.) Tzvelev*, etc. Most of the land is used as pasture.

Using access to the satellite monitoring service "Vega" to monitor the vegetation condition, average NDVI curves were built for the Kharaagol key site for the first 15 years of the twenty-first century [12]. Some nonuniformity of the NDVI distribution was observed throughout the years, caused by different climate conditions. It should be pointed out that in 2007, 2008 and 2012, the NDVI values were rather high, 0.6, which is related to the higher values of temperature and moisture regime during the summer months of these years. The important feature of the system «Vega» is a possibility of the research with meteorological information. It gives the chance to receive spatial distribution of meteorological parameters for any date of the vegetation period. The average summer value of the NDVI in 2011 was 0.45 for this site, which proves sparse vegetation during this year (figure 1).

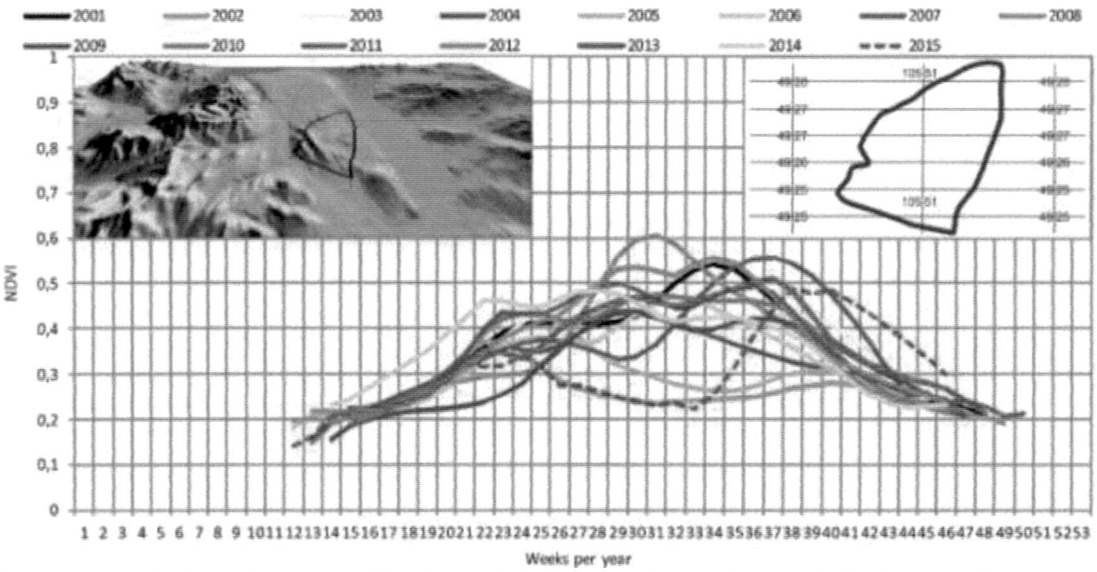


**Figure 1.** Variations of averaged NDVI on 2001-2015 for the north area of the Kharaagol polygon. The upper-right inserted image is the scheme of the site; the upper left inserted image is a 3D-view on the base of DEM SRTM.

NDVI indices were calculated and visualized for the territory of the key site, based on Landsat TM multispectral space images made for September data of a 25-year difference: 1990 and 2015 (figure 2, table 1). While conducting field measurements, GPS-tracks were prepared for the site's borders, which were later converted into vector shape files, functioning as a mask for marking the borders of the sites. The NDVI areal mapping with a 0.1 pitch was conducted only within the limits of marked closed polygons.

**Table 1.** Classification on NDVI graduation for the north area of the Kharaagol polygon.

| NDVI | 17.09.1990 | | 12.09.1994 | | 30.08.2001 | | 08.09.2010 | | 03.09.2014 | |
|---|---|---|---|---|---|---|---|---|---|---|
| | S, km$^2$ | S, % | S, km$^2$ | S, % | S, km$^2$ | S, % | S, km$^2$ | S, % | S, km$^2$ | S, % |
| 0.0-0.1 | 0 | 0 | 0 | 0 | 0.00 | <1 | 0.01 | <1 | 0,00 | <1 |
| 0.1-0.2 | 0.12 | 1 | 0.03 | <1 | 2.19 | 11 | 1.58 | 8 | 16.73 | 87 |
| 0.2-0.3 | 5.44 | 28 | 1.16 | 6 | 15.52 | 81 | 16.45 | 86 | 2.41 | 13 |
| 0.3-0.4 | 10.19 | 53 | 15.37 | 80 | 1.41 | 7 | 1.11 | 6 | 0.00 | <1 |
| 0.4-0.5 | 3.30 | 17 | 2.56 | 13 | 0.01 | <1 | 0.01 | <1 | 0 | 0 |
| 0.5-0.6 | 0.10 | 1 | 0.02 | <1 | 0 | 0 | 0 | 0 | 0 | 0 |
| Total | 19.15 | 100 | 19.15 | 100 | 19.15 | 100 | 19.15 | 100 | 19.15 | 100 |

Comparing the NDVI images, there is a digression of vegetation conditioned mainly by a high pasture load. In September of 1990, most of the site (53 %) had 0.3-0.4 NDVI values, while in September of 2010, 86 % of its area was 0.2-0.3.

In the field expedition studies of 2011-2015, a natural landscape indication of NDVI areas with different values was conducted. The natural and anthropogenic factors of the dynamics of their changes were studied. The territories of plowed fields, cleared by harvests, and the areas of steppe slope pastures in the stage of high digression, caused by cattle overgrazing and characterized by a projective vegetation cover of under 50 %, correspond to NDVI values of 0.1-0.2 dated September 8, 2010, taking in 8 % of the territory.

The most common NDVI value for September is 0.2-0.3. These are the prevailing split slope erosion denudation steppe landscapes with motley grass vegetation (*Leymus chinesis – Artemisia frigida + Potentilla acaulis – Caragana microphylla*). Because of cattle overgrazing, the vegetation of these landscapes experiences average digression; the value of its projective cover is mostly 65-70 %. Of the total projective vegetation cover, the small-leaved pea tree *Caragana microphylla (Pall.) Lam.* takes 17-20 %. Another 15-16 % is taken by the Chinese lyme grass *Leymus chinensis (Trin.) Tzvelev*, 12-13 % is taken by frigid sagebush *Artemisia frigida Willd*, and 9-10 % is taken by the stemless bloodroot *Potentilla acaulis L.*

NDVI values of 0.3-0.4 can be observed for only 6 % of the site area, in hollows and the far western and upper (absolute height marks of 810-870 m) parts of the slopes. The projective cover of the phytocenoses vegetation reaches 80-85 %. Among popular species, there are *Carex duriuscula C.A. Mey.*, which takes 27-33 % of the projective vegetation cover, the small-leaved caragana *Caragana microphylla (Pall.) Lam.* takes 30-33 %, the *Chinese lyme Leymus chinensis (Trin.) Tzvelev* takes 10 %, and the sage-leaf mullein *Phlomis tuberosa L.* takes 10 %.

In September 1990, some of these phytocenoses located in the western part of the area had NDVI values of 0.4-0.55. To all appearances, these phytocenoses had considerable projective cover (around 90-97 %); the highest share belonged to ling, feather grass, and mixed grasses, *Phlomis tuberose L.* in particular. It was proved by studies conducted on territories adjacent to the key site. The reason for phytocenosis digression is also cattle overgrazing.

A different picture is observed in the southern area of Kharaagol model polygon located on the sand massif of the Kharaa River left bank, 18 km south of steppe key area. In 2010, areas of the territories with NDVI values of 0.2-0.3 and 0.3-0.4 decreased somewhat compared to 1990, but areas with 0.1-0.2 values increased sufficiently (twice). This proves a tendency of vegetation cover degradation, similar to that in the northern part of the polygon.

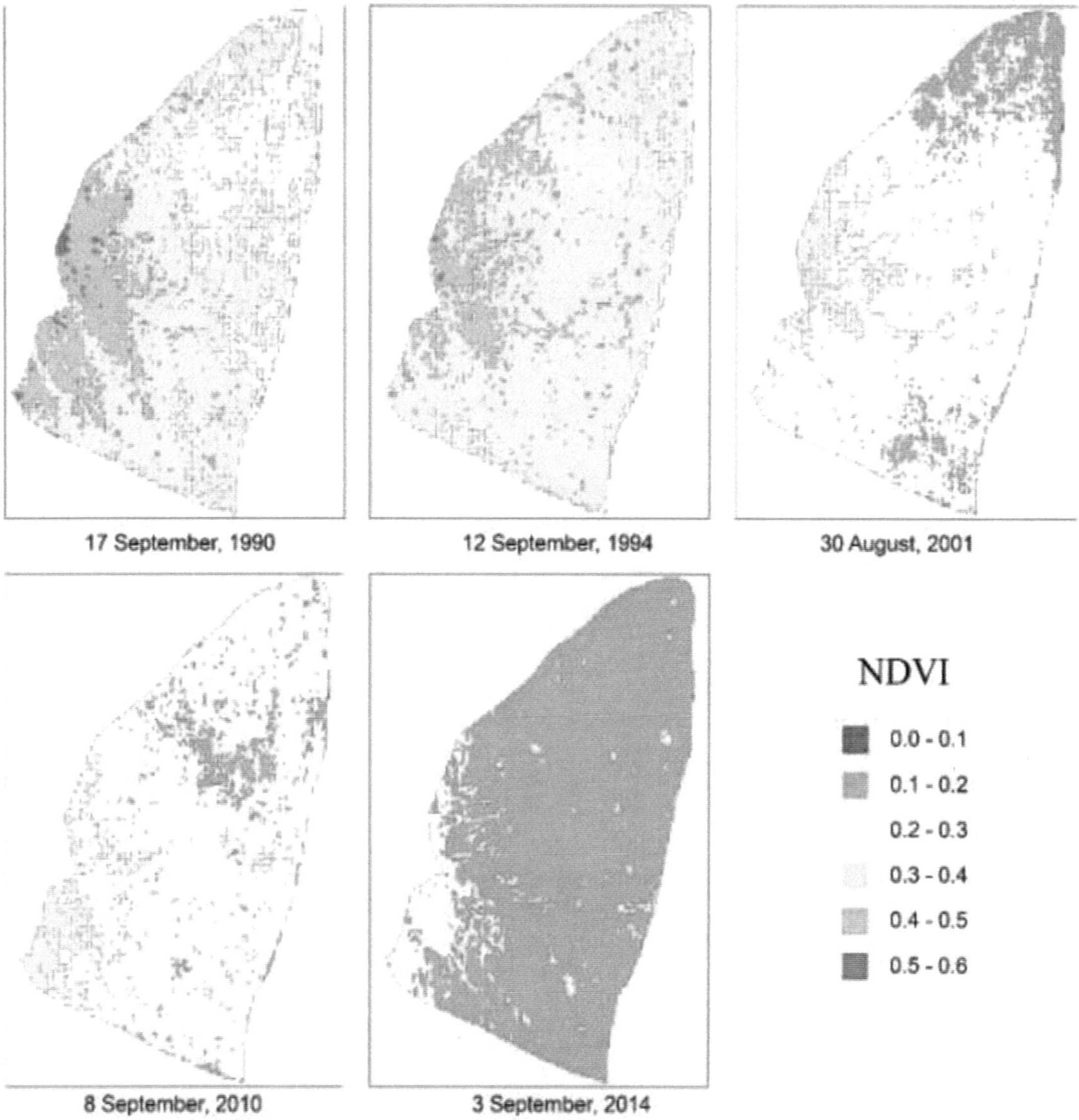


**Figure 2.** NDVI maps for the north area of the Kharaagol polygon.

Natural landscape studies conducted in 2008-2015 on the Kharaagol sandy area and its surrounding areas showed that the main factor of degradation of the vegetation covering it is the grazing and drift of farm animals, primarily sheep and goats, as well as horses and cattle to a lesser extent.

The natural landscape indication of NDVI areas with different values on this massif showed that the NDVI values of 0.1-0.2 appear mostly on barchans, dunes, and desert inter-barchan and inter-dune reductions with traces of intensive cattle grazing with a projective vegetation cover of 3-15 %. Sand areas with a projective vegetation cover up to 3 % have NDVI values of 0-0.1. The prevailing NDVI values for most of the massif, 0.2-0.3, correspond to those for the cover of pine (*Pinus sylvestris L.*) and small elms (*Ulmus pumila L.*). The current decrease in area as compared to 1990 is conditioned not only by farm animal grazing and drift but also by illegal deforestation in some parts and weak reforestation. Denser pine forests with a lot of shrubs make up 14-15 % of the sand massif area. They are located in the top southern part (absolute height marks are 780-815 m), with the NDVI values of 0.3-0.4.

According to the Mongolian State National Statistics Committee [16], a tendency to increase the number of farm animals was observed on the territories of Selenge and Darkhan-Uul aimags in 2000s. Express surveys conducted for arats having a household within the territories of the Kharaagol model polygon showed that nearly half of them immigrated here (to be closer to main market of animal products) with their herds in the 2000s from other periphery Mongolian aimags, the western (mostly) and Gobi.

## 4. Arid climate zone

On the territory of Dundgov aimag (Mid-Gobi), moving from north to south, there is a gradual landscape substitute: desertified steppes of semi-deserts are changed for deserts with saxaul (*Haloxylon sp.*).

Studies showed that in conditions of insufficient moisture (an annual sum of atmospheric condensation less than 100 mm on the territory of Ulziit somon of the southern part of Dundgov aimag, a maximal value of 150 mm in the north, or a complete absence of fresh surface water), Gobi landscapes are affected by degradation processes easily, especially by the physical weathering, deflation, and degradation of vegetation. Even though the general pasture load is low on the territory of Dundgov aimag because of the rather low density of the animal base, the soil surface is trodden around bases within a 0.5-0.7 km radius, and there is hardly any vegetation cover (projective cover from 0 to 1 %).

Up north from the aimag center Mandalgovi, southern dry steppe landscapes dominate. They are located on 70-80 % of the territory. Desertified steppes take up 10-20 %, while saline and alkali saline soil comprise under 8 %. South from Mandalgovi, within the limits of the Mid-Gobi model polygon, desertified steppe landscapes prevail (65-75 %). The southern steppes take up 20-30 %, and the saline ones comprise less than 9 % of the territory.

The average NDVI value for the last decade aggregated for the model Mid-Gobi polygon is 0.12, which proves poor vegetation (figure 3).

Comparison of natural geobotanical studies conducted in 2011-2015 and the results for ecological groups of the Gobi areas of 2010 proves this conclusion, representing the increased share of euxerophyte plants in projective cover for different areas at 12-15 %, indicating continuous invasion of euxeropyte desert species of plants in the north. Thus, the share of euxerophytes has increased in the projective cover from 77 % in August 2010 to 96 % in August 2012 on the key site situated in the lowland, stony, desertified steppes 20 km south of Mandalgovi.

The tendency of xerophytization of Gobi vegetation shows NDVI index analysis of the Mid-Gobi territory polygon from Landsat TM images made in September 1990 and 2011. Thus, if in September 1990, 75 % of the polygon area had a 0.0-0.1 NDVI value, in September 2011, 92 % of the areas had values of 0.0-0.1 values. The natural landscape indication of areas conducted with different NDVI values showed that most (92 % of the area) values 0.0-0.1 can be observed in stony desertified steppes with predominant *Allium polyrrhizum Turcz. ex Regel*, which was popular in the southern Gobi areas in the 1990s. Areas with NDVI values of 0.1-0.2 currently comprising only 8 % of the Mid-Gobi polygon area are observed for ling and needle grass (*Achnaterum splendens – Carex duriuscula*) genuses of saline, stone-free valleys.

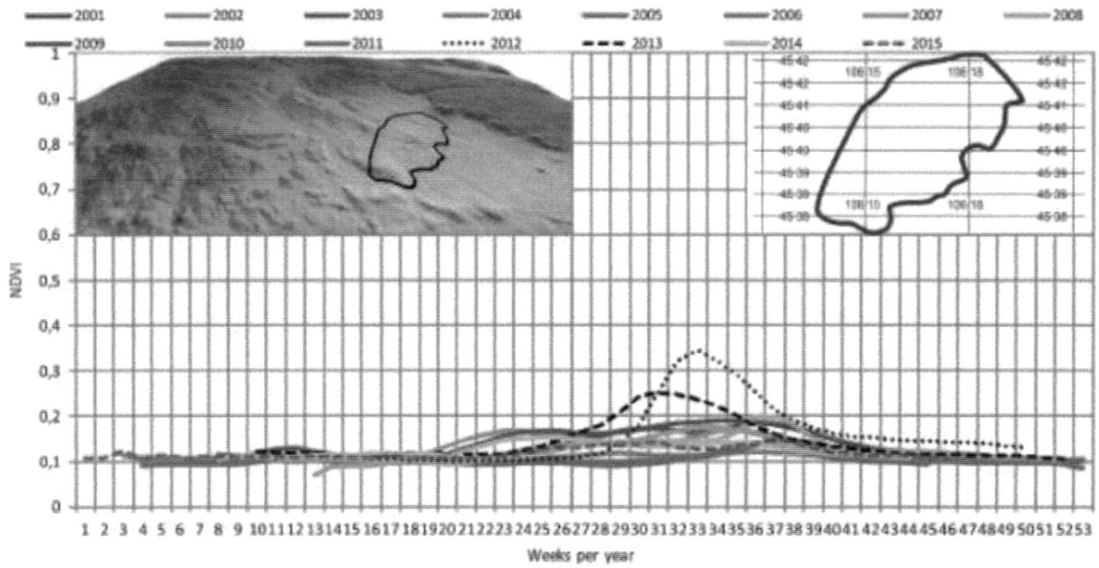


**Figure 3.** Variations of averaged NDVI on 2001-2015 for the Mandalgovi polygon.

Meteorological data analysis proved that in the 2000s there was more severe drought during the summer periods than in the previous century, which, in our opinion, is the main factor for the digression tendency in the Gobi vegetation.

## 5. Conclusions

The performed studies for different channels of optical range, using methods based on the consideration of differences in spectral reflection, together with field natural changes, prove that the degradation of the vegetation cover is increasing for the last two decades and causes of land degradation can be natural or man-made.

Remote sensing methods, based on the analysis of spatial and temporal differentiation of biophysical vegetation parameters, quickly allow a determination of areas of ecological intensity conditioned by the degradation of vegetation cover and desertification.

Cattle overgrazing is observed on the studied areas. It conditions overload for pastures and leads to the digression of their vegetation cover. Therefore, livestock farmers and other land users have to use pasture rationally, changing grazing areas from time to time and thus regulating the grazing rotation.

The vegetation digression during 25-year period is traced across almost the entire territory of the selected polygons. These climatic zones of Central Asia experience different trends of desertification processes associated with both climatic variability (mainly aridization) and anthropogenic influence.

## 6. Acknowledgments

These investigations were partially supported by the Russian Foundation for Basic Research (grant No. 13-05-41378).

We thank Prof. Chimeddorj from the Mongolian State University of Agriculture (Ulaanbaatar) and Dr. Udval from the Research Institute of Animal Husbandry (Ulaanbaatar) for assistance during field investigations.

## References

[1] UNCCD 1994 *United Nations Convention to Combat Desertification in Countries Experiencing Serious Drought And/or Desertification, Particularly in Africa* A/AC.241/27 Paris.
[2] UNCCD 2012 *United Nations Convention to Combat Desertification* www.unccd.int
[3] Hill J, Megier J and Mehl W 1995 Land degradation, soil erosion and desertification monitoring

in Mediterranean ecosystems *Rem. Sen. Rev.* 12 107–130
[4] Reynolds JF 2001 Desertification *Encyclopedia of Biodiversity* ed S Levin (San Diego: Academic Press) pp 61–78
[5] Reynolds J F *et al* 2007 Global desertification: building a science for dryland development *Science* **316** 847–51
[6] Hennemann R. 2001 *Elective on Land Degradation, Assessment, Monitoring and Modelling* (Enschede: Soil Science Division ITC The Netherlands)
[7] Hennemann R 2001 *Some Notes About Soil and Land Degradation in the Naivasha Rangeland Zone* (Enschede: Soil Science Division ITC The Netherlands) p 3
[8] NPACD 1997 *National Plan of Action to Combat Desertification in Mongolia* Ministry of Nature and Environment Government of Mongolia Ulaanbaatar http://www.unccd.int/ActionProgrammes/mongolia-eng2000.pdf.
[9] UNCCD 2004 *Fact Sheets on UNCCD* http://www.unccd.int/fr/resources/publication/Pages/factsheets.aspx
[10] WMO 2006 *World Meteorological Organization World Day to Combat Desertification, Climate and Land Degradation* (Geneva: WMO- No 989)
[11] Gunin P D *et al* 2003 Peculiarities of degradation and desertification of the herbal communities of forest steppe and steppe ecosystems of Southern Transbaikalia *Arid ecosystems* No 9 (19-20) 7–21(*In Russian*)
[12] Lupyan E A *et al* 2011 Satellite service "Vega" for monitoring of vegetation conditions *Modern Problems of Earth's remote sensing* No 8 (1) 190–98 (*In Russian*)
[13] Grubov V I 1982 *Key to Identification of Vascular Plants of Mongolia* (Leningrad: Nauka) (*In Russian*)
[14] Anenkhonov O A *et al* 2001 *Key to Identification of the Plants in Buryatia* (Ulan-Ude: Buryat Science Center) (*In Russian*)
[15] FAO/UNEP 1981 *Provisional Methodology for Desertification Assessment and Mapping* (Rome)
[16] *Mongolian Statistical Yearbook 2009* 2010 (Ulaanbaatar: Mongolian National Statistical Office)

# Linking trace gas measurements and molecular tracers of organic matter in aerosols for identification of ecosystem sources and types of wildfires in Central Siberia

**A V Panov[1], A S Prokushkin[1], M A Korets[2], A V Bryukhanov[3], A N Myers-Pigg[4], P Louchouarn[4,5], N V Sidenko[1], R Amon[4,5], M O Andreae[6] and M Heimann[7]**

[1]Laboratory of biogeochemical cycles in forest ecosystems, V.N. Sukachev Institute of Forest SB RAS, 660036, Akademgorodok 50/28, Krasnoyarsk, Russia
[2]International laboratory of permafrost ecosystems, V.N. Sukachev Institute of Forest SB RAS, 660036, Akademgorodok 50/28, Krasnoyarsk, Russia
[3]Laboratory of forest fire research, V.N. Sukachev Institute of Forest SB RAS, 660036, Akademgorodok 50/28, Krasnoyarsk, Russia
[4]Department of Oceanography, Texas A&M University, College Station, 77843-3146, Texas, USA
[5]Department of Marine Sciences, Texas A&M University, 200 Seawolf Parkway, Galveston, 77554, Texas, USA
[6]Max Planck Institute for Chemistry, Hahn-Meitner-Weg 1, 55128, Mainz, Germany
[7]Max Planck Institute for Biogeochemistry, Hans Knoell 10 07745, Jena, Germany

Email: alexey.v.panov@gmail.com

**Abstract**. Summer 2012 was one of the extreme wildfire years in Siberia. At the surface air monitoring station "ZOTTO" (60°48′N, 89°21′E, 114 m a.s.l.) in Central Siberia we observed biomass burning (BB) influence on the ongoing atmospheric measurements within more than 50 % of the time in June-July 2012 that indicates a 30 times greater wildfire signal compared to previously reported ordinary biomass burning signature for the study area. While previous studies thoroughly estimated a relative input of BB into aerosol composition (i.e. size distribution, physical and optical parameters etc.) at ZOTTO, in this paper we characterize the source apportionment of the smoke aerosols with molecular tracer techniques from large-scale wildfires occurred in 2012 in the two prevailing types of Central Siberian ecosystems: complexes of pine forests and bogs and dark coniferous forests. Wildfires in the selected ecosystems are highly differed by their combustion phase (flaming/smoldering), the type of fire (crown/ground), biomass fuel, and nature of soil that greatly determines the smoke particle composition. Anhydrosugars (levoglucosan and its isomers) and lignin phenols taken as indicators of the sources and the state of particulate matter (PM) inputs in the specific fire plumes were used as powerful tools to compare wildfires in different environmental conditions and follow the role and contribution of different sources of terrestrial organic matter in the transport of BB pollutants into the pristine atmosphere of boreal zone in Central Siberia.

## 1. Introduction

Biomass burning aerosols are one of the least understood aspects of the modern climate system. Biomass burning (including landscape fires) emits 2-4 versus 7.2 Pg C yr-1 released by fossil-fuel combustion and thus produces up to 50% as much carbon dioxide ($CO_2$) as fossil-fuel combustion [1,

2] and caused up to 65% of the variability in the $CO_2$ growth rate between 1997 and 2001 [3]. Consisting of a complex mixture of inorganic and organic compounds organic aerosols comprise up to 90% of the submicron particulate mass in BB emissions [4] and their physicochemical and optical properties vary within fuel types and combustion conditions. Biomass burning aerosols released from Siberian wildfires dominating other BB sources at high latitudes may have significant impacts on regional direct radiative forcing and also may be transported deep into the Arctic regions. Previous studies found that wildfires in Russia and Siberia substantially contributed to the Arctic aerosols [5]. From modeled trajectory analyses [6], Stohl estimated that in summer, during years with average biomass combustion rates, wildfires in boreal forests are able to contribute more aerosol black carbon (BC) to the Arctic compared to anthropogenic sources. Despite the concerns of environmental significance of PM emitted by Siberian wildfires, systematic observations were performed only in some places in Western and North-Eastern Siberia [7, 8]. Therefore, the assessments of BB emissions from Siberian boreal forests and their attribution to potential adverse effects are very sparse.

This gap is to be partially filled with continuous trace gas and aerosol measurements at the background surface air monitoring station "ZOTTO" (Zotino Tall Tower Observatory, 60°48′N, 89°21′E, 114 m a.s.l.) in Central Siberia that operational since late 2006. While previous studies made by Vasileva et al. [9], Heintzenberg et al. [10] and Chi et al. [11] thoroughly estimated a relative input of BB into annual variability of carbon monoxide (CO) mixing ratios and aerosol composition (i.e. size distribution, physical and optical parameters etc.) at ZOTTO, in this study we concentrate mainly on the source apportionment of the smoke aerosols with molecular tracer techniques that we studied in large-scale wildfires occurred in 2012 in the two prevailing types of Central Siberian ecosystems: complexes of pine forests and bogs and dark coniferous forests. Wildfires in the selected ecosystems are highly differed by their combustion phase (flaming/smoldering), the type of fire (crown/ground), biomass fuel, and soil type that highly determines the smoke particle composition. Anhydrosugars (levoglucosan and its isomers) and lignin phenols taken as indicators of the sources and the state of particulate matter inputs in the specific fire plumes were used as powerful tools to compare wildfires in different environmental conditions and follow the role and contribution of different sources of terrestrial organic matter (OM) in the transport of BB pollutants into the pristine atmosphere of boreal zone in Central Siberia.

### *1.1. The monitoring site*

The monitoring site (figure 1) is located in Krasnoyarsk krai, 20 km west of the Yenisei River at the eastern edge of the West Siberian Lowland and lies in a vast region of forests and bogs, relatively undisturbed by anthropogenic influences and relatively inhospitable because of its continental climate. The surrounding area is a flat with small hills and ridges. The climatic conditions of a mean annual temperature of 3.8 °C and a total precipitation of 536 mm derived from the weather station at Bor about 100 km northeast of ZOTTO (61°36' N, 90°01' E, 58 m a.s.l.) indicate a rather cold continental climate [12]. A temperature trend of around +3°C/100 yr was observed within the period of 1950-1999, which is representative for the larger territories of

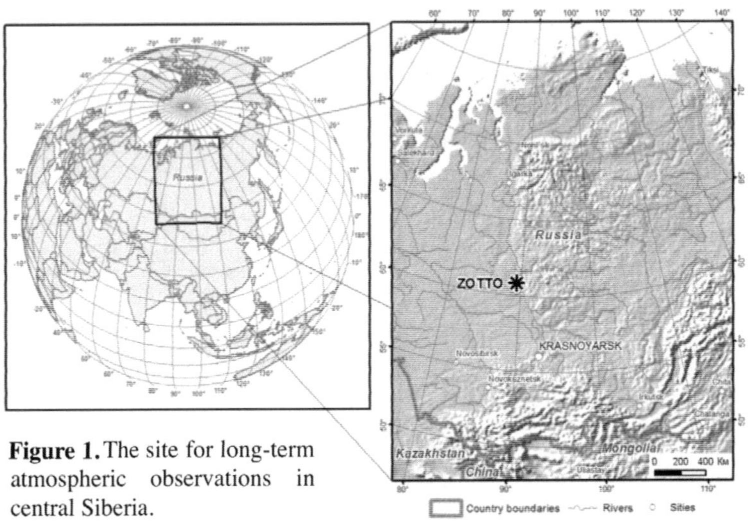


**Figure 1.** The site for long-term atmospheric observations in central Siberia.

Siberia [13]. The region is mostly influenced by Atlantic air masses traveling across the European part of the continent and a cold polar air from North-Eastern Siberia and the Arctic regions [14]. The vegetation of the study area consists of a mosaic of ecosystems that are representative for Central Siberia: large complexes of pine forests (18%) and bogs (16%), dark coniferous forests (26%), mixed forests (27%), broadleaf (5%) and larch (around 2%) forests. The monitoring station "ZOTTO" comprises a 300-m tall mast with inlets installed at the profile of heights (300 m, 227 m, 158 m, 92 m, 52 m and 4 m a.g.l.), where air is sampled and streamed to the analyzers located inside a partly underground laboratory bunker. The tall tower was developed for long-term atmospheric boundary layer (ABL) observations in Siberia that are representative for a very large spatial area. A detailed description of the site is given by Kozlova et al. [15] and Winderlich et al. [16].

### *1.2. Wildfires in Siberia within the last 20 years (1992-2012)*

Within the last 20 years boreal forests in Siberia several times (1996, 2003 and 2012 y.) were greatly influenced by wildfires. In 1996 the Federal statistical service (EMISS; http://www.fedstat.ru) estimated up to 1.03 millions hectares of forests disturbed by wildfires (figure 2) with the most fires occurred in Krasnoyarsk krai (0.3 million hectares) and Irkutsk region (0.4 million hectares). In 2003 extensive wildfires in Irkutsk region (near Baikal lake) contributed around 50% of the BC deposited north of 75°N in spring and summer of that year [17]. Additionally to atmospheric radiative effect, the deposition of such aerosols onto snow and ice surfaces in the Arctic may reduce the albedo and lead to increased melting rates [18]. In the extremely dry summer of 2012 specific wildfire spots could exceed tens thousands of hectares. While the Federal statistical service estimated the total wildfire areas in Russia achieved as much as 2.1 million hectares with up to 50 % of wildfires (1.3 million hectares) occurred in Siberia, the satellite-derived data might 2-3 times exceed the official statistics [19]. These large-scale fires within 2 months (June-July 2012) overlapped with the background surface atmospheric measurements at monitoring station "ZOTTO" in Central Siberia and demonstrated a 30 times stronger wildfire signal compared to previously reported ordinary BB signature for study area [20].

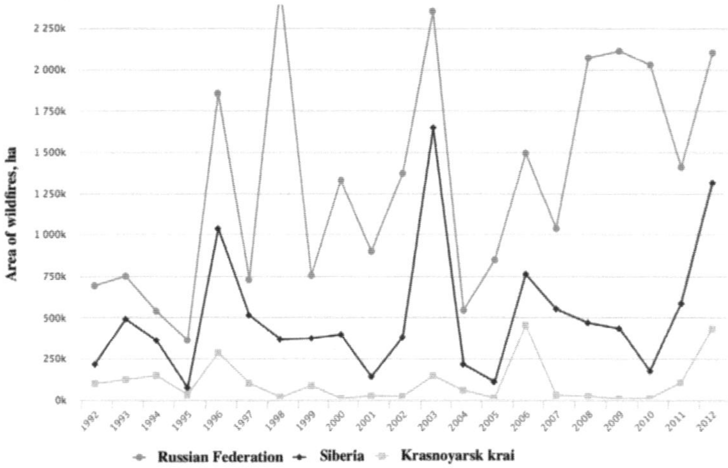


**Figure 2.** Areas of wildfires in Russia and Siberia within the last 20 yrs (1992-2012).

### *1.3. The instrumental setup and methods*

Analysis of air composition in the specific fire plumes was based on the time series of $CO_2/CH_4/CO$ mixing ratios measured at 300 m a.g.l. at monitoring station "ZOTTO". The $CO_2$, $CH_4$, and $H_2O$ measurements were performed by an EnviroSense 3000i analyzer (Picarro Inc., USA, CFADS-17) based on the cavity ring-down spectroscopy technique (CRDS) [21]. Carbon monoxide mixing ratios were measured by an APMA-370 ambient carbon monoxide monitor (Horriba Inc., Japan). The number of meteorological sensors recorded meteorological variables in a vertical profile on the tower. The Hybrid Single-Particle Lagrangian Integrated Trajectory (HYSPLIT) model [22], developed by the National Oceanic and Atmospheric Administration (NOAA), was used to compute 24-hrs backward trajectories for air probes sampled from the specific wildfires. The driving meteorological data fields for the model were obtained from the National Center for Environmental Prediction (NCEP) Global Data Assimilation System (GDAS, global, 2006 – present). Active fire spots were detected from Terra/Aqua MODIS satellite data. The fire behavior associated with emissions was characterized using the Combustion Efficiency (CE) - index describing the relative amount of flaming and smoldering combustion in a wildland fire. As the molar ratio of $CO_2$ emitted to the total moles of carbon emitted, CE was calculated as the ratio of excess moles of carbon emitted as $CO_2$ to the molar sum of carbon released [23]. Chemical analysis of samples of atmospheric particulate matter at 300 m a.g.l. was performed in the specific fire plumes for identifying biomarkers, or compounds indicative of a unique biological source, and thus served as a powerful tool to trace the origin and transformations of organic matter. Anhydrosugars (levoglucosan and its isomers) were analyzed based on the gas chromatography-mass spectrometry (GC/MS) [24]. Lignin analysis was carried out according to the CuO oxidation method suggested by Hedges and Ertel [25] with modifications as described by Kaiser and Benner [26]. The Lignin Phenol Vegetation Index (LPVI) as a quantitative parameter representing the entire characteristics of the vegetation was applied as an additional tool to partition OM among end-member sources [27].

## 2. Results

Levoglucosan (1,6-anhydro-b-D-glucopyranose) and its isomers (mannosan and galactosan) are dehydro-monosaccharide derivatives formed exclusively from the thermal breakdown of cellulose and hemicellulose, respectively, during biomass burning [28]. Thus, they are source specific and can be used as a qualitative marker of biomass combustion in atmospheric particles [24] and as a tracer allowing quantitative apportionment of biomass combustion in the atmosphere. In the aerosol filter probes we found out concentration of levoglucosan in the specific fire plumes ranged from 1.19 up to 33.3 $\mu$g m-3 (figure 3a), with a mean value of 9.22±9.66 $\mu$g m-3.

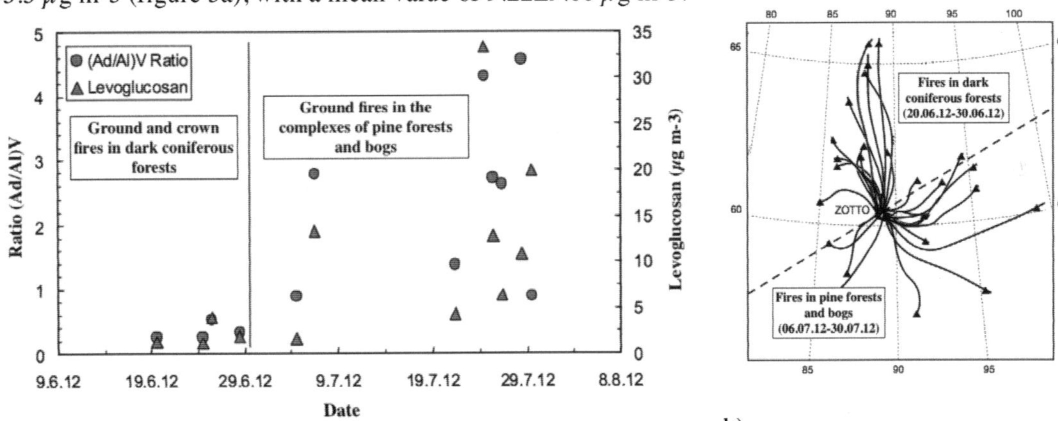


a) b)

**Figure 3.** Time series of levoglucosan and the aldehyde to acid derivatives ratio versus vanillin ((Ad/Al)V) in the period of large-scale wildfires (20.06.2012-30.07.2012) occurred near "ZOTTO" monitoring station (a) and 24-hrs backward trajectories computed for the period of BB detection (b).

The backward trajectory analysis (figure 3b) demonstrates levoglucosan concentration considerably increased from 2,06±1,29 $\mu$g m-3 when air mass travelled from highly intensive ground and crown fires occurred in dark coniferous forests (20.06.2012-30.06.2012) 15-20 km northwest of the measurement site up to 14,59±10,67 $\mu$g m-3 when wind direction changed and we observed a BB signal produced by less severe ground fires occurred in the complexes of pine forests and bogs (06.07.2012-30.07.2012) 12-15 km southeast of the monitoring station. Mannosan and galactosan showed emission trends similar to levoglucosan.

Thermal degradation of fuels occurs ahead and along the fireline, while pockets of intermittent open flame often persist well behind the flaming front. To characterize emissions, wildfire behavior can be described based on the presence or absence of an open flame: "flaming" or "smoldering" combustion [29, 30]. It provides a basis for objectively describing the fire behavior associated with emission measurements. The relative amount of flaming and smoldering combustion in the observed wildland fires we described using the combustion efficiency calculated from trace gas measurements in the ABL at "ZOTTO" monitoring station. The strong feedback (figure 4) of the levoglucosan concentration to CE illustrates stronger thermal degradation of lignocellulosic biomass within highly intensive ground and crown fires occurred in dark coniferous forests with greater values of CE (0.96-0.99) when mainly flaming combustion dominates and fine fuels completely engulfed in flame compared to a signal derived from ground fires in complexes of pine forests and bogs with CE values ranged from 0.93 (smoldering) to 0.95 (flaming) that finally indicates a stronger contribution of a smoldering duff in the fuel bed of the bogs into wildfire smoke.

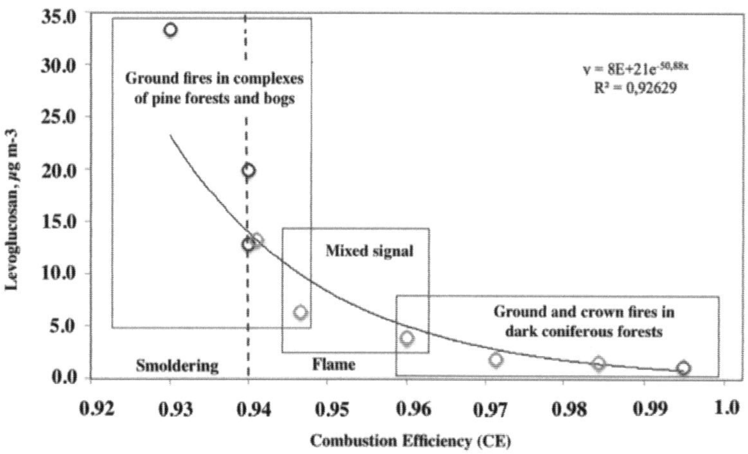


**Figure 4.** Levoglucosan versus Combustion Efficiency (CE) in the period of large-scale wildfires (20.06.2012-30.07.2012) occurred near «ZOTTO» monitoring station.

In turn, the fuel characteristics (type, structure, loading, chemistry etc.) are primarily ecosystem vegetation specific properties that are strongly influenced by environmental conditions. Among molecular properties of organic compounds, lignin biomarkers are a unique proxy for distinction of the vegetation type, because they are specific to individual or a restricted group of plants [27]. To distinguish a relative input of different taxonomical plant groups and tissue types in the OM appeared in smokes throughout the BB we used the differences in abundance of vanillyl (V), syringyl (S) and cinnamyl (C) structural groups [31, 32]. Vanillyl-based phenols are produced by all vascular plants, being the exclusive component in gymnosperm woody tissue. Syringyl-based phenols are produced by angiosperms and found in both woody and nonwoody tissue, although trace levels are found in gymnosperm nonwoody tissue. Cinnamyl phenols are present in the nonwoody tissues of gymnosperms and angiosperms. We used the composition of these three phenol groups to estimate contributions of terrestrial plant tissues. The lignin index-phenols (V, S and C) were defined as total concentration of three V-phenols, three S-phenols and two C-

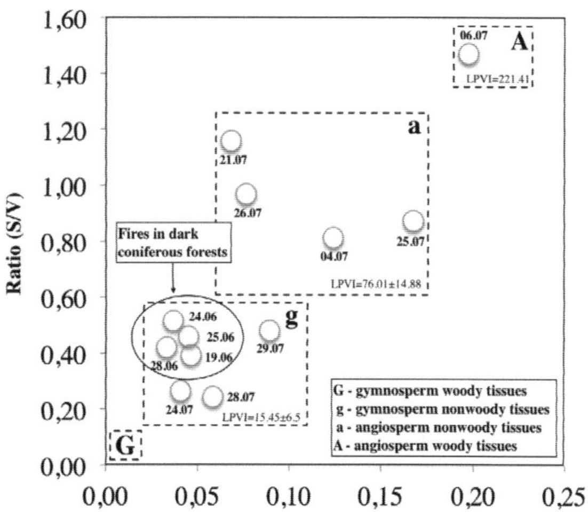


**Figure 5.** Relationship between lignin index-phenols (V, S and C): syringyl-based (S) to vanillyl-based (V) phenols ratio (S/V) versus cinnamyl-based (C) to vanillyl-based (V) phenols ratio (C/V).

phenols normalized to 100 mg of OC, respectively. The S/V value, a measure of relative contributions made by gymnosperm (S/V<0.6; LPVI=15.45±6.5) and angiosperm (S/V>0.6; LPVI=103.29±62.3) plants, demonstrates an average value of 0.67±0.38 throughout the wildfires (figure 5). Generally this value indicates nearly the equal input of both taxonomical plant groups, however applied backward trajectory analysis reflects the contribution of solely gymnosperms (S/V=0.45±0.05) during wildfires occurred in dark coniferous forests. The C/V ratio that differentiates woody and non-woody vegetation [31] is relatively small and values are concentrated in a narrow range (0.04±0.01) that simply illustrates the main contribution into particulate matter during BB in dark coniferous forests is made by gymnosperm nonwoody tissues. In turn the C/V ratio within wildland fires occurred in highly mosaic complexes of pine forests and bogs ranged from 0.04 up to 0.20. Besides of the gymnosperm woody tissues, whose contribution was not found, this range indicates a wide spectrum of terrestrial OM appeared in wildfire smokes contributed by both vascular tissue types (woody/nonwoody) and taxonomical (gymnosperm/angiosperm) plant groups. We found out a relatively equal input of gymnosperm nonwoody (C/V=0.06±0.02; LPVI=15.45±6.5) and angiosperm woody tissues (C/V=0.11±0.04; LPVI=76.01±14.88) in wildfire smokes, while the contribution of angiosperm nonwoody tissues (C/V=0.20; LPVI=212.4) is rather negligible.

Finally, we observed the changes in lignin phenol decomposition state in the OM appeared in wildfire smokes. Lignin decomposition state was estimated using the acid-to-aldehyde (Ad/Al) ratios of lignin phenols, as during microbial decomposition of plant-derived soil organic matter aldehyde forms of lignin phenols are oxidized to their acid derivatives [33]. Thus, increasing Ad/Al ratios reflects an enhancement of decomposition state of organic matter. For observed BB signals we found out sharply increased Ad/Al ratios of vanillin derivatives (Figure 3a) from 0.33±0.27 detected during ground and crown fires in dark coniferous forests up to 2.52±1.36 within ground fires occurred in complexes of pine forests and bogs. These changes reflect distinct sources of burned OM. Particularly there was "fresh" lignin input under intensive flaming combustion in dark coniferous forests: the domination of live biomass combustion. Highly decomposed OM was released to the atmosphere by the smoldering combustion of organic matter accumulated on the soil surface of pine forests and in the bogs: the domination of degraded soil organic matter combustion.

## 3. Conclusions

Anhydrosugars (levoglucosan and its isomers) and lignin phenols taken as indicators of the sources and the state of particulate matter inputs in the specific fire plumes were used as powerful tools to compare wildfires in different environmental conditions and follow the role and contribution of different sources of terrestrial organic matter (OM) in the transport of BB pollutants into the pristine atmosphere of boreal zone in Central Siberia.

In the observed wildfires we found out a considerable growth of the levoglucosan concentration when air masses travelled upwind to the measurement site from highly intensive ground and crown fires occurred in dark coniferous forests to less severe ground fires in complexes of pine forests and bogs. Mannosan and galactosan showed emission trends similar to levoglucosan. The strong feedback of the levoglucosan concentration to combustion efficiency illustrates stronger thermal degradation of lignocellulosic biomass within intensive ground and crown fires in dark coniferous forests with greater values of CE when mainly flaming combustion dominates and fine fuels completely engulfed in flame compared to a signal derived from ground fires in complexes of pine forests and bogs with CE values ranged from 0.93 (smoldering) to 0.95 (flaming) that finally indicates a stronger contribution of smoldering duff in the fuel bed of the bogs into wildfire smoke.

The ratio of syringyl-based phenols to vanillyl-based phenols (S/V) generally indicates nearly the equal input of both taxonomical plant groups into particulate matter in the wildfire smokes, however applied backward trajectory analysis reflects the contribution of solely gymnosperms under wildfires occurred in dark coniferous forests. The ratio of cinnamyl-based phenols to vanillyl-based phenols (C/V) illustrates the main contribution during BB in dark coniferous forests is made by gymnosperm nonwoody tissues. In turn the C/V ratio within ground fires in highly mosaic complexes of pine forests and bogs indicates a wide spectrum of terrestrial OM appeared in wildfire smokes contributed by both vascular tissue types (woody/nonwoody) and taxonomical (gymnosperm/angiosperm) plant groups. We found out a relatively equal input of gymnosperm nonwoody and angiosperm woody tissues while the contribution of angiosperm nonwoody tissues into particulate matter is rather negligible.

The growth of the acid-to-aldehyde (Ad/Al) ratios of lignin phenols that coincides with change of air mass transport from wildfires occurred in dark coniferous forests to BB signal from complexes of pine forests and bogs upwind to the measurement site reflects an enhancement of decomposition state of organic matter. These changes reflect distinct sources of burned OM. Particularly there was "fresh" lignin input under intensive flaming combustion in dark coniferous forests: the domination of live biomass combustion. Highly decomposed organic matter was released to the atmosphere by the smoldering combustion of organic matter accumulated on the soil surface of pine forests and in the bogs: the domination of degraded soil organic matter combustion.

**Acknowledgments**
This study was supported financially by the Russian Scientific Foundation, project # 14-24-00113, the Russian Foundation for Basic Research, grant # 15-45-04423, Krasnoyarsk Region Science and Technology Support Fund, and by the Max Planck Society (Germany).

**References**
[1] Andreae M O and Merlet P 2001 *Global Biogeochem. Cycles* **15** 955–966
[2] Van der Werf G R, Randerson J T, Giglio L, Collatz G J, Kasibhatla P S and Arellano A F 2006 *Atmos. Chem. Phys.* **6** 3423–41
[3] Bowman D M, Balch J K, Artaxo P, Bond W J, Carlson J M, Cochrane M A, D'Antonio C M, DeFries R S, Doyle J C, Harrison S P, Johnston F H, Keeley J E, Krawchuk M A, Kull C A, Marston J B, Moritz M A, Prentice I C, Roos C I, Scott A C, Swetnam T W, van der Werf G R and Pyne S J 2009 *Science* **324** 481–484
[4] Zhang Q, Jimenez J L, Canagaratna M R, Allan J D, Coe H, Ulbrich I, Alfarra M R, Takami A, Middlebrook A M, Sun Y L, Dzepina K, Dunlea E, Docherty K, De-Carlo P F, Salcedo D, Onasch T, Jayne J T, Miyoshi T, Shimono A, Hatakeyama S, Takegawa N, Kondo Y, Schneider J, Drewnick F, Borrmann S, Weimer S, Demer-jian K, Williams P, Bower K, Bahreini R, Cottrell L, Griffin R J, Rautiainen J, Sun J Y, Zhang Y M and Worsnop D R 2007 *Geophys. Res. Lett.* **34** L13801
[5] Eck T F, Holben B N, Reid J S, Sinyuk A, Hyer E J, O'Neill N T, Shaw G E, Vande Castle J R, Chapin F S, Dubovik O, Smirnov A, Vermote E, Schafer J S, Giles D, Slutsker I, Sorokine M, Newcomb W W 2009 *J. Geophys. Res.* **114** D11201
[6] Stohl A 2006 *J Geophys. Res.* **111** D11306
[7] Kozlov V S, Panchenko M V and Yausheva E P 2008 *Atmos. Environ.* **42** 2611–20
[8] Paris J D, Stohl A, Nedelec P, Arshinov M Yu, Panchenko M V, Shmargunov V P, Law K S, Belan B D and Ciais P 2009 *Atmos. Chem. Phys.* **9** 9315–27
[9] Vasileva A V, Moiseenko K B, Mayer J –C, Jurgens N, Panov A, Heimann M and Andreae M O 2011 *J Geophys. Res.* **116** D07301
[10] Heintzenberg J, Birmili W, Seifert P, Panov A, Chi X and Andreae M O 2013 *Tellus B* **65** 1-13
[11] Chi X, Winderlich J, Mayer J -C, Panov A V, Heimann M, Birmili W, Heintzenberg J, Cheng Y and Andreae M O 2013 *Atmos. Chem. Phys.* **13** 12271–98
[12] Schulze E D, Vygodskaya N N, Tchebakova N M, Czimczik C I, Kozlov D N, Lloyd J, Mollicone D, Parfenova E, Sidorov K N, Varlagin A V and Wirth C 2002 *Tellus B* **54** 421–428
[13] Groisman P and Soja A J 2009 *Environ. Res. Lett.* **4** 045002
[14] Eneroth K, Kjellström E and Holmén K 2003 *J. Geophys. Res.* **108** D002730
[15] Kozlova E A, Manning A C, Kisilyakhov Y, Seifert T and Heimann M 2009 *Atmos. Meas. Tech.* **2(1)** 205–220

[16] Winderlich J, Chen H, Gerbig C, Seifert T, Kolle O, Lavrič J V, Kaiser C, Höfer A and Heimann M 2010 *Atmos. Meas. Tech.* **3(4)** 1113–28
[17] Generoso S, Bey I, Attie I –L and Breon F –M 2007 *J. Geophys. Res.* **12** D15302
[18] Hansen J and Nazarenko L 2004 *Proc. Natl. Acad. Sci.* **111(2)** 423–428
[19] Kukavskaya E A, Soja A J, Petkov A P, Ponomarev E I, Ivanova G A and Conard S G 2013 *Can. J. For. Res.* **43** 493–506
[20] Panov A, Chi X, Winderlich J, Bryukhanov A, Ponomarev E, Timokhina A, Birmili W, Heimann M and Andreae M O 2013 *Proc. of 4th Fire Behavior and Fuel Conf.* 46–47
[21] Crosson E R *Applied Physics B-Lasers and Optics* 2008 **92(3)** 403–408
[22] Draxler R R and Hess G D 1998 *Aust. Meteor. Mag.* **47** 295–308
[23] Urbanski S P, Hao W M, Baker S 2009 *Wildland fires and air pollution* (The Netherlands: Elsevier) pp 79–107
[24] Louchouarn P, Kuo L, Wade T L and Schantz M 2009 *Atmos. Environ.* **43** 5630–5636
[25] Hedges J I and Ertel J R 1982 *Anal. Chem.* **54** 174–178
[26] Kaiser K and Benner R 2012 *Anal. Chem.* **84(1)** 459–464
[27] Tareq S M, Tanaka N, Ohta K 2004 *Sci. of the Total Environ.* **324** 91–103
[28] Simoneit B R T, Schauer J J, Nolte C G, Oros D R, Elias V O, Fraser M P, Rogge W F, Cass G R 1999 *Atmos. Environ.* **33** 173–182
[29] Yokelson R J, Griffith D W T and Ward D E 1996 *J. Geophys. Res.* **101** 21067–80
[30] Chen L –W A, Moosmuller H, Arnott W P, Chow J C, Watson J G, Susott R A, Babbitt R E, Wold C E, Lincoln E N and Hao W M 2007 *Environ. Sci. Technol.* **41** 4317–25
[31] Hedges J I and Mann D C 1979 *Geochimica et Cosmochimica Acta* **43** 1803–07
[32] Otto A. and Simpson J M 2006 *Biogeochemistry* **80** 121–142
[33] Rezende C E, Pfeiffer W C, Martinelli L A, Tsamakis E, Hedges J I and Keil R G 2010 *Brasil. Estuarine, Coastal and Shelf Science* **87** 497–486

# Major anion and cation fluxes from the Central Siberian Plateau watersheds with underlying permafrost

**Roman R Kolosov[1], Anatoly S Prokushkin[1], Oleg S Pokrovsky[2]**
[1] V N Sukachev Institute of Forest SB RAS, Akademgorodok 50/28, Krasnoyarsk 660036, Russia
[2] Tomsk State University, 36 Lenin Ave., Tomsk 634050, Russia

kolosov.phd@gmail.com (R. Kolosov)

**Abstract.** The subarctic rivers of the Central Siberian Plateau have specific fed-characteristics due to the permafrost distribution and the active layer thawing dynamics. Two watersheds with different types of permafrost (from insular to continuous) are studied. Different data sources (Roshydromet and our own observations) are used for receiving anions' ($HCO_3^-$, $SO_4^{2-}$, $Cl^-$) and cations' ($Ca^{2+}$, $Mg^{2+}$) fluxes from the Nizhnyaya Tunguska river (1960-2011) and the Tembenchi river (1970-2011). The annual discharge of N. Tunguska for 1939-2011 is characterized by an increase of 0.3 $km^3$/year/year, and for Tembenchi, 0.04 $km^3$/year/year. The major part of the increase (about 60%) is due to spring flooding (May - June). The volume-weighted mean concentrations of major anions and cations in the N. Tunguska river water increased three times in the spring period (40.7 – 116.8 mg/l) and in the summer-fall period (74-212.9 mg/l). On the contrary, such concentrations decreased sharply during the winter mean water period. Due to these results, the total export of main anions and cations from the N. Tunguska river basin rose more than 4,5 times. Two possible reasons can be identified: 1) a water discharge increase of the Subarctic rivers (Peterson et al., 2002); 2) permafrost degradation induced by global climate warming (Frey and McClelland 2009).

## 1. Introduction

Chemical composition of rivers water of Central Siberian Plateau determined by permafrost underlying watersheds. Precisely, permafrost distribution on river basins and active layer thawing dynamics of soil define the fed-characteristics of Subarctic rivers. During the winter mean water period the greatest contribution in river feeding is made by ground water discharged through taliks, during the frost-free period – by precipitation, infiltrating in soil (Prokushkin et al 2011).

Climate data analysis indicates increase of average annual temperature, precipitation and water discharge in Central Siberia rivers since 1950. Concerning, study of chemical elements' fluxes from Central Siberian Plateau watersheds can measure the changes in intensity of biogeochemical processes according to global climate changes.

Our purpose was to analyze main anions and cations export from Central Siberian Plateau watersheds, our objects of study – Nizhnyaya Tunguska and Tembenchi river basins, characterized by different distribution of permafrost beneath (figure 1).

## 2. Materials and methods

### *1.1. River water sampling and hydrochemical analysis*

Hydrochemical composition of Nizhnyaya Tunguska and Tembenchi rivers investigated ourselves from 2004 and 2006, respectively (Evenki Station, Tura, Krasnoyarskii Kray, Russia). There are different methods of water sampling during each period of hydrological year: winter mean water (October – April) – sampling 1 time a month, spring flood (May – June) – 2-3 times a week, summer mean water

(July – August) – 1 time a week. Collected samples are filtered immediately after sampling using 0.45 μm filters (Millipore).

Anions' concentrations ($Cl^-$, $SO_4^{2-}$) obtained by ion chromatography (Dionex ICS 2000), hydrocarbonate concentration analysis based on DIC content, measured on Vario TOC Elementar (conversion factor – 5.08). Cations concentration ($Ca^{2+}$, $Mg^{2+}$) determined using spectrometer Perkin-Elmer 5100 PC, or by ICP-MS method (Agilent 7500).

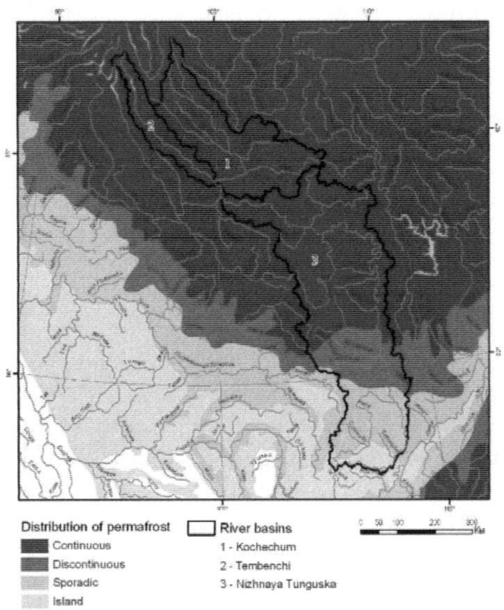


**Figure 1.** Permafrost distribution on studied permafrost (Brown et al 2002 based).

### *1.2. Main anions and cations fluxes analysis*

For hydrochemical composition and water discharges of Nizhnyaya Tunguska and Tembenchi rivers from 1960 archive data of ROSHYDROMET was used. There are two studied gauging stations: 1) #4008 Tura, river basin square – 268000 km$^2$ (studied period – 1960-2011)"; 2) #4046 Tembenchi, 21600 km$^2$ (1970-1993). Due to a few measuring during each year, Roshydromet archive data divided into decades for the analysis of studied anions and cations fluxes' tendency (for N. Tunguska river – 1960-1969, 1970-1979, 1980-1989, 1990-1999, 2000-2009 and 2010-2011; for Tembenchi river – 1970-1979, 1980-1989, 1990-1993).

For calculating daily fluxes, we applied the following stages: 1) elements' concentrations multiplied by water discharges for all available dates; 2) then, we analyzed fluxes (F) function on water discharges (Q) for each anion and cation and we found that this dependence could be described by the function $F = a \times Q^{b}$ most significantly (*a* and *b* – coefficients). Confidence levels (p) for *a* and *b* obtained with using STATISTICA 10 and all coefficients for entire period and all anions had necessary reliability ($p \leq 0.05$, or $p \geq 95\%$). Then, using this function daily and annual (the sum of daily values) fluxes from N. Tunguska and Tembenchi watersheds calculated for the period 1960-2011.

### *1.3. Volume-weighted mean concentrations analysis*

Studied time divided by periods (1955-1959, 1960-1964, 1965-1969 etc.). For volume-weighted mean concentrations calculated using the following method: 1) Main anions and cations concentrations (mg/l) multiplied by water discharge values (m$^3$/s); 2) Then, received values summed up and divided by summed water discharge values for each month, according to selected periods.

## 2. Results and their discussions

### *2.1. Water chemical composition*

In general, investigated rivers water chemical composition is homogeneous. Such factor indicates similar geological structure of N. Tunguska and Tembenchi river basins. In winter mean water period Tembenchi river water is chloride-calcium, N. Tunguska river water is chloride-sodium. During the "open" water period Tembenchi river water becomes hydrocarbonate-calcium, N. Tunguska river water – hydrocarbonate-sodium-calcium. During dry years, N. Tunguska and Tembenchi refer to transitional types of chemical compositions – hydrocarbonate-chloride-calcium and hydrocarbonate-chloride-sodium-calcium types, respectively (by Alekin classification, 1946). Studied main anions and cations have inverse relationship to the flow of their concentration during annual cycle (Bagard et al. 2011, Prokushkin et al. 2011). During the spring flood period, caused by snowmelt, river water are weakly mineralized and belong to ultra fresh type. In summer-fall period, as soil stratum thawing, the outcome of basalt weathering has great significance in river flow (Bagard et al. 2011) and inorganic ions flux, naturally, increases. During winter period, inorganic ions concentrations increase as active layer thawing, and taliks become the main source for river water (Kadamtseva et al. 2005). The chemical compositions of taliks caused by evaporate occurrence and determine the great amount of chloride, sodium and calcium in river water (Kadamtseva et al. 2005).

The average annual mineralization of Tembenchi river significantly lower than N. Tunguska river (67-82 mg/l vs. 301-498 mg/l). It conditions by different types of permafrost distribution beneath studied watersheds. For "northern" part of Central Siberian Plateau (Tembenchi) permafrost is spread continuously, determining the less square of through taliks, which are the source of solutions from subpermafrost highly mineralized brines for river. Also, lower temperatures provide the slowdown of paedogenesis and parent rock (basalts) weathering processes. In "southern" part of Central Siberian Plateau (studied N. Tunguska watershed) continuous permafrost distribution has less than 60%, and average annual temperature is higher than Tembenchi watershed (-7.3 and -10.9 $^{o}$C respectively). Thus, temperatures increase and taliks expansion by permafrost degradation may cause significant mineralization growth of Central Siberian Plateau rivers runoff.

A great increase of volume-weighted concentrations of main ions during the open water period (May – October) revealed by long-term analysis (1955-2015, May – October, Figure 2). Volume-weighted mean concentrations of chloride show a large growth (5.5 to 42.6 mg/l in spring flood time and 9.4 to 67.2 mg/l in summer-fall time). The same is for hydrocarbonate ion – 22.8 to 45.5 mg/l in spring flood time and 43.3 to 82.9 in summer-fall time. Such tendency observed for all studied ions. Wherein increase of chloride, calcium and sodium concentrations indicates increase of evaporate input, assuming permafrost degradation, and taliks expansion in N. Tunguska watershed. During winter mean water reverse pattern is obtained – significant decrease of inorganic ions concentration (April, figure 2). It supposed smaller depth of seasonally freezing part of soil stratum, which determines increase the amount of soil solutions with lower mineralization in winter runoff.

Archive data of main ions concentrations for Tembenchi River limited by period 1970-1993. Volume-weighted mean concentrations of main anions and cations for each month have insignificant changes for entire period. It may be distinguished only two variations - increase of hydrocarbonate concentration (28.16 – 53.45 mg/l) in April and in October (18.21 – 29 mg/l). Such phenomenon may indicates higher intensity of weathering processes on Tembenchi watersheds.

Besides the concentrations increase, described previously, the river flow has the important effect on annual flux of main anions and cations. For the period 1939 – 2011 on N. Tunguska watershed flow increased of 0.31 $km^3$/year/year, on Tembenchi watershed – 0.04 $km^3$/year/year. The spring-flood time has the most important impact on flow growth – about 60%.

### *2.2. Annual fluxes of the main anions and cations*

The increase of ions fluxes observed for N. Tunguska river basin from 1960 to 2011: hydrocarbonate – 2.92-12.77 Mg/$km^2$/year, chloride – 0.95 – 11.07 Mg/$km^2$/year, sulfate – 1.13-2.79 Mg/$km^2$/year, calcium – 1.48-4.64 Mg/$km^2$/year, magnesium – 0.4-1.44 Mg/$km^2$/year (figure 2). It was the most substantial "jump" since 1980: rise of hydrocarbonate from 2.88 to 7.21 Mg/km$^2$/year, chloride – 3.79-7.45 Mg/km$^2$/year. Total flux of the main anions and cations from N. Tunguska watershed increased more than 4.5 times (6.88-32.72 Mg/km$^2$/year). But there was no reliable tendency for flux from Tembenchi watershed. It could be noticed some slight decrease of hydrocarbonate and chloride fluxes and increase of sulfate and calcium fluxes.

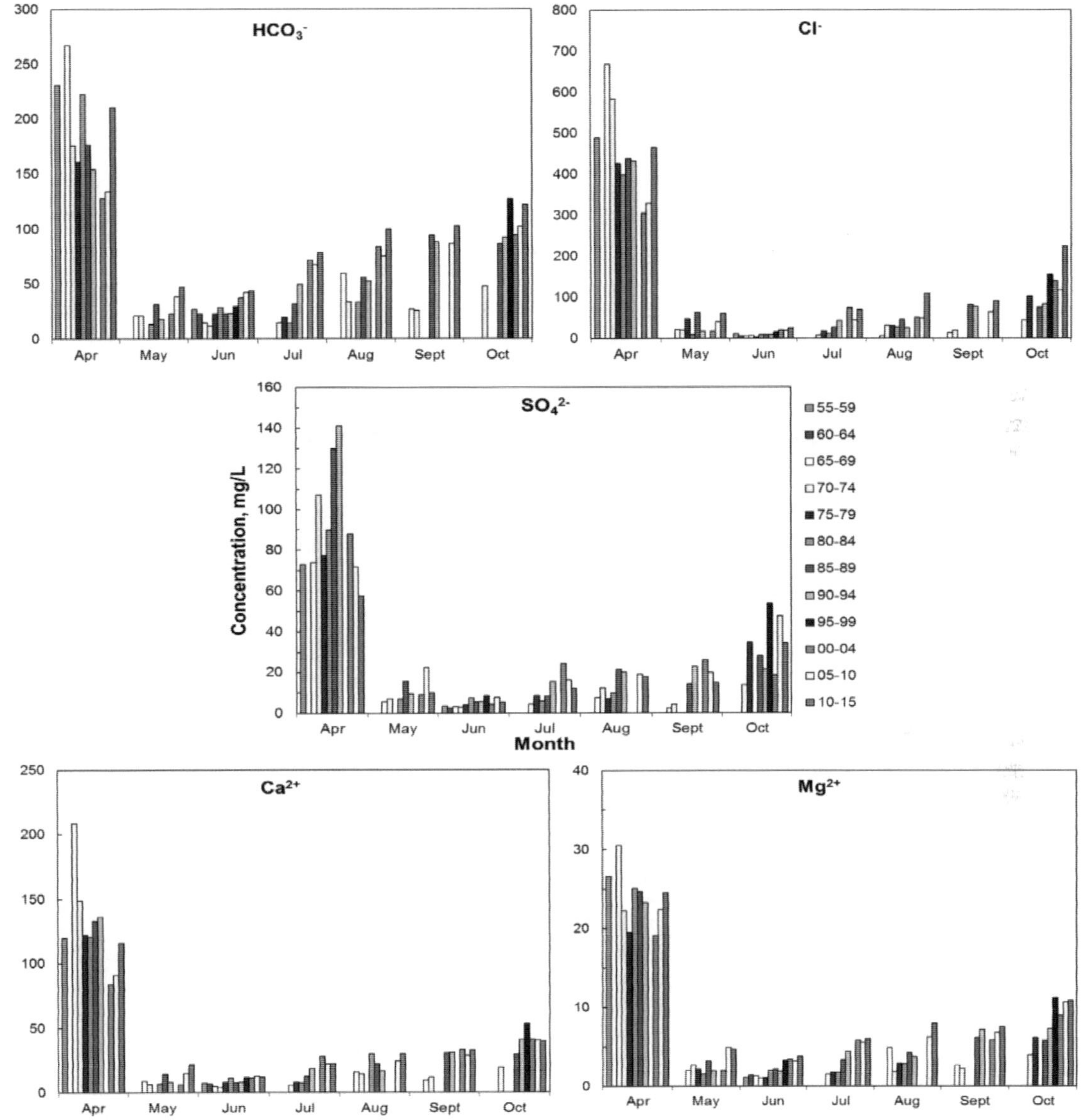


**Figure 2.** Volume-weighted mean concentrations for studied anions and cations in N. Tunguska.

Nevertheless, estimation of ions fluxes from N. Tunguska watershed based on our own investigations fit with estimation based on Roshydromet data. However, some differences in ions flux values observed. In particular, there are essential distinction in sulfate and calcium fluxes values. For Tembenchi, values measured on our instrumental base, have the significant increase of the main ions fluxes in comparison with values, based on Roshydromet. The only exception is sulfate, which flux is higher by Roshydromet data. Such discrepancy may be the result in different methods of sampling and hydrochemical analysis.

At the same time, according to our last observations, it was a significant increase of hydrocarbonate, calcium and magnesium fluxes from Tembenchi river basin for the last decade (2005-2015).

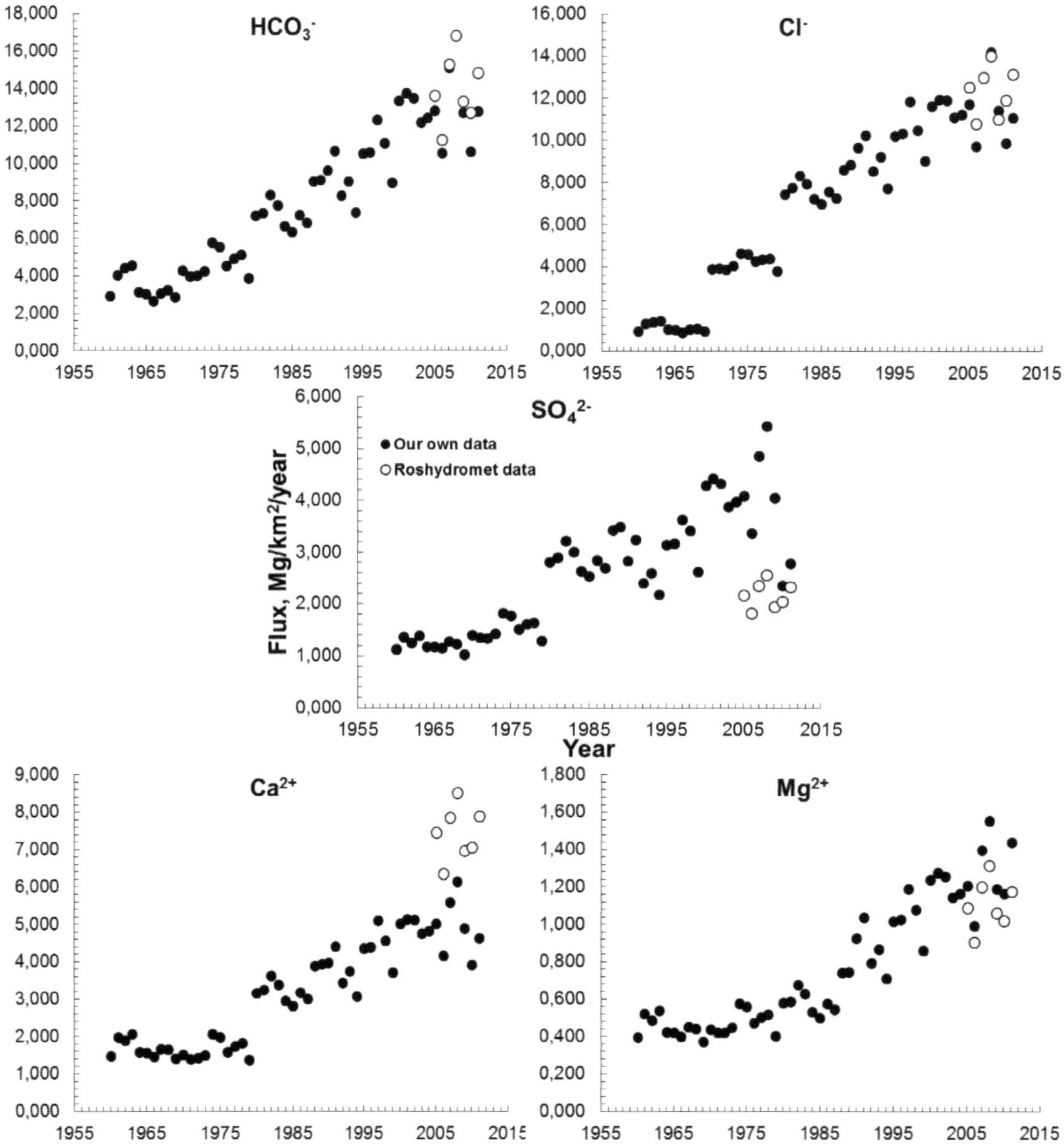


**Figure 3.** Cations and anions fluxes from N/ Tunguska watershed.

In this way, as a result of comparative analysis of Central Siberian Plateau watersheds it can be concluded that the most clear evidence of permafrost degradation and taliks expansion are registered in the zone of discontinuous distribution of permafrost. Such processes, probably, are in initial stage on the central and northern parts of Central Siberian Plateau. Nevertheless, the growth of paedogenic ions (hydrocarbonate, calcium and magnesium) concentrations and fluxes form watersheds draining this territory indicates the increase of active layer and weathering processes intensification. Thus, the main ions concentrations and fluxes analysis from other Central Siberian Plateau watersheds may be the indicators of permafrost degradation and weathering processes intensification.

**Acknowledgements**
This study was funded by RSF Grant #14-24-00113. We thank our laboratory staff for hydrochemistry analysis. Also, we thank Roshydromet for archive data from N. Tunguska and Tembenchi gauging stations.

**References**
[1] Bagard M-L, Chabaux F, Pokrovsky O S, Viers J, Prokushkin A S, Stille P, Rihs S, Schmitt A D and Dupre B 2011 Seasonal variability of element fluxes in two Central Siberian rivers draining high latitude permafrost dominated areas *Geochim. Cosmochim. Acta* **75** 3335-57
[2] Frey K E and McClelland J W 2009 Impacts of permafrost degradation on arctic river biogeochemistry *Hydrol. Process.* **23** 169–82
[3] Gordeev V 2012 *Geochemistry of river-sea system* [in Russian] ed Matushkina Y Y Moscow 452
[4] Kadamtseva T, Kozak S, Sidorkin V, Fedulova A, Yazvin A 2005 The main results of groundwater searches for Tura water needs *Exploration and conservation of bowels of the earth* [in Russian] **11** 30-33
[5] Peterson B J, Holmes R M, McClelland J W, Vorosmarty C J, Lammers R B, Shiklomanov A I, Shiklomanov I A and Rahmstorf S 2002 Increasing river discharge to the Arctic Ocean *Science* **298** 2171-3
[6] Pokrovsky O S, Schott J, Kudryavtsev D I and Dupre B 2005 Basalt weathering in Central Siberia under permafrost conditions *Geochim. Cosmochim. Acta* **69** 5659-80
[7] Prokushkin, A S et al. 2011 Sources and the flux pattern of dissolved carbon in rivers of the Yenisey basin draining the Central Siberian Plateau *Environmental Research Letters* **6** 045212.

# Geochemical characteristics of peat from two raised bogs of Germany

**A M Mezhibor**

Department of Natural Resources, Tomsk Polytechnic University, 30 Lenin Avenue, Tomsk, Russia

amezhibor@gmail.com

**Abstract**. Peat has a wide range of applications in different spheres of human activity, and this is a reason for a comprehensive study. This research represents the results of an ICP-MS study of moss and peat samples from two raised bogs of Germany. Because of the wide use of sphagnum moss and peat, determining their geochemical characteristics is an important issue. According to the results obtained, we can resume that the moss samples from Germany are rich in Cu, As, Y, Zr, Nb, and REE. The geochemical composition of the bogs reflects the regional environmental features and anthropogenic influence.

## 1. Introduction

Peat is an organic substance forming by weak decomposition of the mosses growing in wetlands. Peat has a wide range of applications in different spheres of human activity, including its use as a fuel. Because of the great importance of peat for humans, is it necessary to determine the chemical composition of peat from different regions. When peat deposits are located in areas under the influence of anthropogenic activity, determining their geochemical features is even more necessary for a subsequent peat use.

The geochemical features of raised peat deposits (ombrotrophic bogs) are defined, first of all, by the mineral composition of the underlying grounds and by the surrounding dry lands. The wind soil erosion is also an important provider of atmospheric dust (clay minerals) in the bogs. Chemical elements, as composers of various substances, are deposited on the bog surface by wet deposition and by dry or cloud (fog) water deposition. Because of weak decomposition, the sphagnum mosses forming the peat deposits can save the falling from the atmosphere particles *in situ*. Usually the chemical elements entering bogs from the atmosphere are not subject to significant post-depositional dislocations because of the properties of sphagnum moss to retain them by its "air cells" and accumulations in the root air spaces [3, 5].

There is lack of data on the bog geochemistry of Germany. The concentrations of some chemical elements in the peat in Germany were estimated in a bog from the Black forest [2] and in sphagnum mosses from 8 bogs [6]. The estimates made in [2] relate to the rare earth elements (REE) and do not include some chemical elements presented in this work. The results for 8 bogs of Germany [6] relate only to living sphagnum moss. The objective of this study is to characterize the geochemical features of peat from two bogs of Germany (central part) with the aim to determine the accumulation of chemical elements by sphagnum mosses and peat. The results of the study can be used in different spheres of human activity and for other studies connected with the research of peat and sphagnum moss.

## 2. Sampling sites and methods

For the present research two raised bogs located in the central part of Germany were sampled in September 2011: Strohner Maarchen (Eifel region) and Rotes Moor (Hassian Rhoen) (Figure 1). The central part of Germany has temperate-oceanic climate with relatively cold winters and warm summers. Strohner Maarchen bog is located in an agricultural and touristic region at the foot of a former volcano within the Eifel volcanic field. The underlying magma rocks are enriched in light REE, Zr, Nb, Rb, Ba, U, and Th [7]. Rotes Moor is located in a mountain region without direct anthropogenic influence. The underlying rocks in this area are generally composed by sandstones [4].

Samples of moss and 5-cm of peat were taken with a stainless knife (to exclude any contamination of peat samples) and put into plastic airproof bags. The samples were transported to a laboratory and dried at room temperature for obtaining a constant weight. After that the samples were milled into powder using an electrical stainless grinder. Peat samples of 100 mg weight were studied by mass-spectrometry with inductively-coupled plasma (ICP-MS) at a laboratory of the Chemistry Department of Vladimir State University (analyst I. Podkozlin) with an equipment called ELAN DRC II (PerkinElmer). The ash yield was determined additionally. The samples were weighted, and then dried in a muffle furnace at a temperature of 500°C during four hours. After stabilization of the ash mass, the percent of the ash yield was calculated.

**Figure 1**. A scheme of location of the bogs:
1 - Strohner Maarchen, 2 - Rotes Moor.

## 3. Results and Discussion

When comparing the results obtained for chemical element concentrations in moss and peat, we observe a similar character of the distributions of chemical elements in the bogs (Figure 2).

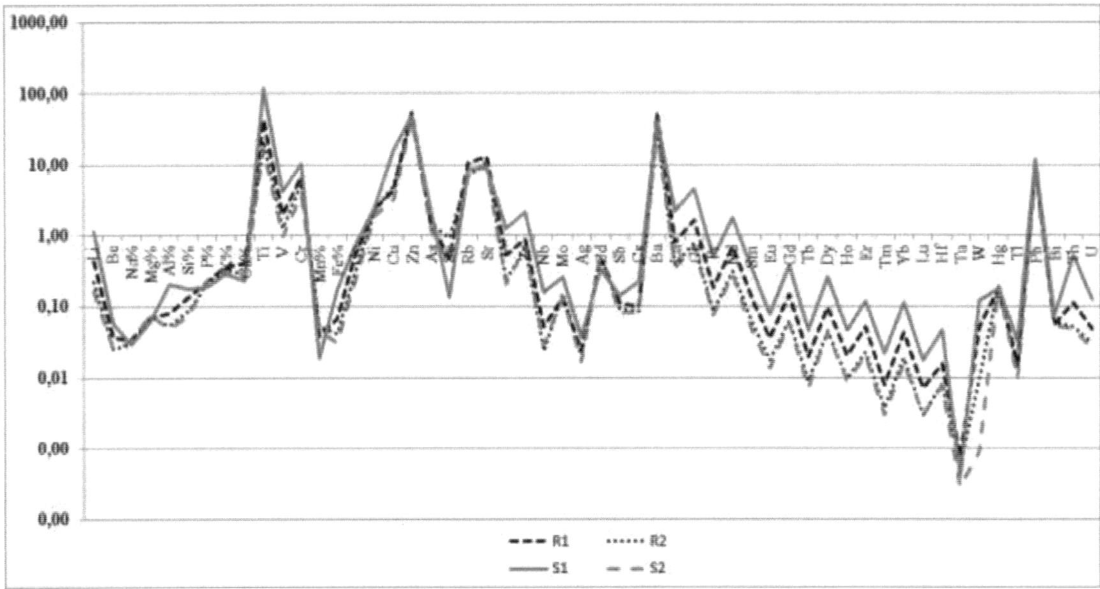


**Figure 2**. Concentrations of chemical elements (mg·kg$^{-1}$) in moss (R1 – Rotes Moor, S1 – Strohner Maarchen) and peat (R2 – Rotes Moor, S2 - Strohner Maarchen) of the bogs.

Higher concentrations of the chemical elements were found in the moss from Strohner Maarchen in comparison with the moss from Rotes Moor. The concentrations of some chemical elements in the moss of Strohner Maarchen are two and more times higher for Li, Al, Ti, V, Fe, Cu, Y, Zr, Nb, Mo, Ag, Cs, REE, Hf, W, Th, and U (Table 1). The results for the mosses of these bogs have shown close values compared with mean estimates of some chemical elements (Ca, Al, Fe, K, Mn, Ni, Pb, Rb, S, Ti, Zn) in 8 bogs of Germany [6], except for Ti in the moss of Strohner Maarchen. The Ti concentration is 2.7-7 times higher than that given in [6]. High Ti contents were also obtained for some moss samples of Germany [2], where Ti concentration in living moss was 90.09 mg·kg-1. The mean concentration of Ti in the mosses of Germany according to [8] is 21.8 mg·kg$^{-1}$.

Comparing the above-obtained results with the mean element concentrations in the moss of Germany (according to [8]), we note that the following chemical elements have higher concentrations: Cu, W, Ti – about 2 times, Li – 3.6 times, Al, U – 4.5 and 4.3 times, Ti, Y, Th, La – 5.3-5.9 times, Cr – 7.4 times, and As – 9.1 times in Strohner Maarchen, and Ti, La, Ce – about 2 times in Rotes Moor. Estimates made in [2] allow comparing the concentrations of REE in the bogs being studied. High concentrations are noted for Er – 4.6 times, Ho, Nd, Pr, Sm, Eu, Ce, La, Dy – 5.0 - 5.9 times, Gd, Tb – 6.6 times in Strohner Maarchen, and La, Ce, Nd, Sm, Er, Tm – 2 times, Eu, Gd, Dy, Ho, Tb – 2.2-2.7 times, Yb, Lu – 1.5 and 1.8 times for Rotes Moor.

The peat samples of the bogs being studied have similar values for the chemical elements except for Se. The concentration of Se in Rotes Moor is 2.2 times higher than the value for Strohner Maarchen. Comparing with REE estimates for a 4-cm peat in [2], there are higher concentrations only for Ce – 3,4 and 2.6 times for Rotes Moor and Strohner Maarchen, respectively.

The ash yield of moss samples in Strohner Maarchen is higher than that in the peat. This may indicate a higher dust load in the recent years of moss growth. Perhaps, higher concentrations of some chemical elements (Ti, Cu, Fe, Cr, As) in the moss samples of Strohner Maarchen are of anthropogenic origin. The region of Strohner Maarchen is influenced by the closely located auto roads and agriculture fields. The location of this bog in the center of an industrially developed territory can also be a reason of the high element concentrations, for example, metals. The great difference in the accumulated REE of the moss of Strohner Maarchen in comparison with the values from [2] allow concluding that the dust load in 2011 was higher than before. The higher ash yield in moss (2.5) than in peat (1.4) also confirms this. The rate of peat accumulation in Europe is quite slow [1], and this explains the similarity of the REE values for the peat samples. Another source of chemical elements (Al, Ti, Zr, Nb, REE, Th, and U) is the dust input from the surrounding ground, whose composition is determined by the composition of the underlying rocks (vulcanites).

**Table 1.** Concentrations of chemical elements (mg·kg$^{-1}$) in moss and peat of the bogs being studied.

| Elements | R1 | S1 | R2 | S2 | Elements | R1 | S1 | R2 | S2 |
|---|---|---|---|---|---|---|---|---|---|
| **Li** | 0.43 | 1.14 | 0.23 | 0.16 | **Cd** | 0.49 | 0.33 | 0.51 | 0.49 |
| **Be** | 0.04 | 0.06 | 0.03 | 0.02 | **Sb** | 0.11 | 0.14 | 0.09 | 0.08 |
| **Na, %** | 0.03 | 0.03 | 0.03 | 0.03 | **Cs** | 0.10 | 0.22 | 0.08 | 0.08 |
| **Mg, %** | 0.07 | 0.06 | 0.07 | 0.08 | **Ba** | 50.6 | 41.3 | 32.0 | 30.4 |
| **Al, %** | 0.08 | 0.21 | 0.05 | 0.05 | **La** | 0.79 | 2.19 | 0.39 | 0.35 |
| **Si, %** | 0.13 | 0.18 | 0.09 | 0.07 | **Ce** | 1.65 | 4.42 | 0.76 | 0.68 |
| **P, %** | 0.23 | 0.19 | 0.22 | 0.24 | **Pr** | 0.19 | 0.49 | 0.08 | 0.08 |
| **K, %** | 0.36 | 0.29 | 0.31 | 0.32 | **Nd** | 0.69 | 1.76 | 0.32 | 0.27 |
| **Ca, %** | 0.41 | 0.23 | 0.29 | 0.28 | **Sm** | 0.13 | 0.34 | 0.06 | 0.05 |
| **Ti** | 42.7 | 117.1 | 21.8 | 16.8 | **Eu** | 0.036 | 0.077 | 0.017 | 0.013 |
| **V** | 1.9 | 4.2 | 1.24 | 0.95 | **Gd** | 0.14 | 0.38 | 0.06 | 0.06 |
| **Cr** | 6.7 | 10.3 | 5.35 | 3.69 | **Tb** | 0.019 | 0.046 | 0.009 | 0.007 |
| **Mn, %** | 0.04 | 0.02 | 0.05 | 0.04 | **Dy** | 0.097 | 0.258 | 0.044 | 0.042 |
| **Fe, %** | 0.07 | 0.17 | 0.04 | 0.03 | **Ho** | 0.020 | 0.045 | 0.009 | 0.009 |
| **Co** | 0.55 | 0.74 | 0.31 | 0.24 | **Er** | 0.052 | 0.119 | 0.023 | 0.020 |
| **Ni** | 2.34 | 2.45 | 2.19 | 1.90 | **Tm** | 0.008 | 0.021 | 0.004 | 0.003 |
| **Cu** | 4.2 | 16.2 | 4.9 | 3.2 | **Yb** | 0.04 | 0.11 | 0.02 | 0.02 |
| **Zn** | 54.0 | 48.32 | 46.8 | 44.3 | **Lu** | 0.007 | 0.017 | 0.003 | 0.003 |
| **As** | 1.5 | 2.24 | 1.45 | 1.21 | **Hf** | 0.02 | 0.05 | 0.01 | 0.01 |
| **Se** | 0.47 | 0.13 | 0.92 | 0.41 | **Ta** | 0.0008 | 0.0004 | 0.0006 | 0.0003 |
| **Rb** | 10.5 | 8.5 | 6.89 | 7.10 | **W** | 0.051 | 0.122 | 0.009 | 0.001 |
| **Sr** | 12.6 | 8.7 | 10.9 | 9.8 | **Hg** | 0.18 | 0.18 | 0.18 | 0.19 |
| **Y** | 0.54 | 1.23 | 0.23 | 0.21 | **Tl** | 0.015 | 0.032 | 0.013 | 0.009 |
| **Zr** | 0.89 | 2.10 | 0.69 | 0.68 | **Pb** | 9.91 | 11.68 | 9.2 | 7.6 |
| **Nb** | 0.05 | 0.16 | 0.02 | 0.03 | **Bi** | 0.06 | 0.07 | 0.06 | 0.05 |
| **Mo** | 0.13 | 0.26 | 0.15 | 0.14 | **Th** | 0.11 | 0.52 | 0.05 | 0.05 |
| **Ag** | 0.02 | 0.04 | 0.02 | 0.02 | **U** | 0.05 | 0.12 | 0.03 | 0.03 |
| **Ash yield, %** | 3.0 | 2.5 | 6.2 | 1.4 | **Th/U** | 2.2 | 4.3 | 1.7 | 1.7 |

Comments: R1, R2 - Rotes Moor moss and peat, respectively, S1, S2 - Strohner Maarchen moss and peat, respectively.

## 4. Conclusions

The results of the present research showed that the regional geochemical features and anthropogenic activity are essential factors influencing the geochemical characteristics of moss and peat in the central part of Germany.

## Acknowledgments
Many thanks to Dr. B. Bonn and W. Haunold from the Institute of Environment of Goethe University, Frankfurt-am-Main and their team for help with sample collection and transportation; to Moritz Schmitt and Ewald Sauer for the peat sampling. This research was partly supported by DAAD and Ministry of Education of Russia (program "Mikhail Lomonosov", 2011).

## References
[1] Amesbury M J, Barber K E, Hughes P D 2011 The methodological basis for fine-resolution, multi-proxy reconstructions of ombrotrophic peat bog surface wetness *Boreas* **40** 161–174.
[2] Aubert D, Le Roux G, Krachler M, Cheburkin A, Kober B, Shotyk W, Stille P 2006 Origin and fluxes of atmospheric REE entering an ombrotrophic peat bog in Black Forest (SW Germany): Evidence from snow, lichens and mosses *Geochimica et Cosmochimica Acta* **70(11)** 2815–26.
[3] Bobbink R, Whigham D F, Beltman B, Verhoeven J T A 2006 Management Wetland functioning in relation to biodiversity conservation and restoration *Wetlands and natural resource* eds R Bobbink, B Beltman, J T A Verhoeven, D F Whigham (Berlin-Heidelberg: Springer-Verlag) pp 2–3.
[4] Brien E 2000 Gewässerlandschaften der Bundesrepublik Deutschland, Maßstab 1 : 1000000.
[5] Dragović S, Mihailović N 2009 Analysis of mosses and topsoils for detecting sources of heavy metal pollution: multivariate and enrichment factor analysis *Environmental Monitoring and Assessment* **157** 383–390.
[6] Kempter H, Frenzel B 2007 The Geochemistry of ombrotrophic Sphagnum species growing in different microhabitats of eight German and Belgian peat bogs and the regional atmospheric deposition *Water, Air and Soil Pollution* **184** 29–48.
[7] McCann T 2008 *The geology of Central Europe* (Mesozoic and Cenozoic vol 2) ed T McCann (London: The Geological SocietyLondon).
[8] Reimann C, de Caritat P 1998 *Chemical elements in the environment: factsheets for the geochemist and environmental scientists* (Berlin-Heidelberg: Springer-Verlag).

# A variational approach to environmental and climatic problems of urban agglomerations

**V V Penenko[1,2] and E A Tsvetova[1]**

[1] Institute of Computational Mathematics and Mathematical Geophysics SB RAS, 6, Akademika Lavrentyeva pr., Novosibirsk 630090, Russia
[2] Novosibirsk State University, 2, Pirogova str., Novosibirsk 630090, Russia

e-mail penenko@sscc.ru

**Abstract.** We discuss some aspects of the development of a variational approach to study the dynamics of climatic and ecological systems under intensive actions of natural and anthropogenic origin. The variational principle essentially represents a versatile tool to create a consistent modeling technology based on models of processes coupled with available observational data. The basic entities included in the formulation of the variational principle are models of processes; data and models of observations; target criteria for forecasting; a priori information about all the required elements of the system. We develop a set of mathematical models combined within the framework of the variational principle. They describe the dynamics of the atmosphere and water bodies in conjunction with a thermally and dynamically heterogeneous surface of the Earth; the hydrological cycle, moisture in the atmosphere and the soil; radiation transfer in the system of the atmosphere and the underlying surface; and transport and transformation of various substances in gaseous and aerosol states in the atmosphere. As an example, we demonstrate the results of calculations performed with a set of numerical models adapted to the conditions of a Novosibirsk city agglomeration. The results of scenario calculations on the formation of mesoclimates and quality of the atmosphere for the typical conditions of Siberian cities are presented.

## 1. Introduction

Throughout the history of mankind, compact residences have been sought in limited areas selected under various criteria. People interact with natural environment, use its resources and, quite frequently, exhaust them. Currently, the results of this interaction are clearly visible, especially in urban agglomerations and industrial regions. Therefore, the studies of problems arising in the interaction with the environment occupy an increasingly prominent place in the research on environmental problems. Now such studies are difficult without the use of mathematical modeling.

Here we discuss some applications of a variational approach to study problems which are critical in the theory of climate and the environment [1-5]. The results of their solutions provide a direct contribution to the sphere of practical applications. These are the assessment of the impact on human health, environmental safety, and the organization of environmental policies. To solve them, mathematical models and technical means for monitoring the parameters of the processes are being developed in many scientific centers of the world [6,7].

Methods for solving environmental problems based on coupled models of dynamics and atmospheric chemistry have been developed at ICM&MG SB RAS since 1980 [2,5]. Descriptions of the variational modeling technologies and some new versions of the models of this class are given in [8,9]. We proceed from the premise that the mathematical models and monitoring systems describe the evolution of the same processes and objects, but by different means. Their union is based on variational data assimilation methods [4, 9-10]. Essentially, the variational principles are a universal tool for creating a consistent modeling technology based on the use of process models coupled with available observational data. Contemporary reviews of works on this subject can be found in [7,11,12] and in the bibliography available there.

## 2. Statement of the problems

To conduct research and to solve practical problems we use a variational approach that combines all the elements of the technology of mathematical modelling. The core objects included in the formulation of the variational principle are: process models; observed data and models of observations; a priori information about all the required elements of the simulation system; target criteria for prediction and design; functional limitations on the quality of the environment and sanitary conditions and environmental standards for the quality of life and health of the population of urban agglomerations.

Consistency is an important factor for the processes of different scales and different nature in conditions of uncertainty, both in the models and in the observational data. To solve this class of problems, we use the following set of mathematical models: hydrodynamics of geophysical fluids (for air and water bodies) in conjunction with a thermally and dynamically inhomogeneous surface of the Earth; the hydrological cycle of moisture transformation in the atmosphere; heat and moisture transport in the soil; radiative transfer in the system of the atmosphere and the underlying surface; transport and transformation of various substances in gaseous and aerosol states in the atmosphere.

Chemically, optically, and thermodynamically active substances, biologically active aerosols and radionuclides are of special interest for practical applications. Secondary products of transformations of primary emissions from the sources affect human health, the quality of air and water, as well as the state of ecosystems. They require special attention.

It should be noted that the whole set of multifunctional factors and different substances manifests itself in an integrated form in the living space of urban agglomerations and may have synergetic effects of their interactions. Therefore, the main issues are to identify the internal mechanisms of the formation and evolution of processes as well as to evaluate possible changes in the quality of the environment for a long period. These changes may occur not only as a result of work of the existing enterprises, but also from new objects of further economic developments of the territories. Moreover, we are interested in possible changes not only in the metropolitan area, but also beyond it, in the areas of risk / vulnerability for the region being studied.

Given these preconditions for the organization of simulation scenarios, we evaluate the extent of interactions in the global climatic system and in the region where the objects being studied are located. For these purposes, we use methods of scale decomposition of the investigated phenomena and evaluation of the characteristic times and intensities of the sources of impact in the region. They allow us to create a set of scripts to reproduce typical meso-regional climates on the background of the global circulation mechanisms [13].

The mathematical models for studying the atmospheric processes mentioned above are represented by systems of equations of hydrodynamics. They are unsteady in time and three-dimensional in space. We study the behaviour of various substances in gas and aerosol states with the help of equations of the convection-diffusion-reaction type. To combine them into a generalized system, we design a main functional of integral identity which acts as restrictions on a class of functions in the generalized formulation of the variational principle. For definiteness, we choose a form of the functional based on the structure of a balance equation for the total energy of the system. Observational data are also included in the system within the framework of a variational principle. This is done by means of special functionals expressing a measure of deviations between the observations and their images, which are calculated by the models of observations with the sought functions of state.

A general structure of the modeling technology is presented in [8-10,14]. The numerical implementation of the variational principles is carried out with functional decomposition and splitting methods for the operators of the process models [1,4,15]. For the construction of numerical schemes we have developed discrete-analytical methods based on a concept of adjoint integration multipliers (factors) [16]. Such methods are convenient for the study of multi-scale and multi-parameter processes with high spatial-temporal resolution, without increasing the dimensions of the grid domains.

## 3. Variational organization of the modelling system

Uncertainties of various kinds are always present in the process models and the observational data. Under an uncertainty we mean the errors of observations, the lack of modern knowledge about the natural processes and their representation in the models. To account for them we introduce additional unknown functions into the model. They represent a quantitative measure of uncertainty in the relevant model equations, in the parameterizations of physical processes, and in the input data and sources. The fundamental importance of these functions is that they provide a quantitative assessment of the degree of conformity of the results of calculations by the models to the results of monitoring the behavior of the real system. In this formulation, the models are not exact. Therefore, we build the variational principle in a weak-constraint formulation. A variety of assessments of the effects of uncertainties for research and practical purposes are being developed in the world (see reviews in [10-12]).

We develop a computationally efficient version of the variational approach with assimilation of observed data. The weak-constraint formulation of the variational principle is equivalent to well-known approaches such as variational methods with weighted least squares of Gauss, including Tikhonov regularization; variational principles of Euler-Lagrange with the technique of Lagrange multipliers, and methods like Kalman filtering.

An advantage of our approach is the use of more efficient algorithms, especially in sequential data assimilation technique with splitting and decomposition methods [9,10]. This allows us to build a forecasting system with non-iterative (direct) data assimilation algorithms allowing a parallel organization of the calculations. The computational efficiency of direct algorithms increases, as compared to the other approaches for solving the problems of data assimilation, with increasing volume of observations in space and time.

The purpose of the study and the sequence of operations when implementing the variational approach define the set of its basic elements:

- the systems of equations in numerical models:
  - ✓ main (forward mode),
  - ✓ adjoint (backward in time),
  - ✓ for calculation of uncertainty functions;
- the set of algorithms:
  - ✓ formation of operators of model sensitivity to variations of all the parameters, sources of influence, and other objects,
  - ✓ calculation of sensitivity functions of the objective functional using the solutions of the main and adjoint problems, and uncertainty functions,
  - ✓ data assimilation and regularization for solving inverse problems,
  - ✓ feedback for the solution of inverse problems for identification of sources of influence and model parameters with respect to specified target criteria and constraints,
  - ✓ estimation of the functions and areas of risk/vulnerability of the object-receptors in relation to existing and potential sources of heat and pollution.

The problem of identifying the active energy centers of action on various spatial-temporal scales is of great interest in the study of critical issues of climatic and ecological systems of the Earth and its specific regions. First of all, on the basis of mathematical models we need to clarify the place of an object in a climatic and ecological system and evaluate its role as a source and receptor of influence in this system. The above-mentioned methods allow us to calculate functions of risk and vulnerability of these objects against the effects of potentially dangerous sources in the region. Therefore, for solving problems of this class we develop a technology of mathematical modeling which involves models of different physical content and spatial-temporal scales. Methods of constructing numerical schemes and algorithms for implementing the mathematical models based on a variational concept of adjoint integrating factors provide sufficient accuracy of approximations for a wide range of spatial-temporal resolutions of the models [16].

## 4. Adaptive targeted monitoring strategy

We develop a scenario approach based on variational methods for the formation of targeted monitoring strategies and planning of observational experiments. In relation to the problem of forecasting the dynamics of the processes and changes in the quality of the atmosphere, this is a versatile research tool for solving a sequence of inverse problems.

Methods of implementing such an information technology are proposed in [13]. We combine methods of scale decomposition of the processes being studied and methods of model sensitivity with assessment of uncertainty in the modeling technique. To this end, we conduct a joint analysis of multidimensional fields of the state functions, adjoint functions, uncertainty functions and a set of sensitivity functions of objective (cost) prognostic functionals.

We use methods of orthogonal decomposition of multidimensional spaces of functions based on singular value decomposition methods to identify the main factors and centers of action with an estimate of characteristic scales of their intensity. The result is a system of informative basis subspaces that are ranked according to some criteria of informativeness of the singular values.

In this way it is possible to separate the subspaces of climatically significant scale from the subspaces of weather noise [13]. Their knowledge is useful when solving practical problems of ecological safety and environmental design for a long period. Such bases are also used in the operational methods of a few-parameter reconstruction of state functions with assimilation of observational data in real time. In particular, if we design new facilities related to emissions of heat, moisture, and pollutants, as well as to changes in the characteristics of the underlying surface, this increases the information content of climate scenarios to obtain estimates of environmental perspectives in the region for a long time.

## 5. A scenario for the Novosibirsk city agglomeration

The seventh in Russia, the city of Novosibirsk (1.7—2.0 million people) is the most important inter-regional center of social and economic development and an attractor for the macro-region of Western Siberia. The cities of Berdsk, Iskitim, Ob, and major scientific centers-satellites of Novosibirsk, Koltsovo, and Krasnoobsk are the main parts of the Novosibirsk agglomeration. Akademgorodok is one of the districts of the city of Novosibirsk.

This industrial area is characterized by a high level of anthropogenic load. The main climatic factor of meso-regional scale is formed as a result of the hydrodynamic and environmental interactions of the objects of Novosibirsk agglomeration with two main water bodies, the Ob river and the reservoir of the Novosibirsk hydroelectric power station. This climatic factor manifests itself continuously in the form of daily and seasonal variations on the background of global processes.

As an example, we present the simulation results of a typical scenario of the formation of mesoclimates in the region under the influence of a background flow, as well as the fields of pollutant concentrations constructed by these mesoclimates.

In the calculations, we use a set of meso-regional models describing the hydrodynamics, transport, and transformation of pollutants adapted to the conditions of the Novosibirsk agglomeration. The system of equations of the models and methods of their realization are described in [8,10,16]. The dynamic model takes into account the processes of hydrological cycle. A parameterization of the processes in the atmospheric surface layer and the main surface heterogeneity are considered as well.

Some land use categories in the region are taken into account: city, river, reservoir, different kinds of forests and farmland.

In the scenario, the circulation of the atmosphere is formed under the influence of a south-western background flow with a wind speed of about 8 m/s given at a height of 3000 m above the ground. Diurnal solar radiation for the summer season is also included. The urban agglomeration is considered as a set of distributed sources of anthropogenic pollution and as a generator of "heat islands" in the atmosphere that interact with the environment. The sources of pollutants and artificial heat fluxes are defined in a parameterized form in view of the intensity of human activity and traffic flows within the metropolitan area.

The results of some scenario calculations on the formation of mesoclimates and air quality for the Novosibirsk agglomeration at two time points are shown in figures 1 and 2. Two-dimensional cross-sections of the fields of concentrations of pollutants and velocity vector at 07:00 a.m. local time are given in figure1. On the left plot (a), a horizontal section at the level of the surface layer is shown. The right plot (b) demonstrates a vertical cross-section through the central part of the city of Novosibirsk along the coordinate line $y = 65\text{km}$. The same cross-sections at 06.00 p.m. local time are shown in figure 2.

Analysis of the results of the scenario calculations has shown that specific local circulation systems are formed in the agglomeration due to the contrasts between the warm and cold areas. The characteristic scales of influence domains of urban heat islands are about 20 km in the horizontal directions and from 500 m to 2 km in the vertical direction.

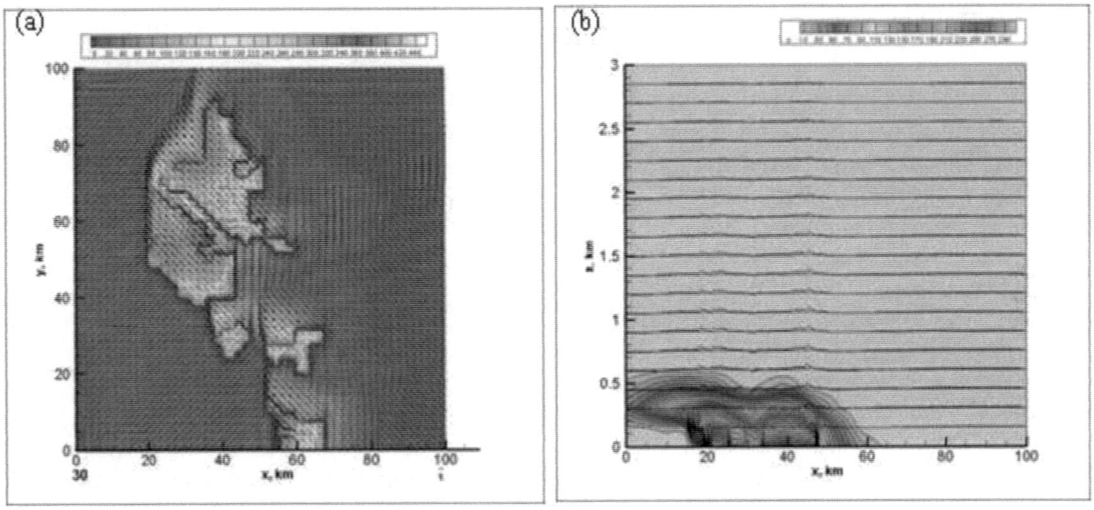


**Figure 1.** Concentration of pollutant (relative units) and velocity vector (m/s) at 7:00 a.m. local time. A 2D horizontal section at the level of the surface layer (a), and a vertical section through the central part of Novosibirsk city (b).

The cold river divides the greatest heat island into two parts. In the daily and seasonal variations, areas of high levels of pollutant concentrations are generated above the urban territories. In the vicinities of the bounds between the city and the surroundings the effects of internal boundary layers manifest themselves in the structure of fields of the state functions. Accordingly, we observe some wave behavior in the space-time dynamics of the fields of wind, temperature, and concentrations of pollutants.

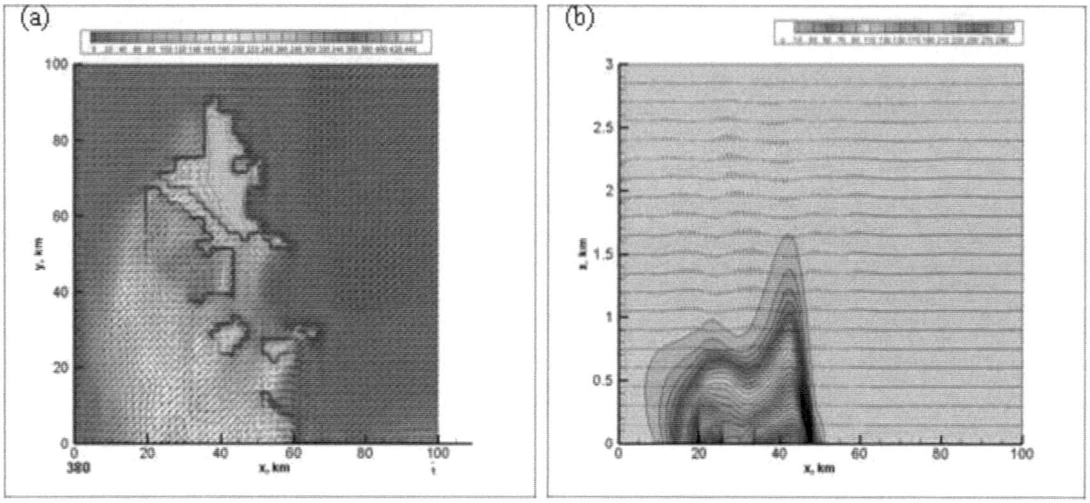

**Figure 2.** The same as in figure1 but for 06.00 p.m. local time.

## 6. Conclusions

Modelling systems of the processes on urban and meso-regional scales are necessary tools for the evaluation of actual status and quality control in normal and especially in emergency situations in urban areas. The scenario approach allows us to explore a variety of climatic and environmental situations in specific conditions of the region under study. With this type of a system, we can solve a variety of inverse problems, including the calculation of risk functions, assessment of the impact of emergency situations with identification of the exposure sources. These systems are of particular importance for a long-term environmental design of the construction and operation modes of objects that represent a potential hazard to the people and the environment.

**Acknowledgements**
The work was supported by the RAS Presidium Programs I.33P and I.II.2P/I.3 and the Russian Foundation for Basic Research (project no. 14-01-00125).

## References

[1] Marchuk G I 1974 *Numerical Solution of Atmospheric and Oceanic Problems* (Leningrad: Gidgometeoizdat) p 303
[2] Marchuk G I 1982 *Mathematical Models in Environmental Problems* (Moscow: Nauka) p 319
[3] Marchuk G I 1992 *Adjoint Equations and Analysis of Complex Systems* (Moscow: Nauka) p 334
[4] Penenko V V 1981 *Methods for Numerical Modeling of Atmospheric Processes* (Leningrad: Gidrometeoizdat) p 352
[5] Penenko V V, Aloyan A E 1985 *Models and Methods for Envronmental Protection Problems* (Novosibirsk:Nauka) p.256
[6] Baklanov A, Machura A, and Sokhi R 2011 *Integrated Systems of Meso-meteorological and Chemical Transport Models*. (Heidelberg Dordrecht: Springer) p 242
[7] *Seamless Prediction of the Earth System: from Minutes to Months* 2015 WMO-No.1156
[8] Penenko V V, Tsvetova E A and Penenko A V 2015 *Izvestiya, Atmospheric and Oceanic Physics* **51** pp 311–19
[9] Penenko V V, Tsvetova E A and Penenko A V 2015 *Russian Meteorology and Hydrology* **40**

pp 365-73
[10] Penenko V V 2009 *Numer. Anal. Appl.* **2** pp 341–51
[11] Navon I M 2009 *Data Assimilation for Atmospheric, Oceanic and Hydrologic Appl.* **1** pp 21-65
[12] Bocquet M et al. 2015, *Atmos. Chem. Phys.*, **15** pp 5325-58
[13] Penenko V and Tsvetova E 2008 *Ecol. Modell.* **217**, 279–291
[14] Penenko V, Baklanov A, Tsvetova E. and Mahura A. 2012 *Pure and Applied Geoph.* **169** pp 447-65
[15] Samarskii A A, Vabishevich P N 2001 *Additive Schemes for Problems of Mathematical Physics* (Moscow: Nauka) p 319
[16] Penenko V V, Tsvetova E A and Penenko A V 2014 *Comp. & Math. with Appl.* **67**, pp 2240–56.

# Numerical study of variational data assimilation algorithms based on decomposition methods in atmospheric chemistry models

**Alexey Penenko**[1,2], **Pavel Antokhin**[3]

[1]Institute of Computational Mathematics and Mathematical Geophysics SB RAS, ICM&MG SB RAS, prospect Akademika Lavrentyeva 6, 630090, Novosibirsk, Russia
[2]Novosibirsk State University, NSU, Pirogova Str. 2, 630090, Novosibirsk, Russia
[3]V.E. Zuev Institute of Atmospheric Optics SB RAS, IAO SB RAS, Academician Zuev square 1, 634055, Tomsk, Russia

aleks@ommgp.sscc.ru

**Abstract**. The performance of a variational data assimilation algorithm for a transport and transformation model of atmospheric chemical composition is studied numerically in the case where the emission inventories are missing while there are additional in situ indirect concentration measurements. The algorithm is based on decomposition and splitting methods with a direct solution of the data assimilation problems at the splitting stages. This design allows avoiding iterative processes and working in real-time. In numerical experiments we study the sensitivity of data assimilation to measurement data quantity and quality.

## 1. Introduction

Forecasting of chemical weather is a challenge for researchers due to the nonlinear, multiscale and dynamic character of the corresponding transport and transformation processes in the atmosphere. It becomes even more challenging when the emission data are not exact. The aim of data assimilation algorithms is to improve the quality of model forecasts with measurement data arriving in the course of calculations. High dimensional atmospheric chemistry problems impose strict requirements on the numerical efficiency of the algorithms. We study the performance of an algorithm that is able to work without any iterative processes. It is achieved by using independent data assimilation procedures at the splitting stages. Overviews of chemical data assimilation can be found in [1, 2, 3].

We consider a scenario where the unknown emission data are estimated with indirect concentration measurements. The algorithm is used to reconstruct the emission effects of a substance with measurements of the other substances. We evaluate the sensitivity of data assimilation to the quality and quantity of measurement data.

## 2. Data assimilation algorithm

### *2.1. Transport and transformation model*

Let us consider a horizontally homogeneous domain

$$z \in \Omega = [0, X_z],\ t \in [0,T],\ \Omega_T = \Omega \times [0,T],$$

bounded by $\partial\Omega_T = \partial\Omega \times [0,T]$. In the domain we consider the atmospheric chemistry transport and transformation model

$$\frac{\partial \varphi_l(z,t)}{\partial t} + div(w\varphi_l(z,t) - \mu(z,t)\, grad\, \varphi_l(z,t)) = S_l(\vec{\varphi}(z,t)) + f_l(z,t) + r_l(z,t),\ (z,t) \in \Omega_T. \quad (1)$$

$$\mu(z,t)\frac{\partial \varphi_l(z,t)}{\partial \vec{n}} + \beta(z,t)\varphi_l(z,t) = g_l(z,t),\ (z,t) \in \partial\Omega_T, \quad (2)$$

$$\varphi_l(z,0) = \varphi_l^0(z), z \in \Omega. \quad (3)$$

Here $\vec{\phi}(z,t)$ is the state function. It has the physical meaning of the field of concentrations at point $(z,t) \in \Omega_T$, e.g. $\phi_l(z,t)$ corresponds to the concentration of $l^{th}$ substance at point $(z,t)$.Here $l = 1,...,N_c$, $N_c$ is the total number of substances considered, $w(z,t)$ denotes the "wind speed", $\mu(z,t)$ is the diffusion coefficient, $S: \mathbb{R}^{N_c} \to \mathbb{R}^{N_c}$ is the transformation operator, $\vec{n}$ is the boundary outer normal direction, $\vec{f}(z,t)$, $\vec{g}(z,t)$, $\vec{\phi}^0(z)$- a priori data from the sources and the initial data, $\vec{r}(z,t)$ is the control function (uncertainty). It is introduced in the perfect model structure to assimilate the data. Each entry of $\vec{f}(z,t)$, $\vec{g}(z,t)$, $\vec{r}(z,t)$, $\vec{\phi}^0(z)$ vectors corresponds to a quantity attributed to l-th substance at point $(z,t) \in \Omega_T$. The transformation operator $S$ is defined by a chemical kinetics system of 22 reacting species from [4,5] augmented with a $SO_2$ reaction taken from a model called CMAQ [6]:

$$NO_2 + hv \to NO + O_3P \qquad hv + O_3 \to O_1D + O_2$$
$$HCHO + hv \to CO + 2H_2O \qquad HCHO + hv \to CO + H_2$$
$$O_2 + O_3P \to O_3 \qquad N_2 + O_1D \to N_2 + O_3P$$
$$O_1D + O_2 \to O_2 + O_3P \qquad H_2O + O_1D \to 2OH$$
$$HO_2 + NO \to NO_2 + OH \qquad NO + O_3 \to NO_2 + O_2$$
$$NO + RO_2 \to HCHO + HO_2 + NO_2 \qquad CO + OH \to CO_2 + HO_2$$
$$HC + OH \to H_2O + RO_2 \qquad HCHO + OH \to CO + H_2O + HO_2$$
$$NO_2 + OH \to HNO_3 \qquad 2HO_2 \to H_2O_2 + O_2$$
$$H_2O + 2HO_2 \to H_2O + H_2O_2 + O_2 \qquad HO_2 + RO_2 \to O_2 + ROOH$$
$$2RO_2 \to \Pr od \qquad OH + SO_2 \to HO_2 + Sulf$$

The reaction rates are taken from [5].They depend on time, i.e. photochemistry is considered. This kinetics system can be presented in production-destruction operator form:

$$S_l(\vec{\varphi}(z,t)) = -P_l(\vec{\varphi}(z,t))\varphi_l(z,t) + \Pi_l(\vec{\varphi}(z,t)),\ l = 1,...,N_c \quad (4)$$

$$P_l, \Pi_l : \mathbb{R}_+^{N_c} \to \mathbb{R}_+, \quad (5)$$

where $P_l$ is the destruction rate functional and $\Pi_l$ is the production functional.

**Direct problem:** Given $\vec{f}(z,t)$, $\vec{g}(z,t)$, $\vec{r}(z,t)$, $\vec{\phi}^0(z)$, determine $\vec{\phi}(z,t)$ from equations (1)-(3). The exact solution $\vec{\phi}_*$ is the solution of the direct problem corresponding to the «unknown» emissions $\vec{r}_*$.

We assume all the functions and model parameters to be smooth enough for the solutions to exist and the transformations to be valid.

For the numerical solution, let us introduce a uniform temporal grid $\omega_t = \{t^j\}_{j=1}^{N_t}$ on $[0,T]$ with step size $\tau$ and $N_t$ points and a uniform spatial grid $\omega_z$ with $N_z$, grid points on $\Omega$. Let $Q(\omega)$ be the space of real grid functions on $\omega$. The direct problem can be efficiently solved with a splitting method. Let us consider an additive-averaged splitting scheme (analogous to [7]) on the intervals $t^j \le t \le t^{j+1}$. The splitting is done with respect to the physical processes of advection-diffusion and transformation. Finally, we have two parallel stages with a step partition $\gamma_z + \gamma_c = 1$ and sources partition $\vec{f} = \vec{f}_c + \vec{f}_z$.

- The convection-diffusion process:

$$\gamma_z \frac{\partial \vec{\phi}_z(z,t)}{\partial t} + \frac{\partial}{\partial z}\left(w(z,t)\vec{\phi}_z(z,t)\right) - \frac{\partial}{\partial z}\left(\mu(z,t)\frac{\partial}{\partial z}\vec{\phi}_z(z,t)\right) = \vec{f}_z(z,t) + \vec{r}_z(z,t),\ (z,t) \in \Omega \times \left[t^{j-1}, t^j\right], \tag{6}$$

$$\mu(z,t)\frac{\partial \vec{\phi}_\beta(z,t)}{\partial \vec{n}} + \beta(z,t)\vec{\phi}_z(z,t) = \vec{g}_a(z,t),\ (z,t) \in \partial\Omega \times \left[t^{j-1}, t^j\right], \tag{7}$$

$$\vec{\phi}_\beta\left(z,t^{j-1}\right) = \vec{\phi}\left(z,t^{j-1}\right),\ \vec{x} \in \Omega, \tag{8}$$

where $\partial\Omega = \{ z = 0, z = X_z \}$. This initial value problem can be approximated in implicit matrix form:

$$\gamma_z \frac{\vec{\phi}_z^j - \vec{\phi}^{j-1}}{\tau} + L_z \vec{\phi}_z^j = \vec{r}_z^j + \vec{f}_z^j, \qquad L_z(\vec{\phi}) := \left\{L_z(\vec{\phi}_l)\right\}_{l=1}^{N_c}. \tag{9}$$

Here $\vec{\phi}^j \in Q(\omega)^{N_c}$ is the solution on the $j$-th time level, $\vec{r}^j \in Q(\omega)^{N_c}$ is the uncertainty on the j-th time level and $L_z : Q(\omega) \to Q(\omega)$ are the approximated advection-diffusion operators from equation (1) corresponding to the spatial dimensions.

- The chemical reaction processes:

$$\gamma_c \frac{\partial \vec{\phi}_c(z,t)}{\partial t} + diag\left(\vec{P}\left(\vec{\phi}_c(z,t)\right)\right)\vec{\phi}_c(\vec{x},t) = \vec{\Pi}\left(\vec{\phi}_c(z,t)\right) + \vec{f}_c(z,t) + \vec{r}_c(z,t),$$

$$(z,t) \in \Omega \times \left[t^{j-1}, t^j\right], \quad \vec{\phi}_c\left(z,t^j\right) = \vec{\phi}\left(z,t^j\right),\ z \in \Omega,$$

Or, in entry form

$$\gamma_c \frac{\partial \phi_{cl}(z,t)}{\partial t} + P_l\left(\vec{\phi}_c(z,t)\right)\phi_{cl}(\vec{x},t) = \Pi_l\left(\vec{\phi}_c(z,t)\right) + f_{cl}(z,t) + r_{cl}(z,t),$$

$$(z,t) \in \Omega \times \left[t^{j-1}, t^j\right],\ l = 1,\dots,N_c.$$

In [8] a family of unconditionally monotonic schemes is constructed, from the first to fourth order of accuracy. One of the single stage schemes is equivalent to the well-known QSSA scheme [9]:

$$\vec{\phi}_c^j = \vec{G}\left(\vec{\phi}^{j-1}\right) + diag\left(\vec{W}\left(\vec{\phi}^{j-1}\right)\right)\vec{r}_c^j. \tag{10}$$

where

$$G_l\left(\vec{\phi}^{\,j-1}(\overline{p})\right)=\phi_l^{j-1}(\overline{p})e^{-P_l\left(\vec{\phi}^{\,j-1}(\overline{p})\right)\tau}+\frac{1-e^{-P_l\left(\vec{\phi}^{\,j-1}(\overline{p})\right)\tau}}{P_l\left(\vec{\phi}^{\,j-1}(\overline{p})\right)\tau}\left(\Pi_l\left(\vec{\phi}^{\,j-1}(\overline{p})\right)+f_{cl}^{j}(\overline{p})\right)\tau,$$

$$W_l\left(\vec{\phi}^{\,j-1}(\overline{p})\right)=\frac{1-e^{-P_l\left(\vec{\phi}^{\,j-1}(\overline{p})\right)\tau}}{P_l\left(\vec{\phi}^{\,j-1}(\overline{p})\right)\tau}\tau.$$

The chemical splitting stage is calculated on a temporal grid $\hat{\omega}_t$ that is $\mathrm{iNt}$ times finer than the original temporal grid, i.e. every interval $\left[t^j,t^{j+1}\right]$ is divided to $\mathrm{iNt}$ subintervals.

- The next step approximation is

$$\vec{\phi}^{\,j}=\vec{\phi}_z^{\,j}+\vec{\phi}_c^{\,j}. \tag{11}$$

An advantage of the scheme is that at each time step the individual processes are evaluated independently (in parallel).

### *2.2. Weakly coupled data assimilation algorithm*

In order to assimilate the measurement data, we connect the measured quantities with the model variables. This is formally done with a measurement operator $H$:

$$\vec{I}(t)=H\left(t,\vec{\phi}^{*}(.,t)\right)+\vec{\eta}(t),\ t\in[0,T], \tag{12}$$

where $\vec{I}(t)$ is the measurement data, $\vec{\phi}^{*}(.,t)$ is the «true» (or exact) solution, $\vec{\eta}(t)$ is the measurement data uncertainty.

**Data assimilation problem:** Determine $\vec{\phi}^{*}(.,t)$ for $t>t^{*}$ with equations (1)-(3), (12) and functions $\vec{f}_a$, $\vec{g}_a$, $\vec{\phi}_a^{0}$, $\vec{I}$ defined on $0<t\le t^{*}$.

We consider $N_M$ *in situ* measurements at the domain grid points $\left\{\left(z_M^m,t_M^m\right)\right\}_{m=1}^{N_M}\subset\omega\times\omega_t$. Hence the m-th measurement is defined by the vector

$$\xi_m=\left\{\left(z_M^m,t_M^m,l_M^m,I_m,\sigma_M^m\right)\right\},\ m=1,...,N_M,$$

where $z_M^m$ is the measurement point location, $t_M^m$ is the measurement time point, $l_M^m$ is the number of substances being measured, $I_m$ is the resulting concentration and $\sigma_M^m$ is the standard variation of the measurement. According to the data assimilation problem statement, at time $t^j$ we can use only the measurements with $t_M^m\le t^j$. Let us define a set of indices

$$\theta^j=\left\{1\le m\le N_M\,|\,t_M^m=t^j\right\}.$$

The corresponding measurement operator and measurement data are defined by

$$H^j\vec{\phi}=\left\{\phi_{l_M^m}\left(z_M^m,t_M^m\right)\right\}_{m\in\theta^j},\ I^j=\left\{\left(1+\left(\sigma_M^m\right)^2\xi\right)\phi_{l_M^m}^{*}\left(z_M^m,t_M^m\right)\right\}_{m\in\theta^j},\sigma^j=\left\{\sigma_M^m\right\}_{m\in\theta^j}.$$

We consider regular measurements in space and time which are defined by temporal $\Delta t_m$ and spatial $\Delta z_m$ gaps between the measurement points, $\xi$ is a normally distributed random variable.

The variational data assimilation method provides a solution to the data assimilation problem as the minimum of a functional with constraints imposed by the model. The functional usually combines the measurement data misfit with the norm of the control variable:

$$J^j\left(\vec{\phi},\vec{r}\right)=\alpha\left\|H^j\vec{\phi}-\vec{I}^j\right\|_{\sigma^j}^2+\left\|\vec{r}\right\|^2,$$

where $\|.\|$ is the norm of a Hilbert space over $Q(\omega)^{N_c}$ and $\langle .,.\rangle$ is the corresponding inner product, $\alpha$ is a regularization (assimilation) parameter which is selected to make the solution closer to the direct model solution or to the measurements. At $t^j$ we update only the control variable $\vec{r}^j$ for this time step. Under a weakly-coupled (or fine-grained) approach [10, 11, 12, 13] the same data are assimilated at the different splitting stages and the results are coupled afterwards. We seek for a minimum of the functional

$$J_f^j\left(\left\{\vec{\phi}_\beta^j,\vec{r}_\beta^j\right\}_{\beta\in\{z,c\}}\right)=\sum_{\beta\in\{z,c\}}J^j\left(\vec{\phi}_\beta^j,\vec{r}_\beta^j\right)$$

under the constraints defined by equations (9), (10) with respect to the independent $\vec{r}_\beta$. Using the method of Lagrange multipliers to solve minimization problems with equality constraints, we can construct the augmented functional

$$\bar{J}_f\left(\left\{\vec{\phi}_\beta^j,\vec{r}_\beta^j\right\}_{\beta\in\{z,c\}}\right)=\sum_{\beta\in\{z,c\}}J^j\left(\vec{\phi}_\beta^j,\vec{r}_\beta^j\right)+\left\langle\gamma_z\frac{\vec{\phi}_z^j-\vec{\phi}^{j-1}}{\tau}-L_z\vec{\phi}^j-\vec{r}_z^j-\vec{f}_z^j,\vec{\psi}_z^j\right\rangle$$
$$+\left\langle\vec{\phi}_c^j-\vec{G}\left(\vec{\phi}^{j-1}\right)-diag\left(\vec{W}\left(\vec{\phi}^{j-1}\right)\right)\vec{r}_c^j,\vec{\psi}_c^j\right\rangle.$$

The components of $\bar{J}_f\left(\left\{\vec{\phi}_\beta^j,\vec{r}_\beta^j\right\}_{\beta\in\{z,c\}}\right)$ corresponding to the different $\beta$ are independent, hence the stationary point coordinates can be found independently.

In order to present an algorithm of finding a stationary point for the convection-diffusion stage, we need further elaboration of the operator $L$. We use approximations of equation (6) that produce the tridiagonal matrix systems

$$-a_i\phi_{i+1}^j+b_i\phi_i^j=\phi_i^{j-1}+\tau r_i^j+\tau f_i^j, i=0, \quad (13)$$
$$-a_i\phi_{i+1}^j+b_i\phi_i^j-c_i\phi_{i-1}^j=\phi_i^{j-1}+\tau r_i^j+\tau f_i^j,\ i=1,\ldots,N-1, \quad (14)$$
$$b_i\phi_i^j-c_i\phi_{i-1}^j=\phi_i^{j-1}+\tau r_i^j+\tau f_i^j,\ i=N. \quad (15)$$

Here the assimilated state is the solution of the minimization problem

$$J\left(\phi^j,r^j\right)\tau=\left(\alpha\sum_{i=0}^{N_z}\left(\frac{\phi_i^j-I_i^j}{\sigma_i}\right)^2M_i^j+\sum_{i=0}^{N_z}\left(r_i^j\right)^2\right)\tau,$$

under the constraints defined by equations (13)-(15) where $M_i^j$ is a spatial-temporal measurement mask (i.e. $M_i^j$ is equal to 1 if there is a measurement at point $\left(z_i,t^j\right)$ and to 0 otherwise), $I_i^j$ is the measurement data at point $\left(z_i,t^j\right)$ (if there is a measurement) and $\sigma_i$ is the measurement device standard deviation of the measurement at point $\left(z_i,t^j\right)$ (if there is a measurement). Introducing Lagrange multipliers, we obtain the augmented functional

$$\bar{J}_f\left(\phi^j,r^j,\psi^j\right)\tau=J\left(\phi^j,r^j\right)\tau+\sum_{i=0}^{N}\left(-a_i\phi_{i+1}^j+b_i\phi_i^j-c_i\phi_{i-1}^j-\phi_i^{j-1}-\tau r_i^j-\tau f_i^j\right)\psi_i^j.$$

Equating the first variations of the augmented functional to zero, we obtain the following tridiagonal matrix equation [10, 11, 12]:

$$-A_i\Phi_{i+1}^j + B_i\Phi_i^j = F_i^j, i=0,$$

$$-A_i\Phi_{i+1}^j + B_i\Phi_i^j - C_i\Phi_{i-1}^j = F_i^j, i=1,...,N-1,$$

$$B_i\Phi_i^j - C_i\Phi_{i-1}^j = F_i^j, i=N,$$

$$A_i = \begin{pmatrix} a_i & 0 \\ 0 & c_{i+1} \end{pmatrix},\ B_i = \begin{pmatrix} b_i & -\dfrac{\tau}{2} \\ \dfrac{2\alpha M_i\tau}{\sigma_i^2} & b_i \end{pmatrix},\ C_i = \begin{pmatrix} c_i & 0 \\ 0 & a_{i-1} \end{pmatrix},\ \Phi_i^j = \begin{pmatrix} \phi_i^j \\ \psi_i^j \end{pmatrix},\ F_i^{j+1} = \begin{pmatrix} \phi_i^{j-1} + \tau f_i^j \\ \dfrac{2\alpha M_i\tau}{\sigma_i^2} I_i^j \end{pmatrix},$$

which can be solved by a direct Gaussian elimination method.

At the transformation stage the algorithm is the same for any grid point $\bar{p}\in\omega$. For brevity let $\vec{\phi}^j = \vec{\phi}_c^j(\bar{p})\in\mathbb{R}^{N_c}$, $\vec{r}^j = \vec{r}_c^j(\bar{p})\in\mathbb{R}^{N_c}$, $\vec{\psi}^j = \vec{\psi}_c^j(\bar{p})\in\mathbb{R}^{N_c}$ and the result is sought-for as a stationary point of the augmented functional

$$\bar{J}(\vec{\phi}^j,\vec{r}^j) = \alpha\sum_{l=1}^{Nc}\left(\frac{\phi_l^j - I_l^j}{\sigma_l}\right)^2 M_l^j + \sum_{l=1}^{Nc}(r_l^j)^2 + \sum_{l=1}^{Nc}\left(\phi_l^j - G_l(\vec{\phi}^{j-1}(\bar{p})) - W_l(\vec{\phi}^{j-1}(\bar{p}))r_l^j\right)\psi_l^j,$$

where $M_l^j$ is equal to 1 if the l-th substance is measured at point $\bar{p}$ at moment $t^j$ and zero otherwise. This minimum is given by the interpolation formula

$$\phi_l^j = \frac{1}{1+Z_l^j}G_l(\vec{\phi}^{j-1}(\bar{p})) + \frac{Z_l^j}{1+Z_l^j}I_l^j,\quad Z_l^j = \frac{\alpha M_l^j}{\sigma_l^2}W_l^2(\vec{\phi}^{j-1}(\bar{p})).$$

The next step is evaluated as in the direct problem:

$$\vec{\phi}^j = \vec{\phi}_z^j + \vec{\phi}_c^j.$$

Note that the resulting algorithm can be implemented without iterations. This allows using it in real-time for large chemical transformation models.

## 3. Numerical experiments

In the numerical experiments, we take a realistic scenario corresponding to August 8, 2013. The turbulence diffusion coefficient $\mu(z,t)$ is calculated from ground heat flux measurements and the friction stress, and from vertical wind profiles based on the k-theory. The vertical speed $w(z,t)$ and $\mu(z,t)$ are presented in figure 1. Realistic initial conditions are considered. The grid parameters are $N_z = 30, X_z = 3000m, T = 75600s, N_t = 281, iNt = 10^4$.

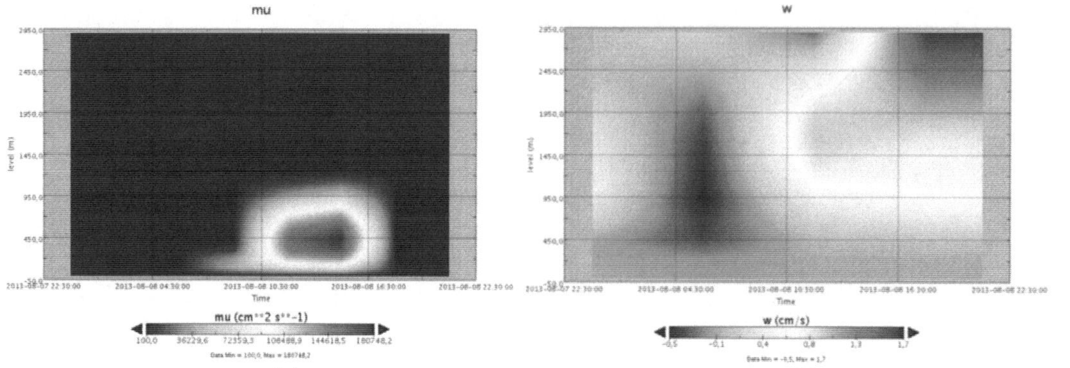


(a) (b)

**Figure 1.** Turbulent diffusion coefficient $\mu(z,t)$ (a) and vertical speed $w(z,t)$ (b).

We compare three configurations for the numerical experiments:

- NoDA (Background solution): Direct problem solution that is driven by initial and boundary conditions only.
- «Truth» (Exact solution): Direct problem solution that is driven by both initial and boundary conditions and by an «unknown» source of $NO_2$ «on the ground» (z=0). The emission rate is chosen to provide a 100% relative difference between «true» and background solutions. The scenario models traffic emissions.
- DA (Data assimilation solution): The result of data assimilation algorithm. In this case, the initial and boundary conditions are the same as for NoDA and «true» solutions. The source is absent as in the NoDA case but there are $NO$ and $0_3$ concentration measurements available for the data assimilation system. In the experiments $\sigma_M^m = 1$, $\alpha = 10^{10}$.

The results are presented for three groups of substances: a substance whose measurement data is assimilated (figure 2), a substance that is emitted in the «true» solution (figure 3) and a substance that is neither emitted nor assimilated (figure 4). In the experiment $\sigma = 0$, $\Delta z_m = 500$ m, $\Delta t_m = 135\text{min}$.

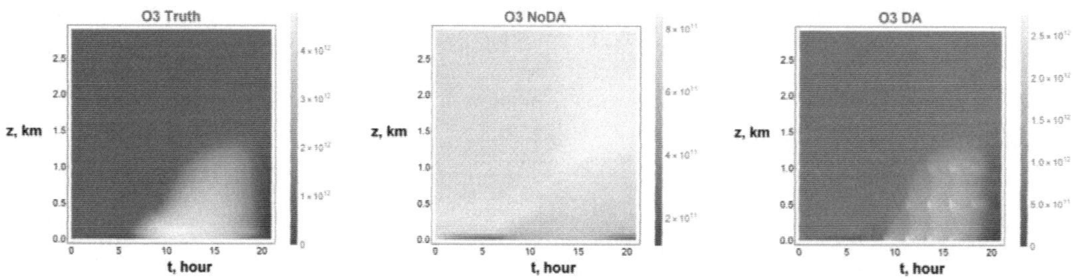


(a) (b) (c)

**Figure 2.** Comparison of «true» solution (a), solution without data assimilation (b) and data assimilation results (c) for a substance whose concentrations are assimilated.

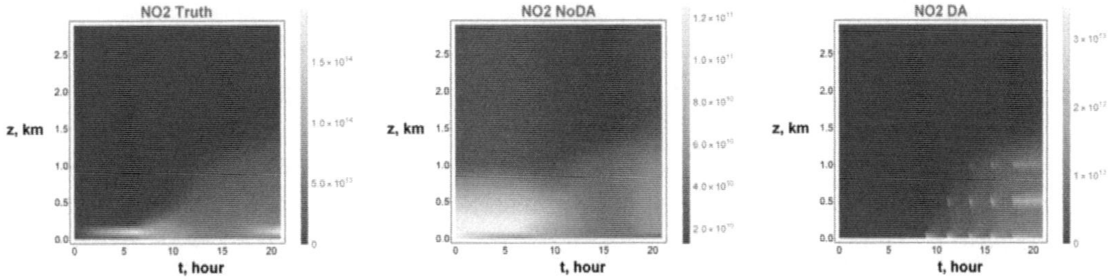


(a) (b) (c)

**Figure 3.** Comparison of «true» solution (a), solution without data assimilation (b) and data assimilation results (c) for a substance that is emitted.

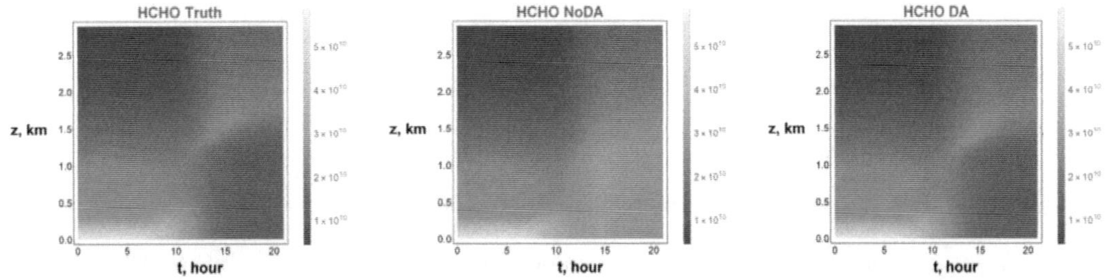


(a) (b) (c)

**Figure 4.** Comparison of «true» solution (a), solution without data assimilation (b) and data assimilation results (c) for a substance that is neither assimilated nor emitted.

From these figures we can conclude that for these substances the data assimilation algorithm was successful at least qualitatively. In the following experiments we will study the sensitivity of the results to measurements data. The results for the substances affected by the emissions (NoDA relative error > 0.5%) are presented in tables 1-3.

Experiment 1: Sensitivity to vertical gaps in data. $\Delta z_m$ is varied with a fixed $\sigma = 0$, $\Delta t_m = 269$ min .

**Table 1.** Relative error of DA versus vertical gaps in measurement points.

| Vertical gap $\Delta z_m$ | H2O2 | HC | HCHO | HNO3 | HO2 | NO | NO2 |
|---|---|---|---|---|---|---|---|
| **500 m** | 0.01 | 0.01 | 0.07 | 0.63 | 0.25 | 0.79 | 0.95 |
| **1000 m** | 0.01 | 0.01 | 0.07 | 0.65 | 0.26 | 0.83 | 0.96 |
| **1500 m** | 0.01 | 0.01 | 0.08 | 0.66 | 0.27 | 0.84 | 0.97 |
| **NoDA** | 0.01 | 0.03 | 0.32 | 0.76 | 0.53 | 0.99 | 1 |

| Vertical gap $\Delta z_m$ | O1D | O3 | O3P | OH | RO2 | ROOH | Prod |
|---|---|---|---|---|---|---|---|
| **500 m** | 0.63 | 0.59 | 0.93 | 0.3 | 0.27 | 0.22 | 0.22 |
| **1000 m** | 0.66 | 0.61 | 0.94 | 0.31 | 0.28 | 0.24 | 0.23 |
| **1500 m** | 0.66 | 0.62 | 0.95 | 0.33 | 0.29 | 0.25 | 0.24 |
| **NoDA** | 0.76 | 0.7 | 1 | 0.87 | 0.57 | 0.5 | 0.48 |

Experiment 2: Sensitivity to temporal gaps in data. $\Delta t_m$ is varied with a fixed $\sigma = 0$, $\Delta z_m = 500$ m .

**Table 2.** Relative error of DA versus temporal gaps in measurements.

| Temporal gap, $\Delta t_m$ | H2O2 | HC | HCHO | HNO3 | HO2 | NO | NO2 |
|---|---|---|---|---|---|---|---|
| **135min** | 0 | 0 | 0.05 | 0.55 | 0.18 | 0.69 | 0.9 |
| **269 min** | 0.01 | 0.01 | 0.07 | 0.63 | 0.25 | 0.79 | 0.95 |
| **404 min** | 0.01 | 0.01 | 0.14 | 0.69 | 0.35 | 0.88 | 0.97 |
| **NoDA** | 0.01 | 0.03 | 0.32 | 0.76 | 0.53 | 0.99 | 1 |

| Temporal gap, $\Delta t_m$ | O1D | O3 | O3P | OH | RO2 | ROOH | Prod |
|---|---|---|---|---|---|---|---|
| **135 min** | 0.55 | 0.52 | 0.87 | 0.22 | 0.2 | 0.15 | 0.15 |
| **269 min** | 0.63 | 0.59 | 0.93 | 0.3 | 0.27 | 0.22 | 0.22 |
| **404 min** | 0.69 | 0.64 | 0.96 | 0.5 | 0.38 | 0.31 | 0.3 |
| **NoDA** | 0.76 | 0.7 | 1 | 0.87 | 0.57 | 0.5 | 0.48 |

Experiment 3: Sensitivity to measurement noise. $\sigma$ is varied with a fixed $\Delta t_m = 269$ min, $\Delta z_m = 500$ m.

**Table 3.** Relative error of DA versus standard deviation of measurement noise.

| Measurement noise, σ | H2O2 | HC | HCHO | HNO3 | HO2 | NO | NO2 |
|---|---|---|---|---|---|---|---|
| **0** | 0.01 | 0.01 | 0.07 | 0.63 | 0.25 | 0.79 | 0.95 |
| **0.1** | 0.01 | 0.01 | 0.07 | 0.64 | 0.26 | 0.81 | 0.95 |
| **0.5** | 0.01 | 0.01 | 0.09 | 0.67 | 0.29 | 0.85 | 0.96 |
| **1** | 0.01 | 0.01 | 0.12 | 0.71 | 0.34 | 0.9 | 0.98 |
| **2** | 0.01 | 0.02 | 0.2 | 0.74 | 0.42 | 0.94 | 1.04 |
| **3** | 0.01 | 0.02 | 0.2 | 0.75 | 0.41 | 0.96 | 1.22 |
| **NoDA** | 0.01 | 0.03 | 0.32 | 0.76 | 0.53 | 0.99 | 1 |

| Measurement noise, σ | O1D | O3 | O3P | OH | RO2 | ROOH | Prod |
|---|---|---|---|---|---|---|---|
| **0** | 0.63 | 0.59 | 0.93 | 0.3 | 0.27 | 0.22 | 0.22 |
| **0.1** | 0.64 | 0.6 | 0.93 | 0.31 | 0.28 | 0.24 | 0.23 |
| **0.5** | 0.68 | 0.63 | 0.95 | 0.36 | 0.31 | 0.28 | 0.27 |
| **1** | 0.72 | 0.67 | 0.97 | 0.45 | 0.37 | 0.34 | 0.33 |
| **2** | 0.75 | 0.71 | 1 | 0.67 | 0.45 | 0.41 | 0.39 |
| **3** | 0.76 | 0.74 | 1.03 | 0.67 | 0.44 | 0.42 | 0.4 |
| **NoDA** | 0.76 | 0.7 | 1 | 0.87 | 0.57 | 0.5 | 0.48 |

The quantitative efficiency of the data assimilation can be studied by comparison of the NoDA results and the results of data assimilation (DA). The small error in both DA and NoDA cases for a substance means that the emissions do not affect the dynamics of its concentrations (as it is for HC and $H_2O_2$). The less error for the DA cases compared to NoDA means that the data assimilation is efficient. In the numerical experiments with high noise levels ($\sigma = 2,3$), the DA becomes inefficient for some substances. In all DA cases the more (smaller gaps) and the better (less noise) data are available the better are the reconstructions.

## 4. Conclusions

The combination of splitting and direct variational data assimilation schemes at the splitting stages allows constructing computationally efficient algorithms for data assimilation of *in situ* concentration measurements in the convection-diffusion-reaction models. The algorithm performance was evaluated in a series of numerical experiments with indirect measurements. In the numerical experiments, the data assimilation algorithm was efficient in all the experiments except for ones with large noise levels. For all the data assimilation cases, the smaller gaps and less noise in data, the better the reconstruction results. We conclude that in the numerical experiments the data assimilation algorithm was able to partially substitute the missing emission inventories with additional concentration measurements.

## Acknowledgments

The work is partially supported by the Presidium of RAS under the Programs I.33P and II.2P/I.3-3; by the projects MK-8214.2016.1 and RFBR 14-01-00125.

## References

[1] Sandu A, Chai T 2011 Chemical data assimilation - an overview *Atmos.* **2** 426–63
[2] Bocquet M, Elbern H, Eskes H et al. 2015 Data assimilation in atmospheric chemistrymodels: current status and future prospects for coupled chemistry meteorology models *Atmos. Chem. Phys.* **15** 5325-58.
[3] Elbern H, Strunk A, Schmidt H, Talagrand O 2007 Emission rate and chemical state estimation by 4-dimensional variational inversion *Atmos. Chem. Phys.* **7** 3749-69
[4] Stockwell W, Goliff W 2002 Comment on Simulation of a reacting pollutant puff using an adaptive grid algorithm by R. K. Srivastava et al *J. Geophys. Res.* **D.107** 4643–50.
[5] Gery M W, Whitten G Z, Killus J P, Dodge M C 1989 A photochemical kinetics mechanism for urban and regional scale computer modeling *J. Geophys. Res.* **94** 12952-6
[6] Byun D W, Schere K L 2006 Review of the governing equations, computational algorithms and other components of the Models-3 Community Multiscale Air Quality (CMAQ) Modeling System *Appl. Mech. Rev.* **59**(2) 51–77
[7] Samarskii A A, Vabishchevich P N 1995 *Computational Heat Transfer Vol.1,2* (Chichester :Wiley) 1-850
[8] Penenko V, Tsvetova E 2013 Variational methods for construction of monotone approximations for atmospheric chemistry models *Numer. Anal. Appl.* **6**(3) 210-20
[9] Hesstvedt E, Hov O, Isaacsen I 1978 Quasi-steady-state-approximation in air pollution modelling: comparison of two numerical schemes for oxidant prediction *Int. J. Chem. Kinet.* **10** 971–94
[10] Penenko A 2006 Some theoretical and applied aspects of sequential variational data assimilation *Comp. tech.* **11**(2) 35-40
[11] Penenko V V 2009 Variational methods of data assimilation and inverse problems for studying the atmosphere, ocean, and environment *Num. Anal. and Appl.* **2**(4) 341-51
[12] Penenko A V, Penenko V V 2014 Direct data assimilation method for convection-diffusion models based on splitting scheme *Comp. tech.* **19**(4) 69-83
[13] Penenko A et al. 2015 Direct variational data assimilation algorithm for atmospheric chemistry data with transport and transformation model *Proc. SPIE* **9680** 968076-1-12

# Assessment of mercury content in poplar leaves of Novokuznetsk agglomeration

E E Lyapina[1,2], D V Yusupov[2], E M Tursunalieva[2] and V V Osipova[2]

[1] Institute of Monitoring of Climatic and Ecological System of SB of RAS
[2] National Research Tomsk Polytechnic University, Tomsk, Russia

**Abstract.** In this paper, the content of mercury in poplar leaves in the Novokuznetsk industrial agglomeration and along the automobile route Novokuznetsk-Mezhdurechensk is assessed. The geoecological indicators are also calculated.

## 1. Introduction

Mercury is one of the most toxic metals polluting the environment. The total amount of mercury as a natural impurity connected with coal, oil and ore concentrates of non-ferrous metals. The main sources of atmospheric mercury emissions are coal and oil products combustion installations, as well as non-ferrous industry. Weighted average mercury content in Russian coal is estimated to be 0.08 mg / kg, and it largely depends on the average mercury concentration in coal from Kemerovo region, for which the half of mercury total amount from the extracted coal of Russia is accounted [4].

The goal of the study is assessment of mercury and its distribution in the territory of Novokuznetsk agglomeration according to the poplar leaves research data.

## 2. Methods and research objects

Novokuznetsk is one of the largest metallurgical and coal-mining centers of Russia. It is the second city in the number of inhabitants (more than 550 thousand of people) and the first one in the area (424 $km^2$)in Kemerovo region [2]. The city is a center of Novokuznetsk (Kuzbass) agglomeration, which has population of over 1.3 million people (figure 1). This agglomeration is the 12th largest in Russian Federation. The industry of agglomeration has a strong specialization and represented by ferrous and nonferrous metallurgy; the coal industry (mines, open cuts), machine-building, construction and heat power industries.

The coal industry in the region amounts 56% of mineral coal production and 83% of coking coal of Russia. Mining and open cuts are situated in all city districts and municipal areas of the agglomeration. Russia's largest coal companies are Kuzbassrazrezugol, Yuzhkuzbassugol, and Raspadskaya.

The metallurgical industry is concentrated mainly in Novokuznetsk and presented by: Novokuznetsk and West-Siberian Metal Plants, ferroalloy and aluminum plants.

The largest enterprises producing construction materials are presented by: Kuznetsk cement plant, Baydaevsk brick factory, large house-building combines, enterprises manufacturing glass and soft roof, extracting non-metallic building materials for the needs of construction companies within Kemerovo region and outside of it.

The power plants are TPP of Kuznetsk Metallurgical Combine, Zapadno-Sibirskaya TPP, Kuznetsk TPP, South Kuzbass GRES, Tomusinsk GRES, Novokuznetsk GTPP.

**Figure 1.** Novokuznetsk industrial agglomeration.

Ecological situation in Novokuznetsk is assessed by state environmental agencies as hard [2]. In 2013-2014, Novokuznetsk was included in the list of Russian cities with the highest level of air pollution by such indicators as soot, suspended matters, nitrogen dioxide, hydrogen fluoride, carbon monoxide, formaldehyde and benzapyrene [1, 5].

In the 63 km eastwards of Novokuznetsk in sublatitudinal basin at the confluence of rivers Tom and Usa, Mezhdurechensk is located, its population is about 100 thousand of people. It is located in the south-eastern part of the Kuznetsk coal basin in the territory of Tomusinsk deposit. The mercury content in coal of this deposits is 0.8 mg / kg [4]. The coal industry is a city-forming industry. The main sources of air pollution are boiler houses, coal mines, open cuts, processing plants. They emit dust, carbon monoxide, nitrogen oxides, sulfur dioxide and heavy metals in the atmosphere.

The research object was the leaves of black poplar (Populus nigra L.). A black poplar reaches 30-35 m in height and 1-2 m in diameter of trunk. Poplar differs from other woody plants with rapid growth rate and more frequent use for landscaping of urban areas in the temperate zone. Poplar leaves are effective air filter. Due to the special structure of poplar surface, leaves are natural geochemical tablet to capture dust aerosols and reflect short-term seasonal (summer) cumulative effect.

In September 2015 in Novokuznetsk in the regular network of 2 × 2 km along the route Novokuznetsk-Mezhdurechensk with a grid of 3-5 km, black poplar leaves were sampled (figure 2). Totally, 45 samples were selected. Leaves were sampled by the average sample method with approximately the same age of the trees at a height of 1.5-2 meters above the ground, according to methodological guidelines [3]. Sample preparation for analysis includes drying by the environmental temperature, mechanical grinding and stirring to achieve the highest homogeneity of the material, sample taking for analysis.

**Figure 2.** The map of poplar leaves sampling in the territory of Novokuznetsk industrial agglomeration.

The sample analysis was conducted in the laboratories of International innovative academic center «Uranium Geology», Geoecology and Geochemistry Department, Tomsk Polytechnic University. Mercury content in the poplar leaves samples was detected on the Atomic Absorption Spectrometer RA-915+ with pyrolyzer PYRO-915 (pyrolysis reaction, mercury detection limit – 5 mkg/kg). For the control of mercury measurements in dry mass of poplar leaves in this devise, a standard sample "birch leaf" was used (GSO 8923-2007, COOMET CRM 0067-2008-RU).

## 3. Results and discussion

The results are statistically processed and presented in Tables 1 and 2. The analysis of these tables allows us to distinguish features of mercury distribution in the territory of Novokuznetsk industrial agglomeration. The average mercury content in dry mass of poplar leaves in the territory of Novokuznetsk agglomeration (26.7 ng / g), which is comparable with the content of mercury in the standard sample composition of poplar leaves (GSV-3) registered abroad (Institute of Geophysical and Geochemical Exploration (China)) - 26.0 ng / g. The maximum mercury content in the sample 1.9 times exceeds the value. By the value of variation coefficient, the homogeneous sample is typical for mercury (<50%), for Novokuznetsk, as well as along the route Novokuznetsk-Mezhdurechensk.

Average mercury content in the territory of Novokuznetsk, as well as along the route Mezhdurechensk-Novokuznetsk is approximately the same and do not differ significantly. The maximum contents in the investigated areas are within the measurement error. However, the minimum concentration in the territory of Novokuznetsk is 2 times higher in comparison with the data for the route Novokuznetsk-Mezhdurechensk.

**Table 1.** Statistical parameter of mercury content (ng/g) in dry mass of poplar leaves in the territory of Novokuznetsk (N=33).

| Index | Value | Index | Value |
|---|---|---|---|
| Average | 23.6±1.5 | Standard deviation | 7.9 |
| Minimum | 7.9 | Excess | 0.5 |
| Maximum | 46.7 | Asymmetry | 0.7 |
| Median | 22.4 | Variation coefficient, % | 35 |

**Table 2.** Statistical parameter of mercury content (ng/g) in dry mass of poplar leaves along the route Novokuznetsk-Mezhdurechensk (N=12).

| Index | Value | Index | Value |
|---|---|---|---|
| Average | 26.4±2 | Standard deviation | 8.3 |
| Minimum | 14.3 | Excess | 0.93 |
| Maximum | 43.9 | Asymmetry | 0.90 |
| Median | 25.3 | Variation coefficient, % | 31 |

Schematic map of the spatial distribution of the concentrations of mercury in Novokuznetsk, built using the programs SURFER and COREL DRAW 10 and 16, is shown in Fig. Four significant halos of gross mercury 1.5-2 times exceeding the average value were discovered. The largest in area and high-contrast halo (38-48 ng / g) is located in the central part of the city and connected with industrial zones of the Central TPP and Metallurgical Combine (site of rail WSMC). Two other local halos (38-44 ng / g) are concentrated in the Zavodskoy district (in the north-eastern part of the city), and refer to the industrial zone of WSMC, including the West Siberian TPP. The fourth, less contrast halo is in the Kuznetsk district (in the eastern part of the city) within the impacted areas of the Kuznetsk thermal power station, aluminum and ferroalloy plants. The prevailing south-western wind direction influences the halos configuration (figure 1).

It has been established that detected halos of mercury are located in industrial areas of the city. Mercury contained in the leaves of poplar is one of the geochemical indicators of industrial specialization of Novokuznetsk. The question of mercury emission sources within the industrial areas of the city is not completely clear. What is the separate contribution of TPP, plants of ferrous and non-ferrous metallurgy in the total emissions of mercury into the environment of Novokuznetsk?

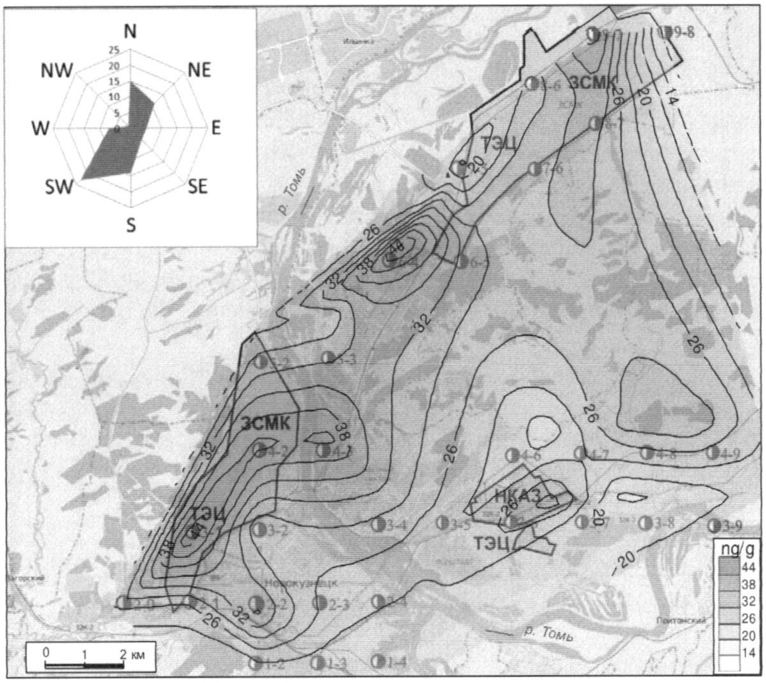


**Figure 3.** Isoconcentrations of mercury in the territory of Novokuznetsk according to poplar leaves sampling. Upper left inset – summer wind rose.

Figure 4 represents a graph showing mercury distribution in dry mass of poplar leaves along the route Novokuznetsk-Mezhdurechensk. Increased mercury concentrations in poplar leaves refer to the western and eastern sections of the route Novokuznetsk - Mezhdurechensk. Western section joins to Novokuznetsk - vil. Atamanovo (sample At-1), the eastern section is located in Mezhdurechensk area(sample Mk-2, 3, 4).

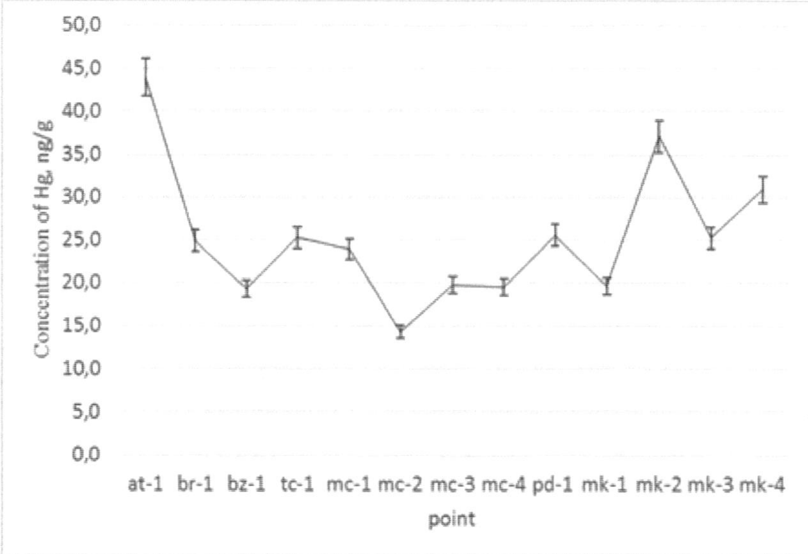


**Figure 4**. Mercury content distribution in dry mass of poplar leaves along the route Novokuznetsk-Mezhdurechensk.

Legend: At-1 (vil. Atamanovo), Br-1 (vil. Borovkovo), Ts-1 (Tom-Usinsk GRES), Pd-1 (vil. Podobas), Ms-1-4 (Myski), Mk-1, 2, 3, 4 (Mezhdurechensk). Upper left inset – summer wind rose.

So, determined areas with the highest mercury content in poplar leaves refer to the territories, which have the greatest technogenic impact in the summer season and reflect the influence of the industrial areas of Novokuznetsk, as well as the impact of coal open cuts near Mezhdurechensk. It is worth paying attention to the fact that the influence of Tom-Usinsk GRES (sample T-1) as a potential source of mercury emissions into the environment is not evident in the chart.

The calculated geoecological indicators of mercury load in the territory of Novokuznetsk industrial agglomeration in the summer have shown the up to 4 times excess of mercury content in poplar leaves in comparison with the mercury Clarke for terrestrial plants (Table 3). The excess over the temporary acceptable concentration is 2 times.

By comparison of the obtained data on the mercury concentration in poplar leaves in the studied areas with the results given in the literature, a significant excess compared to the mercury content in Europe vegetation (Aboal, 2001) and medium for terrestrial plants (according Yanina, 1992) should be mentioned. However, in general, the obtained in this research data are comparable with the background for terrestrial plants (Yanina, 1992) (Table 4).

**Table 3.** Geoecological characteristics of mercury load on the territory of Novokuznetsk agglomeration according to poplar leaves research data.

| Location | C[a], ng/g | Cc[b] | TAC[c] |
|---|---|---|---|
| Novokuznetsk | 8-49/26 | 0.7-4.1/2.2 | 0.1-0.2/0.1 |
| Route Novokuznetsk-Mezhdurechensk. | 14-44/25 | 1.2-3.7/2.1 | 0.6-1.8/1.1 |

[a] C - mercury content;

[b] Cc - concentration coefficient (regarding to Clarke for terrestrial plants after Dobrovolsky, 2003 – 12 ng/g)

[c] TAC - temporarily admissible concentration

**Table 4.** Indicative levels of mercury content in flora of parts of the world.

| Region | Average content, ng/g | Literature source |
|---|---|---|
| Background value in terrestrial plants | 30-700 | Yanin, 1992 |
| Average value in terrestrial plants | 15 | |
| Europe | 8 | Aboal, 2001 |
| Norway | 47-116 | Steinnes, 2005 |

## 4. Conclusions

Mercury containing in poplar is a sensitive indicator of industrial specialization of Novokuznetsk agglomeration area. The results of this research can be used for biogeochemical monitoring in industrialized, and in particular, in coal mining areas. The obtained data are comparable with those mentioned in the literature and do not exceed them. The calculated geoecological indicators of mercury load on the investigated areas in the summer have shown the increased mercury concentration by poplar leaves (up to 4 times), and as a result, the up to 2 times excess of the estimated amount of temporary acceptable concentrations.

*This work was partly financially supported by RSF №15-17-10011*

## References

[1] 2011 *State Report on Environment and Environmental Protection in the Russian Federation in 2014* (Moscow: Ministry of Natural Resources of the Russian Federation)

[2] 2015 *Report on Environment in Novokuznetsk in 2014* (Novokuznetsk: Committee of Environmental Protection and Natural Resources of Novokuznetsk Sity Administration)

[3] 1981 *Guidelines for Conducting of Field and Laboratory Researches of Soil and Plants by the Control of Metal Pollution of Environment* (Moscow: Hydrometeoizdat)

[4] 2005 *Assessment of Mercury Input to Environment from the Territory of the Russian Federation* (Moscow: Federal Service for Ecological, Technological and Nuclear Supervision Service/Copenhagen: Danish Agency for Environmental Protection)

[5] 2014 *Report on Environment and Environmental Protection in the Kemerovo Region in 2013* (Kemerovo: Department of Natural Resources and Environmental Protection of the Kemerovo Region)

# Experimental and numerical study of gas-to-particle conversion in an emission plume from mining and metallurgical industry based on airborne sounding in a polar atmosphere

**Denis V. Simonenkov[1], Vladimir F. Raputa[2], Tatyana V. Yaroslavtseva[3], Boris D. Belan[1]**

[1]V.E. Zuev Institute of Atmospheric Optics SB RAS (IAO SB RAS). 1, pl. akad. Zueva, Tomsk 634055, Russia
[2]Institute of Computational Mathematics and Mathematical Geophysics SB RAS. 6, Prosp.Akad. Lavrentyeva, Novosibirsk 630090, Russia
[3]Novosibirsk Scientific Research Institute of Hygiene of Federal Service for Surveillance on Consumer Rights Protection and Human Wellbeing. 7, ul. Parkhomenko, Novosibirsk 630108, Russia

e-mail: simon@iao.ru

**Abstract**. The results of an airborne survey of plumes from the Norilsk Mining and Metallurgical Plant by an Optik-É AN-30 aircraft laboratory on November 10, 2002 are discussed. Most pollutants are blown out of the city in the gas phase in the form of acidic oxides (mainly sulfur). Mapping of the substances is performed along the main trajectory of air mass transport at a distance of 20-140 km from the city. Horizontal flights were performed at 400, 600, 800, and 1200 m above sea level at equidistant traverses (from 3 to 6 at each height) normally to the main flow direction. Most pollution was concentrated above the 400-m level. An active gas-to-particle conversion was observed at a distance of 60-100 km from the emission source. In the plume areas distant from the source there was a sulfate anion increase from 4% to 51% in aerosol composition weight and a calcium decrease from 64% to 9%. Under the conditions of low humidity in the polar atmosphere in winter, $SO_2$ is apparently removed from the air mainly due to dry heterogeneous condensation with calcium oxide as the main counteragent of industrial origin. The concentrations of these active pollutants in the plume are well approximated by a two-parameter transformation model.

## 1. Introduction

A comprehensive airborne survey of the atmosphere over the Norilsk industrial region was performed by an aircraft laboratory called Optik-É AN-30 [1] in November 2002. The purpose of the complex experimental study was to evaluate the impact of the Norilsk Mining and Metallurgical Plant on air quality of the adjacent and remote areas of this region.

The main source of air pollution is the emission of contaminants from nonferrous metallurgical enterprises. As a result of ore roasting, sulfide sulfur is oxidized to gaseous sulfur dioxide, which accounts for up to 96% of the total emission mass; dust, carbon monoxide, and nitrogen oxides only account for 1-1.5% [2]. Thus, most emissions are blown out of the city in the gas phase in the form of acidic oxides (mainly sulfur). The chemical processes of sulfur transformation, according to the theory [3, 4], are its oxidation to higher valence, VI, and neutralization: $H_2S \rightarrow SO_2 \rightarrow H_2SO_3 \rightarrow H_2SO_4 \rightarrow M(HSO_4)_m \rightarrow M_2(SO_4)_m$, where $M^{m+}$ is any metal cation. In fact, the major part of the emissions is sulfur dioxide. In accordance with calculations of [4, 5], the maximum amount of sulfuric acid in the atmosphere during the transport is reached in about 20 hours, and the average life time of sulfur dioxide due to dry deposition and chemical transformations is about 40 hours. The rate of removal of sulfur dioxide abruptly increases by precipitation. In addition, our first experiments [6] showed that the aerosol fraction of the emissions should be sampled at a distance of more than 20 km from the source.

## 2. Experimental studies

Based on the above information, the mapping of pollutants was performed along the main trajectory of air mass transport at distances of 20-140 km from the city. The flight was carried out on November 10, 2002 in daytime according to the scheme shown in figure 1.

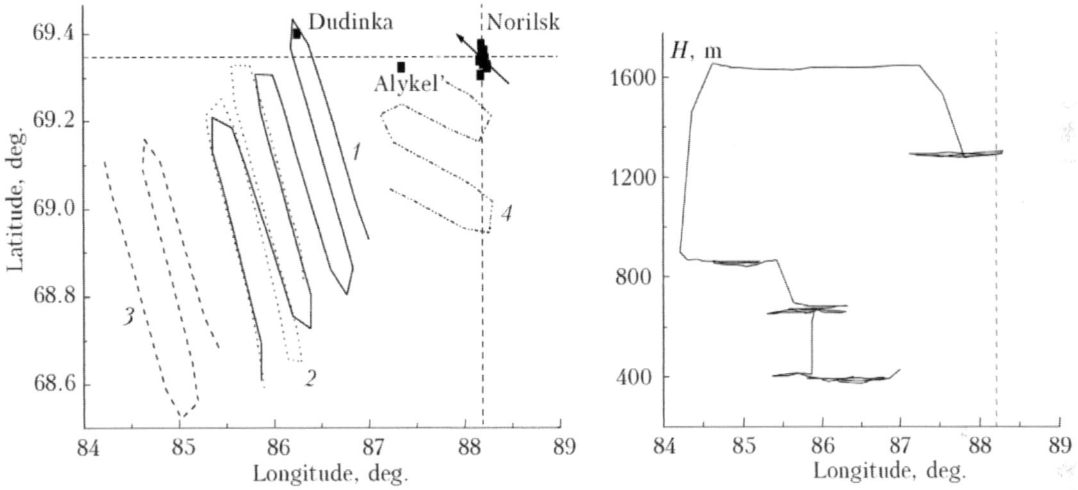


**Figure 1.** Flight route. November 10, 2002 (*1* – 400 m, *2* – 600 m, *3* – 800 m, *4* – 1200 m).

Horizontal flights were performed at heights of 400, 600, 800, and 1200 m above sea level at equidistant traverses, from 3 to 6 at each height. The morning weather in the Norilsk area was determined by the central part of a filling cyclone with surface fronts, and then by a low-gradient field of low pressure. The air mass was of arctic origin.

Maps of horizontal pollutant distributions were constructed with the measurement data at all height levels. Since the profiles of trace gas concentrations from 600 m to 1200 m showed a tendency to decrease with altitude down to the background values, we will focus our attention on the data of the 400-m level (figure 2), where the most part of pollution was concentrated.

The $SO_2$ concentration in the air from the city reached 450 $\mu g/m^3$. The nearest edge of the square was at a distance of 58 km, and the far edge, at 100 km from the city. A significant difference was observed between the urban plume concentrations and the background conditions, where the sulfur dioxide concentration decreased to 50-60 $\mu g/m^3$. This confirms the theory of Israel and Ryaboshapko that $SO_2$ may remain in a plume for tens of hours [4, 5].

Most of the other components of the plume (carbon monoxide, carbon dioxide, and fine aerosol particles) have similar distribution profiles. The data on ultrafine aerosols also confirm that there is an active process of aerosol formation via gas-phase chemical reactions in the plume and at its boundaries.

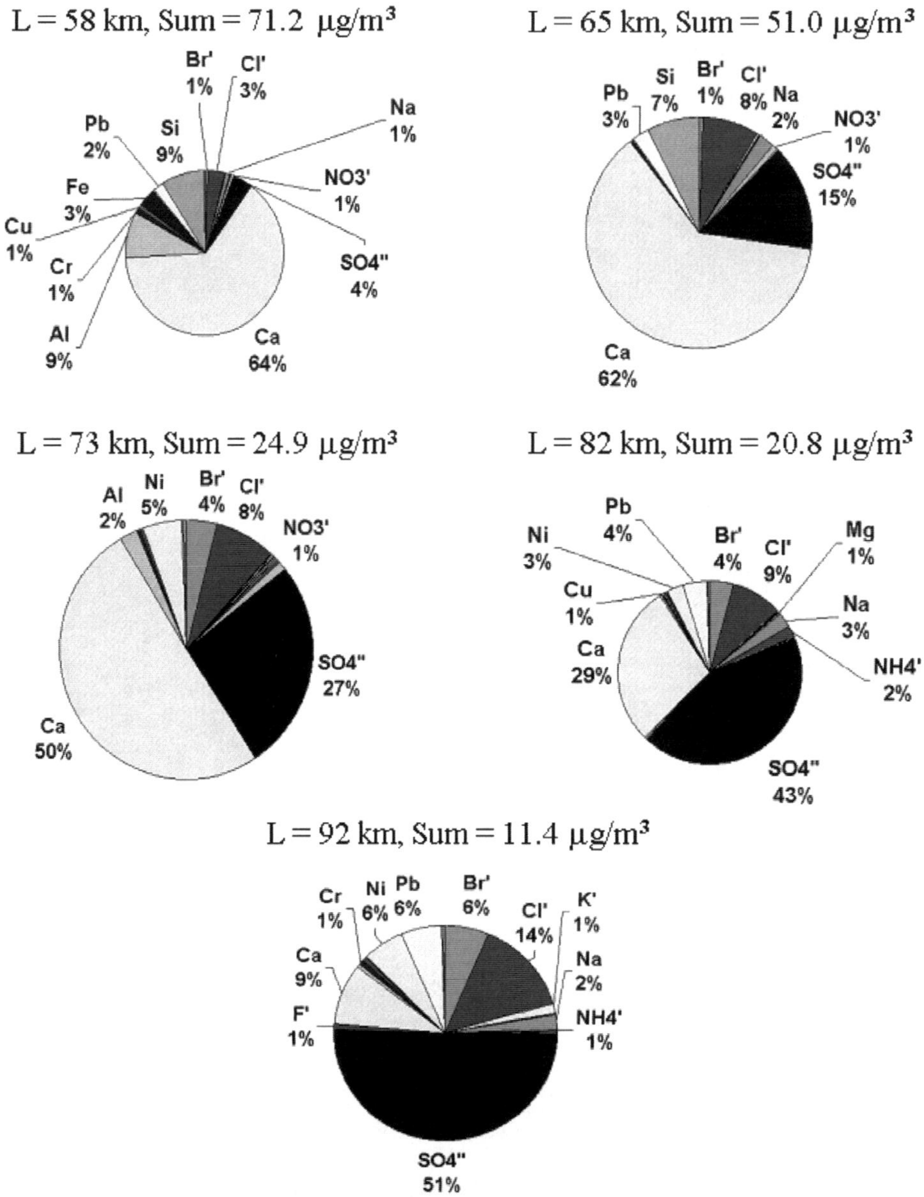


**Figure 2.** Comparative chemical composition of aerosols sampled at 400-m level at various distances from Norilsk (L; Sum - total concentration of all aerosol elements and ions).

The validity of all these processes is confirmed by the evolution of the chemical composition of the aerosols. Figure 2 shows that in plume areas distant from the source there is a sulfate anion increase from 4% to 51% in aerosol composition weight and a calcium decrease from 64% to 9%.

The source of calcium is, apparently, the lime added during roasting to enrich the copper-nickel ore (separation of iron-bearing pyrite) [7]. The calcium is emitted into the atmosphere during combustion as ultrafine and fine particles of 20-180 nm [8], and apparently reacts directly with sulfur dioxide in such reactions as $CaO + SO_2 \{+ 2H_2O\} = CaSO_3\{*2H_2O\}$, $CaSO_3*2H_2O + 1/2O_2 = CaSO_4*2H_2O$ [9].

IOP Conf. Series: Earth and Environmental Science **48** (2016) 012023 doi:10.1088/1755-1315/48/1/012023

## 3. Models to reconstruct the concentration fields in the plume

Based on the characteristics of passive substance transport at large distances from a plane source, the following model of reconstruction was used to describe the concentration field along the direction of emission [10]:

$$Q(x, y, \theta) = \theta \Big/ \sqrt{(x-\lambda)^2 + (y-\mu)^2} \tag{1}$$

$$\theta = M/2\pi u H \ , \qquad M = \iint\limits_S m(\xi, \eta) d\xi d\eta \ , \tag{2}$$

where $M$ is the total mass of pollutants from the territory *S*. $m(\xi, \eta)$ is the emission of pollutants from a point $(\xi, \eta)$ belonging to *S*, and $u$ and $H$ are the average wind speed and height of the mixing layer, respectively. The point *(x, y)* is assumed to be at a distance of more than 7-10 km from *S*. The function $m(\xi, \eta)$ is usually unknown.

The following two-parameter relationship was used to estimate the concentration of active pollutants in the plume [5]:

$$C(r, \vec{\theta}) = \theta_1 \exp(\theta_2 r) \ , \tag{3}$$

where $r$ is the distance from the source.

The parameters $\theta$, $\theta_1$, $\theta_2$ in the relations (1) and (3) are estimated by a least squares method using the observational data.

## 4. Numerical analysis of the transport processes of passive and active pollutants

The results of numerical reconstruction, using the relation (1), of the concentrations of lead and chlorides in the plume at a reference point located at a distance of 58 km from Norilsk are presented in figure 3. The distribution of the passive pollutants in the atmosphere is consistent with the physical theories.

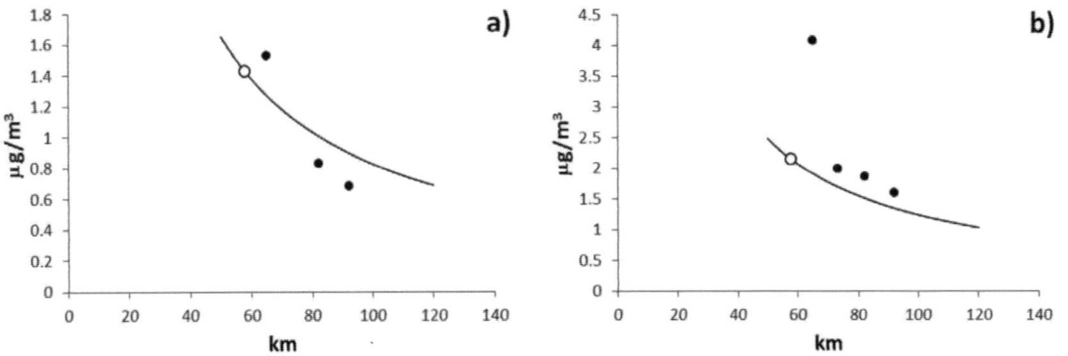


**Figure 3.** Measured and reconstructed concentrations of lead (a) and chlorides (b) in the plume. ○, ●: Reference and control measurement points.

The results of the estimation of calcium and sulfate concentrations performed with the relation (1) are presented in figures 4a and 4b. It is evident that the passive tracer propagation model is of little use in this case. The model (3) reproduces the transformation process quite adequately in the distance interval under analysis (figure 4c).

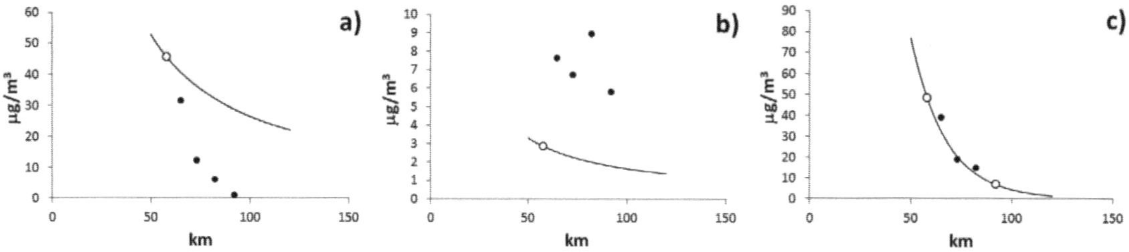


**Figure 4.** Concentrations of calcium (a) and sulfates (b) measured and reconstructed by model (1). Total concentration of calcium and sulfates in the emission plume (c) estimated by model (3).

## 5. Conclusions

Talking into account the air transport velocity, an active gas-to-particle conversion was observed at a distance of 60-100 km from the emission source. It confirms the average estimates of the sulfur dioxide lifetime in real atmosphere [4, 5]. Under the conditions of low humidity in the polar atmosphere in winter, $SO_2$ is apparently removed from the air mainly due to dry heterogeneous condensation, with calcium oxide as the main counteragent of industrial origin. The concentrations of these active pollutants in the plume are well approximated by a two-parameter model of transformation.

## Acknowledgments

This work was supported by the Presidium of RAS (program no.15), the Department of Earth Sciences RAS; the Russian Foundation for Basic Research (grants no. 14-05-00526 and 14-05-00590), the Foundation for Global Environmental Research of the National Institutes of the Ministry of Environment of Japan.

## References

[1] Antokhin P N, Arshinov M Yu, Belan B D, Davydov D K, Zhidovkin E V, Ivlev G A, Kozlov A V, Kozlov V S, Panchenko M V, Penner I E, Pestunov D A, Simonenkov D V, Tolmachev G N, Fofonov A V, Shamanaev V S and Shmargunov V P 2012 Optik-É AN-30 aircraft laboratory: 20 years of environmental research *Journal of Atmospheric and Oceanic Technology* vol 29 N 1 pp 64-75

[2] Bezuglaya E Yu, Rastorgueva G P and Smirnova I V 1991 *By what an Industrial City Breathes* (Leningrad: Gidrometeoizdat) p 256

[3] Ryaboshapko A G 2002 *Remote atmospheric transfer and atmospheric budgets of sulfur, nitrogen, and mercury compounds* Author's Abstract of Doct. Phys.-Math. Sci. Dissert. (Moscow: IKGE) p 29

[4] Ryaboshapko A G 1988 *Trudy Instituta Prikladnoy Geofiziki* vol 71 pp 25–32

[5] Israel Yu A, Nazarov I M, Pressman A Ya, Rovinskii F Ya and Ryaboshapko A G 1989 *Acid Rains* (Leningrad: Gidrometeoizdat) p 269

[6] Belan B D, Zuev V E and Panchenko M V 1995 Main results of airborne sounding of aerosol conducted at the Institute of Atmospheric Optics from 1981 till 1991 *Atmospheric and oceanic optics* (Tomsk: IAO SB RAS) vol 8 N 1-2 pp 131-156

[7] *Popular Library of Chemical Elements* 1983 vol 1 (Moscow: Nauka) p 576

[8] Savenko S V 1991 *Natural and Anthropogenic Sources of Atmospheric Pollution* (Moscow: VINITI) p 212

[9] Garshin A P 2000 *Inorganic chemistry in schemes, figures, tables...* (St-Peterburg: Lan') p 288

[10] Raputa V F, Ol'kin S E and Reznikova I K 2008 Numerical analysis of observation data of regional pollution of an area source *Atmospheric and oceanic optics* (Tomsk: IAO SB RAS) vol 21 N 6 pp 486-489

# A numerical study of gravity waves in the atmosphere: smooth and steep orography effects

**M S Yudin**

Institute of Computational Mathematics and Mathematical Geophysics SB RAS, 6, Akademika Lavrentieva pr., Novosibirsk 630090 , Russia

e-mail m.yudin@ommgp.sscc.ru

**Abstract**. Two types of nonhydrostatic meteorological models are applied to modeling of front wave propagation: a finite-difference model based on a terrain-following coordinate transformation and a finite-element model based on triangular elements. The former is used for a relatively smooth orography, and the latter for steep surfaces. The front surface is described in both models by an equation for advection of a scalar substance, which is solved with a third-order semi-Lagrangian procedure. Various meteorological effects, the influence of stratification and the introduction of an inversion layer above an orographic obstacle are investigated and discussed.

## 1. Introduction

Orographic retardation of gravity flows in the atmosphere is a subject of great practical and theoretical interest [1, 4, 5]. Atmospheric phenomena take place on a wide range of horizontal length scales. The scales of the flows of interest range from micro to macroscales. Flows ranging from several to thousands of kilometers are called mesoscale ones.

Atmospheric fronts over complex terrain are examples of mesoscale gravity flows. A terrain-following coordinate system is most often used to describe numerically a local topography of complicated shape. The domain becomes a rectangular one that can be easily discretized into a finite-difference grid. However, the transformed equations are more complicated than the original ones. In addition, it can be shown that the transformation function must satisfy some smoothness restrictions. In the present paper a finite-element model is used as an alternative to the above approach. Specifically, a 3D non-hydrostatic finite-difference meteorological model and a 2D finite-element model are used to simulate the effects of atmospheric front propagation over a hill.

The propagation of an atmospheric front over steep terrain is a phenomenon of great practical importance in meteorology [1-5]. This is also a subject matter of interest for numerical modelers, since atmospheric fronts can be considered as surfaces of discontinuity in the atmosphere. To simulate the deformation of these surfaces by spatial obstacles like mountains and valleys with good accuracy, efficient numerical methods are needed. Two distinct approaches can be recognized in the numerical simulation of front propagation. In one approach, the front to be calculated is considered as a gravity current driven by a cold air source [7-8]. In the other, the front surface is considered as a passive scalar, a tracer to distinguish between warm and cold air masses [10].

In the present paper, a preliminary investigation is carried out to simulate cold front propagation over a steep hill in two dimensions with a finite-element model and in three dimensions with a finite-difference model. Specifically, 3D non-hydrostatic meteorological model is used. The model is based on spatial discretizations that conserve some important quantities of the phenomena under study like momentum and scalars. Also, an efficient procedure is used to calculate the advection of scalars. A 2D finite-element model based on triangular elements is used to simulate the same phenomenon of cold front propagation over an idealized orographic obstacle.

## 2. Governing equations

For the simulations we use a small-scale non-hydrostatic model, which is essentially a form of the Navier-Stokes equations in the Boussinesq approximation. A version of the equations relevant for a compressible fluid is used here. Since sound waves in the atmosphere have small effects on gravity flows like fronts, the compressible version of the Navier-Stokes equations is taken here for pure technical reasons. The equations are discretized with a splitting method in time, and the compressible form of the equations is suitable for an efficient splitting. Sound waves are then eliminated by the use of an Asselin-type filter [4]. The governing equations may be written as follows:

$$\frac{dU}{dt}+\frac{\partial P}{\partial x}=f_1(V-V_g)-f_2W+R_u,$$

$$\frac{dV}{dt}+\frac{\partial P}{\partial y}=-f_1(U-U_g)+R_\upsilon,$$

$$\frac{dW}{dt}+\frac{\partial P}{\partial z}+\frac{gP}{C_s^2}=f_2U+g\frac{G^{1/2}\bar{\rho}\theta'}{\theta}+R_\varpi$$

$$\frac{d\theta}{dt}=R_\theta,$$

$$\frac{ds}{dt}=R_s,$$

$$\frac{1}{C_s^2}\frac{\partial P}{\partial t}+\frac{\partial U}{\partial x}+\frac{\partial V}{\partial y}+\frac{\partial W}{\partial z}=\frac{\partial}{\partial t}\left(\frac{\bar{\rho}\theta'}{\theta}\right)$$

$$U=\bar{\rho}u, V=\bar{\rho}\upsilon, P=\bar{\rho}p',\ W=\bar{\rho}\omega$$

where p', θ' are deviations from a reference state for pressure $\bar{p}$ and potential temperature $\bar{\theta}$, respectively, $s$ is the specific humidity, $C_s$ is the sound wave speed, $u_g$ and $v_g$ are geostrophic wind components representing the synoptic part of the pressure, $f_1$ and $f_2$ are the Coriolis parameters, $g$ is the gravity constant. $R_u$, $R_v$, $R_\omega$, $R_\theta$, and $R_s$ are terms describing the subgrid-scale processes in the K-theory for turbulent diffusion in meteorology [4].

## 3. Advection of a substance

The problem examined in this study has very small physical diffusion, and the governing equations can be considered as almost inviscid and adiabatic. Special operators of space discretization are used to provide conservation of momentum and scalars in the finite-difference model (for a detailed description see [11-13]. Triangular elements are used in the finite-element model [12]. In this section we consider a suitable scheme for the advection of a scalar, e.g. temperature.

For fully hyperbolic problems explicit schemes have been widely used in the past and will probably continue to be used in the future. Traditionally, problems for which the governing equations are hyperbolic everywhere have been solved by the method of characteristics. Sometimes traditional methods of characteristics are slower than competitive finite difference methods. However, some finite difference schemes use a form of the governing equations such that knowledge of the characteristic locations can be exploited in an efficient way [2].

A flux-corrected transport scheme developed by Smolarkiewicz [4] was successfully applied to simulation of the propagation of an idealized atmospheric front by Schumann [10]. This is an example of a general technique of the predictor-corrector type in which large diffusion is introduced at the predictor stage and an almost equal amount of anti-diffusion is introduced at the corrector stage.

In this paper, the advection of a scalar like temperature is treated with an efficient semi-Lagrangian finite difference scheme [6]. Here the advection of a scalar is calculated in two steps:

1. Determination of the so-called departure point. This is the point from which the point under consideration is reached at the next time step.
2. Interpolation of values of the advected scalar from grid points to the departure point.

$$x_D = x - \int u\,dt$$
$$f(x,t+\Delta t) = f(x_D,t).$$

Here $u$ is the velocity vector and $\Delta t$ is the time step size.

At the first step, interpolation of the velocity vector is performed. The second step is devoted to interpolation of the advected scalar. The procedure of interpolation determines the accuracy of the method. For velocity, linear interpolation is used. At the second step, a third-order accuracy scheme will be used in this paper for reasons to be discussed below.

The total advection of temperature is split into horizontal and vertical components. For simplicity, we present here only a one-dimensional scheme. The scheme is designed as follows: an arbitrary function $f$ is expanded into a Taylor series up to terms of the fourth order. The free coefficients of this expansion are determined through values of the function at grid points. Denote $\lambda = (x_D - x_i)/\Delta x$. Here $\Delta x$ is the grid size. By solving the system of linear equations for the free coefficients, we finally obtain:

$$\begin{aligned} f(t+\Delta t) &= f_i(1-\lambda/2-\lambda^2+\lambda^3/2) \\ &+ f_{i+1}(\lambda+\lambda^2/2-\lambda^3/2) \\ &+ f_{i+2}(-\lambda/6-\lambda^2+\lambda^3/6) \\ &+ f_{i-1}(-\lambda/3+\lambda^2/2-\lambda^3/6) \end{aligned}$$

Numerical experiments have been performed to compare various semi-Lagrangian schemes. Some conclusions follow:

1. The first-order schemes have large numerical diffusion.
2. The second-order schemes are non-monotonic and have a small-scale wavelike structure.
3. In the third-order schemes, the above two effects are essentially reduced.
4. The schemes of order higher than 3 have a significant increase in the operation count but only a small increase in the solution quality.

The above third-order semi-Lagrangian scheme will be used as a reasonable compromise between cost and accuracy.

## 4. Smooth surface orography simulation

In this Section, we apply the above finite-difference model to simulating the propagation of a cold atmospheric front over a hill-type obstacle in three dimensions. The input parameters are taken from [10].

The obstacle is a circular mountain with an axially symmetric sinusoidal height profile of 2 km and a surface diameter of 200 km. The computational domain is 600 x 400 x 6 km.

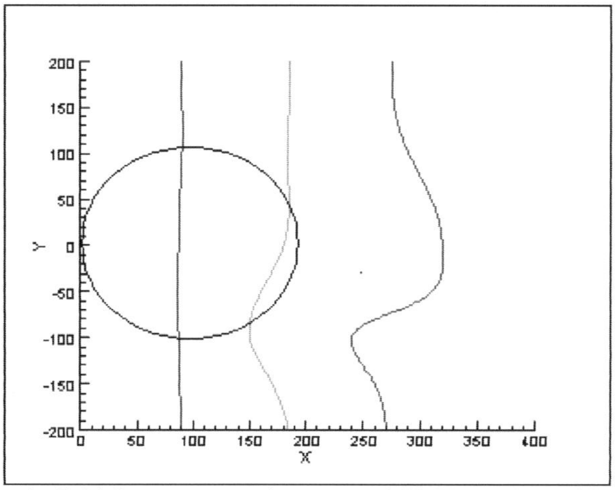


**Figure 1.** Cold front propagation over a hill. Neutral stratification.

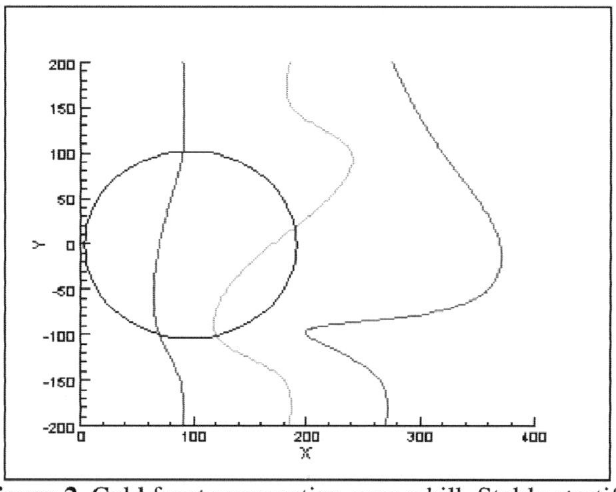


**Figure 2.** Cold front propagation over a hill. Stable stratification.

Figure 1 shows the results of simulations on the propagation of a cold atmospheric front with an asymptotic height of 4 km and a temperature jump of 3K for neutral stratification. Surface isochrones at a sequence of non-dimensional time varying from 0 to 1 are shown. In Figure 2 the same is shown for a stably stratified atmosphere.

The results show acceleration of the front on the northern side and retardation near the mountain center. Examination of the flow also shows strong anticyclonic motion over the mountain, which causes significant front deformation. It follows from Figure 2 that stable stratification greatly increases the effects of acceleration and retardation. A qualitative agreement is observed between the above results and similar temperature patterns obtained in [6] with different schemes.

## 5. Steep surface orography simulation

In this section the effects of introducing an inversion layer into a stably stratified atmosphere over a steep orographic obstacle are studied with the finite-element model described above. The model parameters are taken as in papers [7-8]. In these papers, a cold front propagating over an axially-symmetric obstacle in the form of a bell-shaped hill is simulated with a finite-difference model. The height of the hill is 600 m. The calculation domain is 25x2 km. In contrast to [7-8], where the front is generated by a volume of cold air, in the present study the front is initially given in the form of a step-function of 400 m in height.

An inversion layer of 300 m in height is placed over the isolated hill described above. The atmosphere has the standard stable stratification of 3.5K/100 m, whereas the stratification of the inversion layer is 1K/100 m. The results of a series of calculations of cold front propagation over the hill and a plain are shown in Tables 1 and 2. The calculated values of windward and leeward speeds are given in the last columns.

It can be seen from these tables that the introduction of the inversion layer produces a significant decrease in the front speed both for the currents over the obstacle and those over flat orography. This is in agreement with the results of papers [7-8], where the front was generated by a volume of cold air. As in [7-8], the introduction of the elevated inversion layer causes an intensification of the vertical air motion and increases the mesoscale heat flux. This results in a more intensive entrainment.

**Table1.** Calculated windward and leeward speeds of cold front propagation over an orographic obstacle. Stable stratification.

| obstacle height (m) | initial front height (m) | stratification ( K / 100 m) | windward speed (m/s) | leeward speed (m/s) |
|---|---|---|---|---|
| 600 | 400 | no inversion | 4.9 | 2.7 |
| 600 | 400 | inversion | 4.4 | 2.2 |

**Table2.** Calculated windward and leeward speeds of cold front propagation over a plain. Stable stratification.

| obstacle height (m) | initial front height (m) | stratification ( K/100 m) | windward speed (m/s) | leeward speed (m /s) |
|---|---|---|---|---|
| 0 | 400 | no inversion | 5.1 | 5.1 |
| 0 | 400 | inversion | 4.6 | 4.6 |

## 6. Conclusions

The above results have been obtained with two types of models: a 3D non-hydrostatic finite-difference meteorological model and a 2D finite-element model. Both models are dicretized versions of the the Navier-Stokes equations in the Boussinesq approximation written in a compressible form.

The 3D model is based on spatial discretizations that conserve some important quantities of the phenomena under study, atmospheric gravity currents. Also, an efficient procedure is used to calculate the advection of scalars. Section 4 presented an application of the non-hydrostatic version of the small-scale meteorological model to simulating flows that occur due to the retardation of a cold front by a steep mountain in three dimensions. The simulations described above have been carried out to test the specific discretizations for the advection operators mentioned above (see[11-13]). The preliminary simulation results are in qualitative agreement with the existing theoretical insight, observations, and calculations performed by other authors [10].

The 2D finite-element model is based on triangular elements. Section 5 showed an application of the model to simulating cold front propagation over an idealized hill-type obstacle in a stratified atmosphere with an inversion layer over the isolated hill. The study was performed under stable stratification in and beyond the inversion layer. It has been shown that the introduction of the inversion later produces a significant decrease in the front speed both for the currents over the obstacle and those over flat orography.

Although the present studies are of limited scope, these preliminary results show that both the finite-difference and the finite-element model can be used for the simulation of the propagation of gravity flows like atmospheric fronts over steep orographic obstacles. These studies will be extended to more realistic situations in forthcoming papers.

**Acknowledgements**

This work was supported by the Presidium of RAS under program 1.33P and the Russian Foundation for Basic Research under grant 14-01-00125-a.

**References**

[1] Marchuk G I 1974 *Numerical Solution of Atmospheric and Oceanic Problems* (Leningrad: Gidgometeoizdat) p 303
[2] Marchuk G I 1982 *Mathematical Models in Environmental Problems* (Moscow: Nauka) p 319
[3] Penenko V V and Aloyan A E 1985 *Models and Methods for Envronmental Protection Problems* (Novosibirsk: Nauka) p 256
[4] Pielke R A 2002 *Mesoscale Meteorological Modeling* (San Diego: Academic Press) p 676
[5] Davies H C 1984 On the Orographic Retardation of a Cold Front *Beitr. Phys. Atmos.* **57** 409-418
[6] Ritchie H 1987 Semi-Lagrangian Advection on a Gaussian Grid *Mon. Wea. Rev.* **115** 136-146
[7] Bischoff-Gauss I, Gross G 1989 Numerical Studies on Cold Fronts. Part 1: Gravity Flows in a Neutral and Stratified Atmosphere *Meteorol. Atmos. Phys.* **40** 150-158
[8] Bischoff-Gauss I, Gross G and Wippermann F 1989 Numerical Studies on Cold Fronts. Part 2: Orographic Effects on Gravity Flows *Meteorol. Atmos. Phys.* **40** 159-169
[9] Ikawa M 1988 Comparison of Some Schemes for Non-Hydrostatic Models with Orography *J. Meteor. Soc. Japan* **66** 753-776
[10] Schumann U 1987 Influence of Mesoscale Orography on Idealized Cold Fronts *J. Atmos. Sci.* **44** 3423-41
[11] Yudin M S 2011 Propagation of a Gravity Current in the Atmosphere over a Valley *Bull. Novos. Comput. Center* **14** 65-70
[12] Yudin M S and Wilderotter K 2006 Simulating Atmospheric Flows in the Vicinity of a Water Basin *Computational Technologies* **11** 128-134
[13] Yudin M S 2012 Comparison of FDM and FEM Models for a 2D Gravity Current in the Atmosphere over a Valley *Bull. Novos. Comput. Center* **13** 95-101

# Sodars and their application for investigation of the turbulent structure of the lower atmosphere

**N P Krasnenko**[1, 2], **L G Shamanaeva**[3,4]

[1]Chief Scientific Worker, Institute of Monitoring of Climatic and Ecological Systems SB RAS, Tomsk, Russia
[2]Professor, Tomsk State University of Control Systems and Radioelectronics, Tomsk, Russia
[3]Senior Research Scientist, V. E. Zuev Institute of Atmospheric Optics SB RAS, Tomsk, Russia
[4]Assistant Professor, National Research Tomsk State University, Tomsk, Russia

E-mail: krasnenko@imces.ru, sima@iao.ru

**Abstract.** Possibilities of sodar application for investigation of the spatiotemporal dynamics of three components of wind velocity vector, longitudinal and transverse structural functions of wind velocity field, structural characteristics of temperature and wind velocity, turbulent kinetic energy dissipation rate, and outer scales of temperature and dynamic turbulence in the atmospheric boundary layer are analyzed. The original closed iterative algorithm of sodar data processing taking into account the classical and molecular absorption and the turbulent sound attenuation on the propagation path is used that allows the vertical profiles of the characteristics of temperature and wind velocity field to be reconstructed simultaneously and their interrelations to be investigated. It is demonstrated how the structure of temperature and wind turbulence is visualised in real time.

## 1. Introduction

Sodars or acoustic radars are widely used all over the world to investigate the structure and dynamics of the atmospheric boundary layer (ABL) [1–18]. The principle of their operation is based on the effect of sound scattering by small-scale turbulent inhomogeneities of the atmosphere. Doppler three-component monostatic sodars that employ the backscattered acoustic signal of audio frequencies are the simplest and most widespread means of ground-based remote sounding of the ABL. They visualise the small-scale structure of atmospheric turbulence, allow the ABL temperature stratification to be elucidated, boundaries of surface and elevated temperature inversions and height of the mixing layer to be determined, and turbulence intermittence and wave motions to be detected.

The knowledge of characteristics of atmospheric turbulence is of fundamental importance for investigation of the structure and dynamics of the atmospheric boundary layer and construction of its mathematical models widely used for synoptic forecast, theoretical analysis, and prediction of electromagnetic and acoustic wave propagation as well as diffusion of pollutants. Characteristics of the atmospheric turbulence depend on the altitude over the Earth surface. Sodars typically measure the characteristics of the atmosphere at altitudes from several meters to several hundred meters.

In the present paper, works of authors [12–18] are generalized, and the possibility of sodar application for simultaneous comprehensive measurements of the vertical profiles of turbulence characteristics is demonstrated. To analyze the random wind velocity field in the atmosphere, the structure functions – the average differences of the squared wind velocity components at two points – are used. The application of sodars allows long time series of observations of the vertical profiles of wind velocity vector to be obtained and the structure functions of the wind velocity field to be calculated for separation of the observation points up to several hundred meters.

The spatiotemporal dynamics of the characteristics of atmospheric turbulence, including three components of wind velocity vector, longitudinal and transverse structure functions of wind velocity field, structure characteristics of temperature and wind velocity, turbulent kinetic energy dissipation rate, and outer scales of temperature and dynamic turbulence are analyzed. The above-indicated characteristics are retrieved from time series of vertical profiles of the wind velocity vector and the backscattered signal power measured with the sodar having a working frequency of 1700 Hz and the minisodar having the working frequency of 4900 Hz. The original closed iterative algorithm of sodar data processing considering the classical and molecular absorption and the turbulent sound extinction on the propagation paths to the scattering volume and back to the receiver is used that allows the vertical profiles of the characteristics of temperature and wind velocity field to be reconstructed simultaneously and their interrelation to be investigated. In the literature, including our papers, only results of measurements of individual turbulence characteristics are described. Many papers are devoted to measurements of the temperature and wind velocity structure characteristics [1–7]; results of acoustic sounding of the turbulent kinetic energy dissipation rate are presented, for example, in [8, 9]; the wind velocity structure functions are considered in [10, 11].

## 2. Methods of sounding and data processing

Doppler sodars allow long time series of vertical profiles of the backscattered signal power $P_j(z_k)$ and of the three components of wind velocity vector $V_{i,j}(z_k)$ to be obtained in vertical strobes $z_k$, where $i = x, y, z$, $j = 1, \ldots, N$ designates the serial number of the vertical profile in the series of $N$ measurements, and $k = 1, \ldots, M$ designates the serial strobe number. They can be used to calculate the temporal longitudinal and transverse structure functions for each strobe $z_k$:

$$D_{rr}\left(z_k, n\Delta t\langle V(z_k)\rangle\right) = \frac{1}{N-n-1} \times \sum_{j=1}^{N-n-1}\left[V'_{j+n}(z_k) - V'_j(z_k)\right]^2, \; n = 1, \ldots, N/5, \tag{1}$$

$$D_{tt}\left(z_k, n\Delta t\langle V_z(z_k)\rangle\right) = \frac{1}{N-n-1} \sum_{j=1}^{N-n-1}\left[V_{z,j+n}(z_k) - V_{z,j}(z_k)\right]^2, \; n = 1, \ldots, N/5. \tag{2}$$

Here

$$|V(z_k)| = \sqrt{\langle V_x(z_k)\rangle^2 + \langle V_y(z_k)\rangle^2 + \langle V_z(z_k)\rangle^2}, \tag{3}$$

$$\langle V_i(z_k)\rangle = \frac{1}{N}\sum_{j=1}^{N} V_{i,j}(z_k), \quad i = x, y, z, \tag{4}$$

$$V'_j(z_k) = \left\{V_{x,j}\langle V_x(z_k)\rangle + V_{y,j}(z_k)\langle V_y(z_k)\rangle + V_{z,j}(z_k)\langle V_z(z_k)\rangle\right\} / \left|\langle V(z_k)\rangle\right| \tag{5}$$

is the longitudinal component of the wind velocity vector retrieved from individual measurement for strobe $z_k$, and $\langle V(z_k)\rangle$ is the wind velocity vector for the strobe $z_k$ averaged over the measurement period $T = Nt$.

The vertical profiles of the structure characteristic of the wind velocity field $C_V^2(z_k)$ were calculated from the formula:

$$C_V^2(z_k) = \left\langle \left[ V_{z,j}(z_k + r) - V_{z,j}(z_k) \right]^2 \right\rangle \cdot r^{-2/3}. \tag{6}$$

The separation of the observation points was chosen either in horizontal, or in vertical direction. In the first case, $C_{Vt}^2(z_k)$ was calculated from the vertical component of the wind velocity measured in the strobe $z_k$ with the time delay $n_t \Delta t$ ($n_t = 1, \ldots, N/5$, $N$ is the number of measurements in the series, and $\Delta t$ is the pulse repetition period; in this case, $r = n_t \Delta t \langle V(z_k) \rangle$). In the second case, $C_{Vz}^2(z_k)$ was calculated from the vertical component of the wind velocity measured in different strobes spaced at the distance $r = n_z \Delta z$ (here $\Delta z$ is the strobe length and $M$ is the number of strobes in the sounding range).

The turbulent kinetic energy dissipation rate $\varepsilon(z_k)$ was calculated from the wind velocity structure characteristic using the formula:

$$\varepsilon(z_k) = \left[ \frac{C_V^2(z_k)}{1.97} \right]^{3/2}. \tag{7}$$

From the backscattered signal power $P_j(z_k)$, using the iterative algorithm, vertical profiles of the temperature structure characteristic and of the outer scale of temperature turbulence were calculated.

Mathematically, the outer scale of temperature turbulence was calculated using the closed iterative algorithm considering the excess turbulent signal attenuation. The expression for the $i$th iteration of the outer scale of temperature turbulence $L_{0T}^{(i)}$ depending on the altitude $z_k$ above the underlying surface has the form

$$L_{0T}^{(i)}(z_k) = \left[ \frac{C_T^{2(i)}(z_k)}{2.8\gamma_a^2} \right]^{3/4}. \tag{8}$$

Here $\gamma_a$ is the adiabatic temperature gradient, $C_T^{2(i)}(z_k)$ is the $i$th iteration of the temperature structure characteristic that for the $j$th measurement was calculated from the formula

$$C_{0Tj}^{2(i)}(z_k) = \frac{2.7 \cdot 10^2 A_c \lambda^{1/3} T^2 z^2 P_j(z_k)}{\gamma_1 \gamma_2 c \tau A_t P_0 L^{(i-1)}(z_k)}, \tag{9}$$

$c$ is the sound velocity, $\lambda = c/f$ is the wavelength, $f$ is the working sodar frequency, $T$ is the surface air temperature, $\tau$ is the sodar pulse length, $A_c$ is the instrumental constant, $A_t$ is the transceiving antenna aperture of the sodar, $P_j(z_k)$ is the signal power backscattered from altitude $z_k$ in the $j$th measurement, $j = 1, \ldots, N$, $P_{0j}$ is the transmitted signal power, $\gamma_1$ and $\gamma_2$ are the coefficients of electric signal conversion into the acoustic one and *vice versa*,

$$L^{(i)}(z_k) = L_{abs}(z_k) \cdot L_{turb}^{(i)}(z_k) \tag{10}$$

is the $i$th iteration of the extinction coefficient, $i = 1, \ldots, M$, caused by the classical and molecular absorption of sound waves and their excess turbulent attenuation during propagation from the transmitter to the scattering volume and back to the receiver.

The dependence of the spatial wind velocity structure function on the separation of the observation points $r$ in the inertial ($l_{0V} \ll r \ll L_{0V}$, where $l_{0V}$ is the inner and $L_{0V}$ is the outer scale of dynamic turbulence, respectively) and energy ranges ($L_{0V} \ll r$) is described by following formula:

$$D(z_k, r) = \begin{cases} C_V^2(z_k)\, r^{2/3}, l_{0V} \ll r \ll L_{0V}(z_k), \\ C_V^2(z_k) \cdot L_{0V}(z_k)^{2/3}, L_{0V}(z_k) \ll r. \end{cases} \quad (11)$$

It is proportional to $r^{2/3}$ in the inertial range and then is saturated at a constant. It is the inflection point of the dependence of the structure function on the separation of the observation points in power 2/3 that determines the outer scale $L_{0V}(z_k)$ of the dynamic turbulence. Formulas (1), (2) and (6)–(9) were implemented in the closed iterative computing algorithm.

Measurements were performed with a three-component monostatic Doppler sodar having a working frequency of 1700 Hz, pulse repetition period of 11.5 s, and pulse duration of 150 ms. One receiving sodar antenna was oriented vertically, and two others were tilted at angles of 20° to the vertical in mutually orthogonal planes. The normalized backscattered signal power was measured with the altitude resolution $\Delta z = 20$, the minimum sounding altitude changed from 48 to 74 m depending on the surrounding noise level during measurements. Iterations stopped at $z_M$ when they started to diverge. The signal/noise ratio was additionally monitored from facsimile records of sodar signals. Series of $N = 53$ sodar measurements were processed, which provided 10-minute averaging period. The instrumental constants of the sodar entering into formula (11) were determined by sodar calibration from results of independent measurements of $C_T^2$ by an ultrasonic thermometer/anemometer under conditions of developed convection.

Results of measurements with the Doppler minisodar having a working frequency of 4900 Hz, pulse duration $\tau = 60$ ms, and pulse repetition period of 4 s were also processed. Radiation was transmitted and received in three directions – vertical and tilted at the angles $\alpha = 14°$ to the vertical in two mutually orthogonal planes. Vertical profiles of the three wind velocity components $V_{i,j}(z_k), i = x, y, z, j = 1, \ldots, N$ were measured in $M = 40$ strobes $z_k$ of length $\Delta z = 5$ m in the altitude range 5–200 m. Processing of series from $N = 150$ profiles provided the wind velocity structure functions averaged over the 10-minute period.

## 3. Measurement results

Doppler monostatic sodars are widely used to measure vertical profiles of the wind velocity vector in the atmospheric boundary layer. From their measurements the vertical profiles of the wind velocity structure characteristics $C_V^2$ are also retrieved. Sodars are also used to retrieve vertical profiles of the temperature structure characteristics $C_T^2$ from the backscattered signal intensity. However, a comparison of sodar data with the data of local measurements showed that the divergence of sodar and *in situ* values of $C_T^2$ is on average equal to 40 % and can reach 2–5 times depending on the atmospheric conditions, which was explained by disregarding of turbulent signal attenuation.

The original method of acoustic sounding of atmospheric turbulence taking into account turbulent signal attenuation allows vertical profiles of the temperature and wind velocity structure characteristics to be retrieved simultaneously.

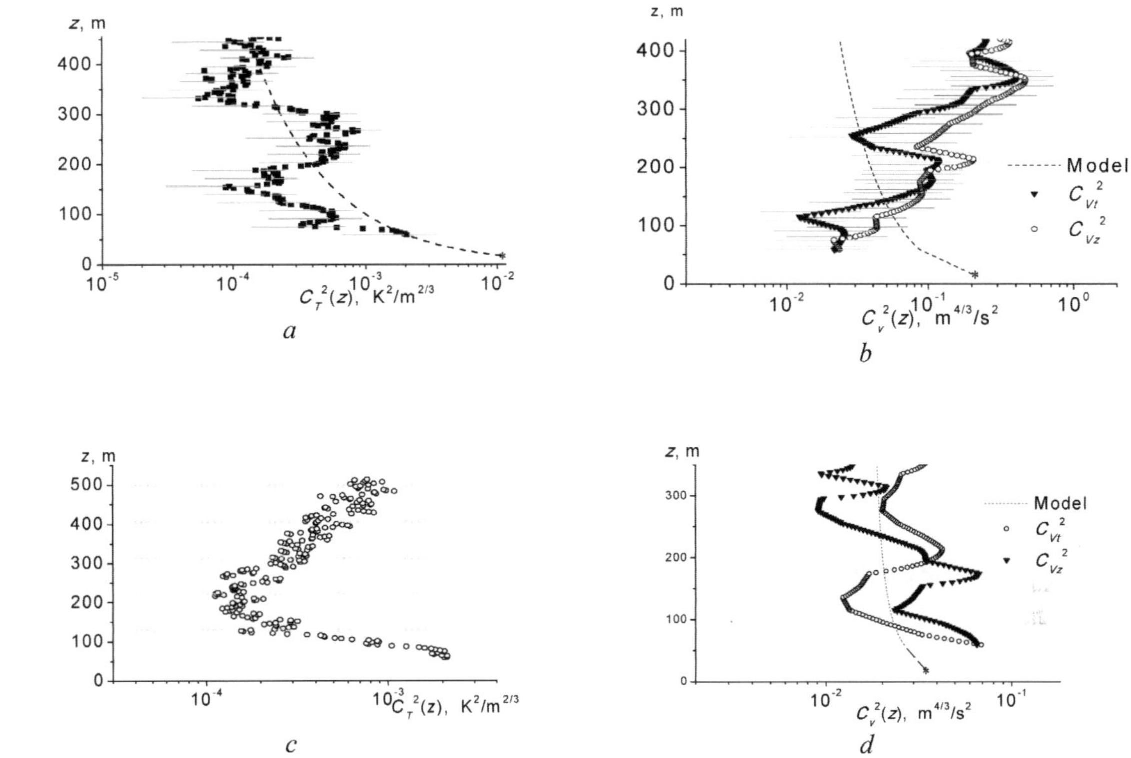


**Figure 1.** Temperature and wind velocity structure characteristics.

Figure 1 shows synchronous vertical profiles of the structure characteristics of temperature, $C_T^2(z)$, and wind velocity for the vertical (designated by $C_{Vz}^2$) and horizontal separation of observation points (designated $C_{Vt}^2$) retrieved from sodar measurements in the morning (10:00, *a* and *b*) and evening hours (16:00, *c* and *d*). The averaging time was 10 minutes. Asterisks show results of *in situ* measurements with micropulsation sensors.

Figure 2,*a–c* illustrates synchronous temporal dynamics of the temperature (facsimile records *a*) and wind turbulence (transverse, $D_{tt}$ (*b*), and longitudinal, $D_{rr}$ (*c*), wind velocity structure functions) in the lower 200-meter layer of the atmosphere in gradation of artificial colors for two 10-minute series of minisodar measurements. The data of continuous minisodar measurements during 6 days were processed. Thermal plumes characteristic for daytime convection are clearly seen on the facsimile records. Their upper boundary is subject to quasi-periodic fluctuations and has a clearly expressed tendency to an increase in altitude during observation period. Values of the transverse (*b*) and longitudinal (*c*) structure functions, in m$^2$/s$^2$, in gradations of artificial colors, are indicated to the right of the figures. It can be seen that the transverse structure function is much less than the longitudinal one that indicates strong anisotropy of atmospheric fluctuations in the longitudinal and transverse directions and squeezing of the small-scale turbulence in the vertical direction. The behavior of the longitudinal structure function also characterizes the dynamics of the mixing layer height.

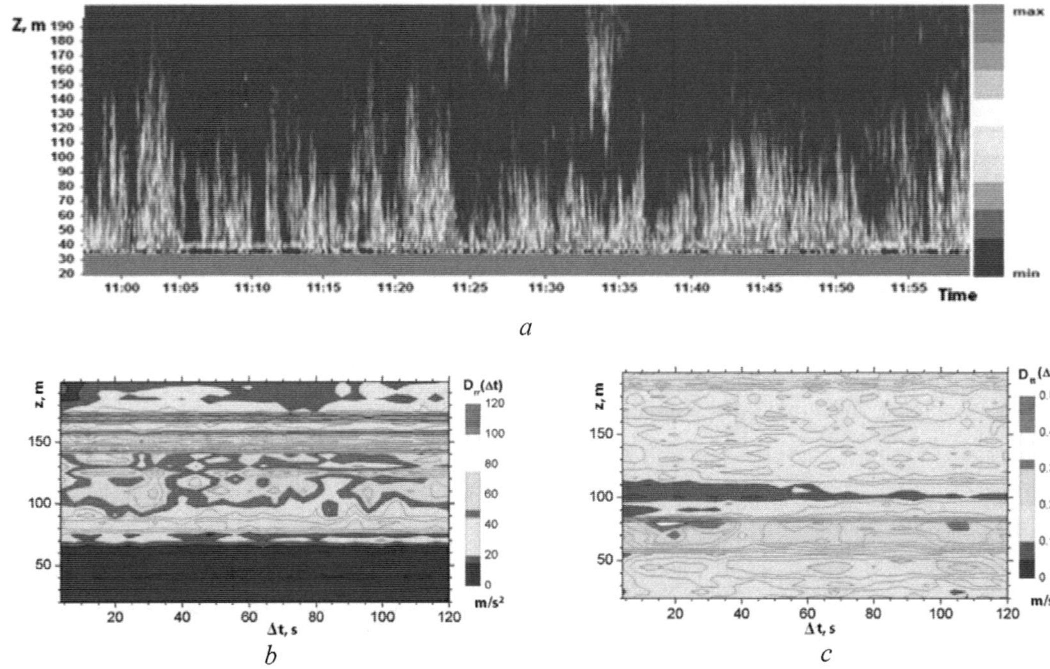


**Figure 2.** Dynamics of the structure of temperature and wind turbulence in the morning at 11:00.

Figure 3,*a* shows a typical example of the temporal longitudinal wind velocity structure function $D_{rr}(z,\Delta t)$ calculated from formula (1) for sodar measurements in summer from 10:00 till 10:10, local time, and Fig. 3,*b* shows the spatial structure function $D_{rr}(z,\Delta r)$ corresponding to it calculated using the hypothesis of the frozen turbulence. It can be seen that the temporal structure functions with increasing time delay of observations remain practically unchanged undergoing only local oscillations. Their values increase with sounding altitude. The spatial structure functions first increase with increasing separation between the observation points and then saturate, as predicted by the theory. An analysis of the results obtained demonstrated that the saturation is observed at smaller separations $\Delta r$ with increasing sounding altitude.

Figure 4 shows the longitudinal temporal wind velocity structure function calculated from formula (1). Vertical profiles of the wind velocity vector were measured with the minisodar in summer from 16:00 till 16:10. It can be seen that the wind velocity field is homogeneous with time; it remains almost invariable with increasing time and grows with altitude of the observation point. In this case, when the altitude increases from 30 to 110 m, the structure function increases from ~10 $m^2/s^2$ to ~50 $m^2/s^2$. For higher altitudes, $D_{rr}(z,\Delta t)$ increases faster, reaching ~100 $m^2/s^2$ for $z = 125$ m. Two layers of intensive turbulence in which $D_{rr}(z,\Delta t)$ reached 135 $m^2/s^2$ were located at altitudes 125–145 m and 170–180 m. For altitudes above 180 m, the structure function was saturated at $D_{rr}^{\mathrm{sat}}(z,\Delta t) \approx 100$ $m^2/s^2$.

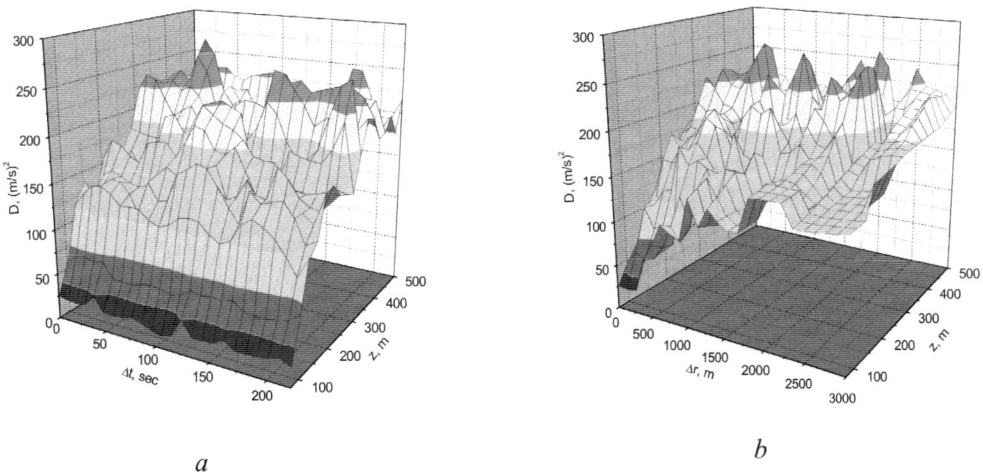


**Figure 3.** Longitudinal temporal (*a*) and spatial (*b*) wind velocity structure functions retrieved from sodar measurements.

The corresponding temporal transverse structure function $D_{tt}(z,\Delta t)$ calculated from formula (2) is shown in Fig. 5. It is much less than $D_{rr}(z,\Delta t)$. Strong anisotropy of the time spectra of amplitude fluctuations of sound wave during its longitudinal and transverse propagation relative to the direction of the wind velocity was also pointed out earlier. Moreover, the correlation in the longitudinal direction was much stronger than in the transverse one. The temporal transverse structure function, unlike the longitudinal one, changed with time. It increased with altitude and saturated at $D_{tt}^{\text{sat}}(z,\Delta t) \approx 0.23$ m$^2$/s$^2$. Small-scale inhomogeneities with lifetimes of the order of 10 s were clearly pronounced.

It should be noted that in addition to the traditional facsimile records of sodars that illustrate the dynamics of temperature turbulence in the atmospheric boundary layer, the results presented here illustrate the dynamics of the wind turbulence indicating layers of more intensive wind turbulence that can be used, for example, to improve aircraft safety.

The diurnal dynamics of the wind velocity structure characteristic is illustrated in Figure 6, where the vertical profiles of $C_V^2$ calculated from the data of hourly minisodar measurements averaged over 10-minute periods (over 150 individual profiles) on September 13, 2006 from 11:00 till 22:00. Here profiles (*a*) were measured from 11:10 till 11:20, profiles (*b*) from 12:10 till 12:20, ..., profiles (*l*) from 22:10 till 22:20. The red curves are for the vertical separation of the observation points, and the blue curves are for their horizontal separation. Horizontal bars indicate 95 % confidence intervals. The dashed curves illustrate the altitude dependence $C_V^{2(theor)}(z) \sim z^{-2/3}$ predicted theoretically for developed convection affixed to the data of surface measurements. From the figures it can be seen that for this series of measurements, $C_V^{2(\text{theor})}(1\ \text{m})$ remains constant and equal to 0.18 m$^{4/3}$·s$^2$ from 11:00 till 15:00, local time, then at 16:00 it decreases to 0.12 m$^{4/3}$·s$^2$, from 17:00 increases from 0.15 m$^{4/3}$·s$^2$ to 0.22 m$^{4/3}$·s$^2$ at 18:00, reaching a maximum value of 0.25 m$^{4/3}$·s$^2$ at 19:00 and 20:00, then decreases to 0.1 m$^{4/3}$·s$^2$ at 21:00, and increases again to 0.22 m$^{4/3}$·s$^2$ at 22:00.

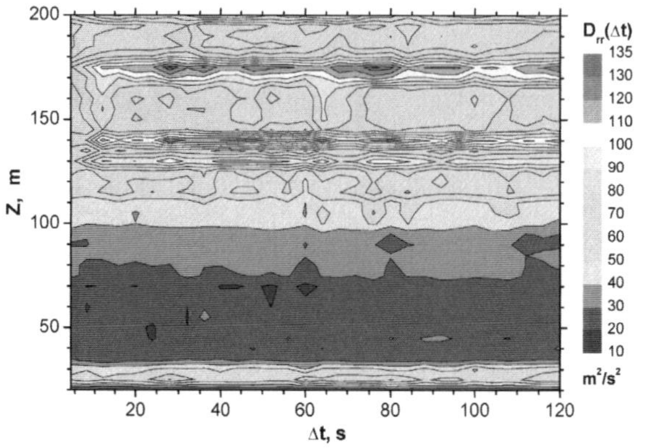


**Figure 4.** Longitudinal temporal wind velocity structure function $D_{rr}(z, \Delta t)$ calculated from minisodar measurements.

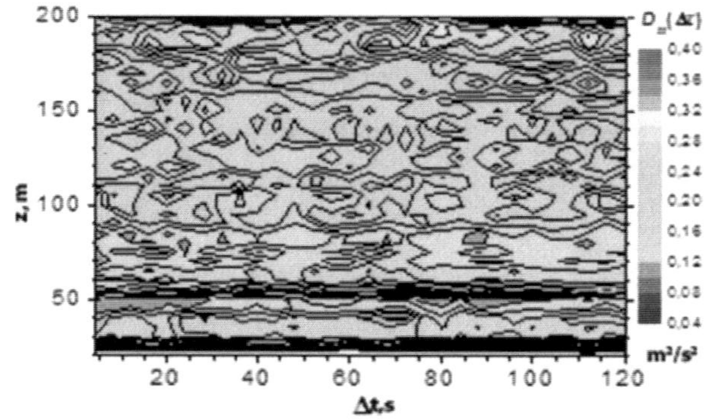


**Figure 5.** Transverse temporal wind velocity structure function $D_{tt}(z, \Delta t)$ calculated from minisodar measurements.

The temporal dynamics of vertical profiles of turbulent kinetic energy dissipation rate calculated from formula (7) from minisodar data taking into account the excess turbulent attenuation is shown in Figure 7. The month, day, year, and time of the beginning of 10-minute series of sodar measurements are indicated on the right of the figures. For a comparison, asterisks show results of ε retrieval from measurements with a 2-micron coherent Doppler lidar.

Good agreement of the results of lidar and acoustic measurements can be seen. Analysis of the results shown in Figure 7 demonstrated that $\varepsilon(z)$ decreases approximately by an order of magnitude with increasing local measurement time from 01:00 till 09:00, which is in agreement with lidar measurements. It reached a minimum at 11:00. The dissipation rate on September 16 increased by approximately one and a half orders of magnitude when the local measurement time increased from 11:00 till 21:00 and then started to decrease when the measurement time further increased till 23:00. For $z = 100$ m on September 17, it decreased from ~$8 \cdot 10^{-4}$ m$^2$/s$^2$ at 01:00 to $1 \cdot 10^{-4}$ m$^2$/s$^2$ at 05:00.

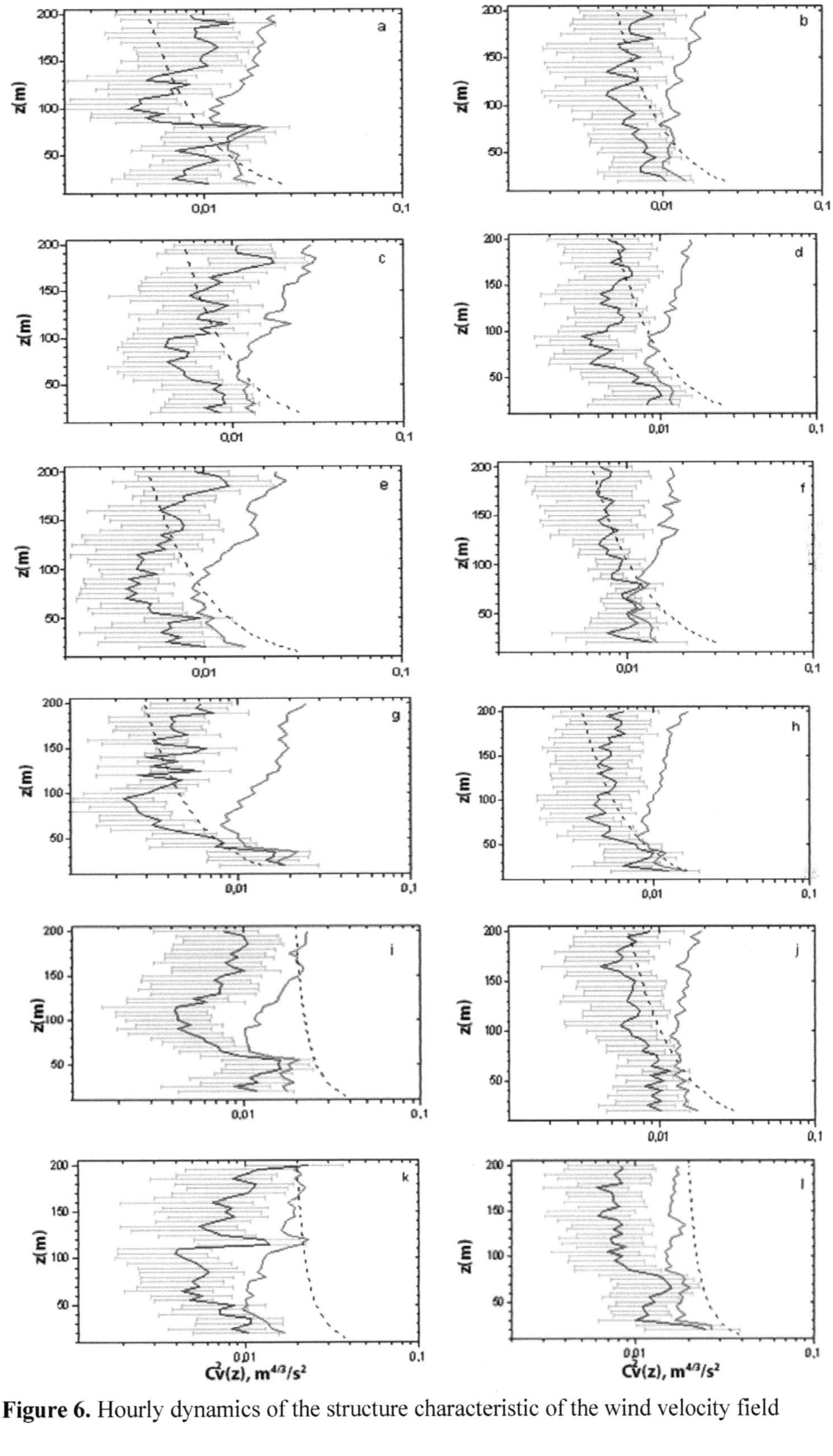


**Figure 6.** Hourly dynamics of the structure characteristic of the wind velocity field

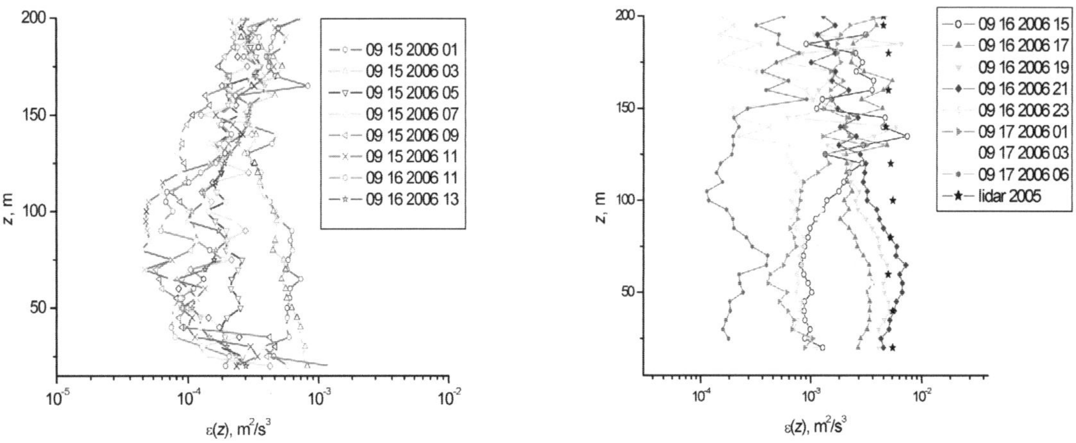


**Figure 7.** Turbulent kinetic energy dissipation rate retrieved from minisodar measurements.

Figure 8 shows the results of measurements of the outer scale of temperature turbulence $L_{0T}(z)$ in developing convection (*a*) in the morning (from 10:00, local time), developed daytime convection (*b*) (from 15:00, local time), and stable stratification (*c* and *d*). It can be seen that the outer scale of temperature turbulence increases with altitude that is in agreement with its existing models.

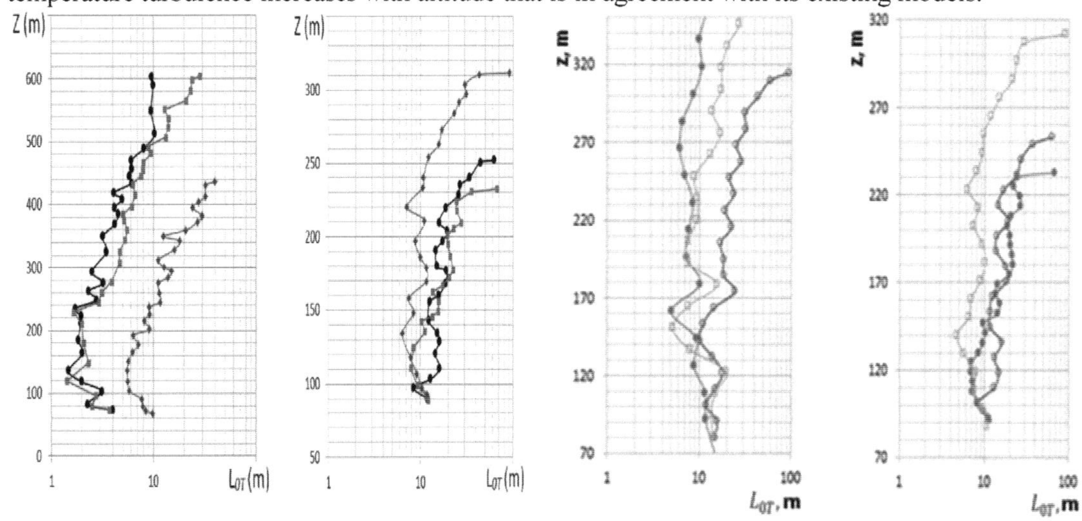


*a* *b* *c* *d*

**Figure 8.** Outer scale of temperature turbulence for 3 successive 10-minute series of sodar measurements (circles, diamonds, and squares) in developing (*a*) and developed convection (*b*) and in stable stratification (open circles, closed circles, open squares) (*c* and *d*). In case *d*, measurements started at 01:00, local time.

Figure 9 shows semidiurnal hourly dynamics of vertical profiles of the outer scale of dynamic turbulence calculated from 10-minute series of minisodar measurements from 11:00 till 23:00, local time. From the figure it can be seen that the outer scale of dynamic turbulence changes in the range from 5 m to 100 m, which is in good agreement with its theoretical estimation and results of aircraft measurements. The general tendency of the outer scale to increase with altitude is also traced. As to the temporal dynamics, it can be seen that maximal outer scales are observed in the morning and in the evening. Moreover, the outer scale in the morning changes greater than in the daytime and in the evening, and its vertical stratification is also more clearly pronounced in the morning. For a comparison, Figure 9,*d* shows the altitude profiles of the outer scale of dynamic turbulence from night measurements with a 2-micron Doppler lidar from 23:23 till 23:40, local time, retrieved from the azimuthal structure function after spatial filtration of lidar data (circles) and from the radial structure function after spatial filtration of lidar data (diamonds). Good agreement of sodar data with lidar measurements is seen from the figure. Attention is also drawn to the clearly expressed layered structure of the outer scale of dynamic turbulence retrieved from both sodar and lidar data.

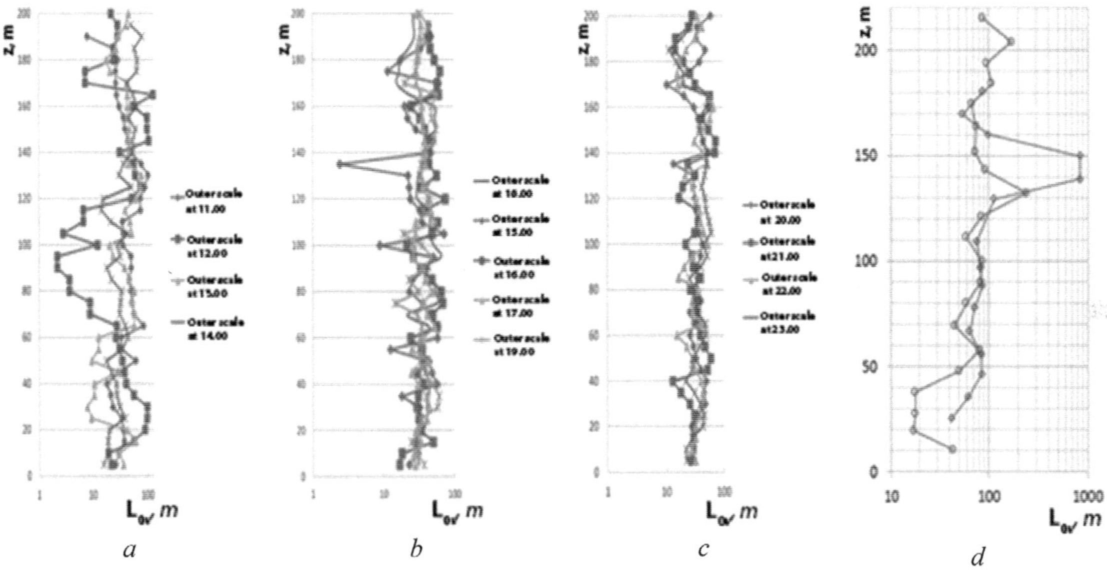


**Figure 9.** Semidiurnal hourly dynamics of vertical profiles of the outer scale of dynamic turbulence retrieved from minisodar measurements. Time of the beginning of10-minute measurementseriesis indicated to the right of the figures.

## 4. Conclusions

The capabilities of sodar application for investigations of the spatiotemporal dynamics of the three components of wind velocity vector, longitudinal and transverse structure functions of wind velocity field, temperature and wind velocity structure characteristics, turbulent kinetic energy dissipation rate, and outer scales of temperature and dynamic turbulence in the atmospheric boundary layer are described. The efficiency of the original closed iterative algorithm of sodar data processing taking into account turbulent sound extinction on the propagation path for simultaneous retrieval of the vertical profiles of the characteristics of temperature and wind velocity fields and investigation of their interrelation is demonstrated. It is demonstrated that sodars are very efficient means for studying and monitoring of the wide range of the above-indicated parameters of the lower atmosphere.

## References

[1] Singal S P 1997 *Acoustic remote-sensing applications* (Berlin: Springer-Verlag)
[2] Kallistratova M A and Kon A I 1985 *Radioacoustic sounding of the atmosphere* (Moscow: Nauka) (In Russian)
[3] Krasnenko N P 1986 *Acoustic Sounding of the Atmosphere* (Novosibirsk: Nauka) (In Russian)
[4] Krasnenko N P 2001 *Acoustic Sounding of the Atmospheric Boundary Layer* (Tomsk: Vodolei) (In Russian)
[5] Bradley S 2007 *Atmospheric Acoustic Remote Sensing: Principles and Applications* (CRC Press

Taylor&Francis Group)

[6] Kallistratova M A, Kuznetsov R D, Petenko I V 2014 *Turbulence, Dynamics of the Atmosphere and Climate. Proc. Int. Conf. Devoted to Memory of Academician A M Obukhov* M: GEOS 90-96. ISBN 978-5-89118-654-5. 593-620

[7] Petenko I.V., Shurygin E.A. 1999 *Boundary-Layer Meteor.* **93** (3) 381-394

[8] Quintarelli F 1993 *J. of Applied Meteorology* **32** 1433-40

[9] Greenhut G K 1989 *J. of Applied Meteorology* **28** 99-106

[10] Coulter R L 1990 *Boundary-Layer Meteorology* **52** 75–91

[11] Coulter R L, Martin T J 1996 *Boundary-Layer Meteorology* **81** 23–33

[12] Kapegesheva O F, Krasnenko N P, Stafeev P G, Shamanaeva L G 2012 *Izv. Vyssh. Uchebn. Zaved. Fiz.***55** 10 1132-36

[13] Kapegesheva O F, Krasnenko N P, Stafeev P G, Shamanaeva L G 2012 *Izv. Vyssh. Uchebn. Zaved. Fiz.***55** (9/2) 228-232

[14] Krasnenko N P, Shamanaeva L G 2012 2012 *Izv. Vyssh. Uchebn. Zaved. Fiz.***55** (12) 1486-92

[15] Krasnenko N P Kapegesheva O F Stafeev P G Shamanaeva L G 2013 *Izv. Vyssh. Uchebn. Zaved. Fiz.***56** (6) 667-673

[16] Kapegesheva O F Krasnenko N P Stafeev P G Shamanaeva L G 2013 2012 *Izv. Vyssh. Uchebn. Zaved. Fiz.***56** (9/2) 254-255

[17] Kapegesheva O F Krasnenko N P Stafeev P G Shamanaeva L G 2013 *Opt. Atmos. Okeana* **26** (11) 964-968

[18] Krasnenko N P Shamanaeva L G 2014 *Memories of Faculty of Physics Lomonosov Moscow State University* 146306

# Detection of geodynamic activity areas based on the Earth's electromagnetic noise parameters

**Vasily F Gordeev, Sergey Yu Malyshkov, Sergey G Shtalin, Vitaly I Polivach and Vladimir A Krutikov**

Institute of Monitoring of Climatic and Ecological Systems, Siberian Branch of the Russian Academy of Sciences, 10/3, Akademichesky pr., Tomsk 634055, Russia

e-mail: gordeev@imces.ru

**Abstract**. In this paper, a method of using the Earth's natural pulsed electromagnetic noise for mapping of anomalies of intensely strained state of the Earth's crust is substantiated. Examples of using the method for mapping of geodynamically dangerous sites and monitoring of processes that pose threats to the operation of industrial facilities are presented.

## 1. Separating space and time variations in a flow of EM signals

Natural pulsed electromagnetic fields are mostly produced by shallow crustal discontinuities, cracks and microcracks. Electromagnetic noise patterns have been employed for mapping faults, monitoring stress and strain in the ground, and for predicting geodynamic movements since the beginning of PEMF studies [1, 2], though they remain of almost no use in geophysical prospecting applications. The lithospheric component of pulsed EM responses becomes prominent only if the instruments are sensitive enough [3, 4] but records of very sensitive acquisition systems are very noisy and abound in atmospheric signals. On the other hand, low sensitive systems record only the heaviest thunderstorms and miss lithospheric responses. Thus, the data acquisition units have to be adjusted to a certain optimum sensitivity range depending on local ground properties. It is convenient to tune instruments against empirical calibration relationships, based on years-long measurements of the Earth's natural pulsed electromagnetic field in different parts of Eurasia. In this it is important that the behavior of the recorded time series were proximal to diurnal ENPEMF rhythms typical of a given season [4, 6]. The precision of ENPEMF measurements is about 10 nT. One has to take into account that about 90 % of responses come from sources outside the target objects and cannot represent their geology. The way of picking spatial ENPEMF anomalies from the flow of recorded signals has been patented (RF Patent No. 2414726) [5].

In the suggested approach, atmospherics and remote responses from objects outside the area of interest are removed from the flow of recorded signals. Filtering from irrelevant components is provided at the stage of data acquisition and processing by:

- using a network of synchronously operated largely spaced fixed and mobile EM stations;
- tuning the stations to the optimum sensitivity and checking their consistency in terms of sensitivity and precision;
- discriminating between remote and local responses.

PEMF signals vary both in space and time. The fixed stations in ENPEMF systems record only variations in time and serve for reference, while the others are mobile and sample both temporal and spatial field patterns along profiles that cross the target areas. The systems include at least two receivers, the greater their number the better the resolution.

Local and remote responses can be discriminated according to their amplitude and time of arrival at distantly spaced stations. Remote responses (e.g., atmospherics) propagate along the Earth-ionosphere waveguide to arrive at proximal stations almost simultaneously, and have the same amplitudes. Responses of large lithospheric objects reach the surface and travel further as ground rays at about the light speed and decay very slowly, which provides their synchronous recording at all stations with similar amplitudes.

Unlike the remote signals, local responses travel mostly through rocks in the vicinity of stations. Signals immediately over the source and those from more distant points have markedly different amplitudes because of rapid decay. In threshold acquisition systems that cut off low signals, more distant receivers can record fewer responses than those located near the anomaly. However, prominent local responses recorded by all distantly spaced stations have different amplitudes depending on the source-receiver distance, which is employed in the instruments and techniques we suggest.

## 2. Mapping stress and strain of rocks and faulting patterns

As it was mentioned above [6], natural Earth's EM noise results from crustal discontinuities, domains of high stress and strain, cracks and microcracks. Mechanic-to-electric energy conversion under the effect of strain waves from the mantle, tides, microseisms, winds, and technological loads produces pulsed EM fields which make up the natural lithospheric EM background. According to long-term measurements in different areas, lithospheric ENPEMF responses have well-defined diurnal and seasonal rhythms because crustal strain waves are associated with Earth's rotation about its axis and about the Sun. The diurnal cycles depend on the calendar dates, geographic coordinates, and geophysical properties. Typical diurnal variations respond to changes in crustal rhythms, for instance, when a consolidated domain of several interlocked fault blocks forms during earthquake nucleation or when stress and strain change rapidly. Thus, ENPEMF measurements can provide a universal tool for geophysical surveys and geodynamic monitoring or for any other application in geosciences.

We applied ENPEMF surveys to assess slope stability in the Kama River right bank in central Russia where the Urengoi – Pomary – Uzhgorod gas pipeline crosses the river. There is a cascade of landslides in the slope, which pose serious hazard to the line. During 1D profile surveys, one reference station was fixed and sampled ENPEMF temporal variations. It was placed ~150 m away from the slope, on a flat surface outside the landslide zone, and the slope stability and stress-strain patterns in the landslide were estimated against its data. Before the acquisition, the stations were synchronized with clocks built into MGR recorders, to fractions of a second. This allowed removing the responses that arrive simultaneously at all stations and thus come from outside the landslide. Information was inferred from the profile-to-reference station ratios of the measured parameters. Sampling at all stations was every 1 s (i.e., at least 300 counts at each point). Statistical processing included filtering the data from diurnal cycles and remote responses. Measurements were run along eleven profiles, both along and across the slope. In addition to measuring stress and strain in the southern part of the landslide, surveys along profile 1, upslope from the river, between pipelines one and two, aimed at estimating whether the detected anomalies remained stable with time. Each point was sampled twice by two different stations, 20 minutes one after another, which allowed testing the measurement quality and checking the anomalies in a while after the first sampling.

Repeated measurements at station 2 were taken 20 min after sampling at station 4; the EM noise patterns used to infer stress and strain in the southern part of the pipeline are shown in Figure 1.

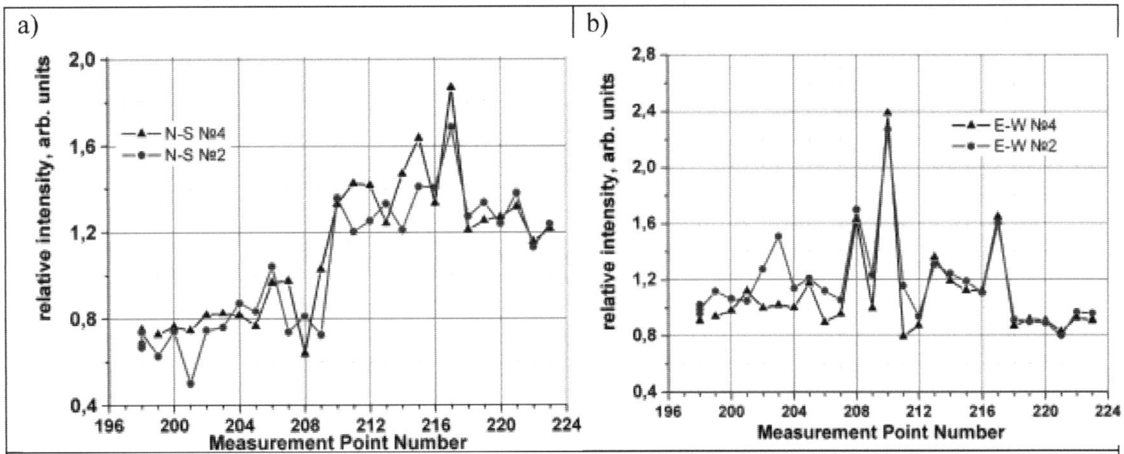


**Figure 1.** Variations of Earth's EM noise. a: N—S channel; b: W—E channel.

The results from different stations obviously remain consistent after filtering out the time component of ENPEMF. The highest landslide activity is recorded in the steepest part of the slope. Spatial variations are quite stable and hold with time. The spatial ENPEMF pattern remains almost invariable at least for 20 min and persists within some region of scatter. However, this stability may be upset either by interior processes inside the landslide or by exterior triggers, such as seasonal changes in ground moisture, snow melting, long rainfalls, technological loads, etc.

Images in Figure 2 are processed 2D data. Alternating zones of extension and compression are restricted mainly to the northern and central parts of the landslide slope. They appear as peaks and troughs in the 3D model above, with their heights and depths proportional to the stress magnitude, and as red or blue zones, respectively, in the map below.

**Figure 2.** Patterns of stress and strain in landslide slope based on radiowave measurements.

Zones of extension, as well as their sharp borders with compression zones, are especially hazardous for pipelines. The most active landslide zones are located on the right, both in the 3D model and in the map (Figure 2), where exactly the compression and extension stresses are the highest and where most of pipeline accidents occurred for the past years. Thus, the ENPEMF method, being cheap and straightforward, turns out to be no less informative than geomorphological analysis, seismic survey, or drilling.

The data from the landslide slope of the Kama right bank demonstrate the applicability of PEMF surveys to estimating stress and strain of rocks. According to evidence of multiple tests, with repeated surveys, synchronous sampling at several stations, and checks against the classical methods, the appropriate use of the ENPEMF techniques ensures precise, well reproducible, and reliable records of slope activity.

Thus records of the Earth's natural pulsed EM field can highlight zones of high and low sliding activity in unstable slopes, as well as zones of extension and relative compression and stress directions.

## 3. Monitoring of sliding activity in unstable slopes, assessment of the degree of their danger

In the end of 2007 an automatic system for slope stability monitoring was launched within the Urengoi – Pomary - Uzhgorod pipeline in the area where it crosses the Kama. The system currently comprises twelve multichannel stations MGR-01 deployed at sites of prominent stress-strain anomalies chosen proceeding from geophysical data. The stations consist of a recorder, a receiving antenna, and a battery, all placed in a tight container transparent for radiowaves. The containers are buried under the ground to avoid damage from vandalism while the transmitter GPS antennas emerge on the ground surface and are naturally camouflaged. The reference station (T4rep) is set up on the flat surface about 80 m away from the slope. Data from the other stations are processed with regard to temporal variations of ENPEMF parameters from the reference station. Furthermore, special smart devices (UB) pick up strain of metal in all nine pipes of the line on the slope top and at its base.

The EM noise data are saved in the recorder's memory and are available for reading on a PC or for sharing via a password-protected FTP web server. The recorded field parameters are collected, processed, and visualized at a specially designed web portal [7] equipped with all necessary tools. In the database created at the portal, the processed data are available in a graphic form or in a mode of hazard monitoring, which allows automatic assessment of slope stability at certain sites according to analysis of data in a data page.

Figure 3 shows a window a portal user sees. The ENPEMF data are visualized as curves in

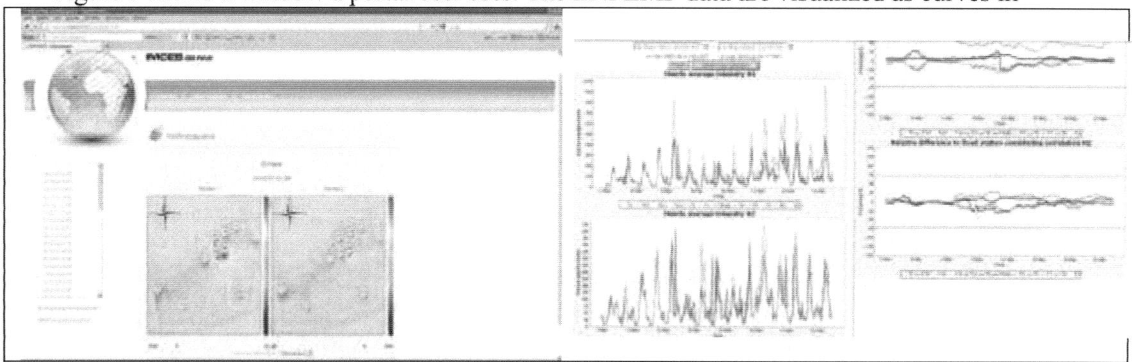

**Figure 3.** The view of a window at the web portal.

two radiowave channels (N—S channel H1 and W—E channel H2). Stress and strain data at the monitored landslide sites are available as color-coded curves and provide information on sliding hazard. The greater the curve departure from zero (green line) the more rapidly stress and strain change at the respective PEMF measurement point.

The algorithm of slope stability monitoring from PEMF parameters employs two criteria: one shows the field recorded on the slope relative to that at the reference station and the other estimates the similarity of data from different stations. The excess flow of ENPEMF signals at each station (T1, T2, T3, T5, T6, T7, T8, T9, T10, T11, T12) relative to the reference field at T4 rep is found as

$$K_{(C-IO)} = \frac{(N1_{T(i)} - N1_{T(r)})}{N1_{T(r)}},$$

$$K_{(3-B)} = \frac{(N2_{T(i)} - N2_{T(r)})}{N2_{T(r)}}$$

- where $N1_{T(i)}$ is the flow of ENPEMF signals in the N—S direction at the i-th station; $N2_{T(i)}$ is the flow of ENPEMF signals in the W—E direction at the i-th station;;
- $N1_{T(r)}$, $N2_{T(r)}$ are the flows of ENPEMF signals in the N—S and W—E directions at the reference station.

The coefficients *K*<0 correspond to compression in the respective direction (N—S or W—E) and *K*>0 correspond to extension, the higher the *K* value the greater the stress. This parameter is a main diagnostic criterion of the geodynamic state of ground.

The second parameter accounts for correlation between ENPEMF diurnal variations at the reference and i-th stations, with Spearman's rank correlation coefficient between the two statistical samples found as

$$r = 1 - \frac{6}{n^3 - n}\sum_{i=1}^{n}(x_i - y_i)^2$$
, where

$n$ is the size of the sample (number of measurements used in the calculation);

$x_i$ is the rank of the i-th element of the first sample in the sample itself (e.g., for the sample {5, 8, 4, 6} $x_1 = 2$, $x_2 = 4$, $x_3 = 1$, $x_4 = 3$);

$y_i$ is the rank of the i-th element of the second sample in the sample.

Then the integral slope stability coefficient is found as

$$R_{об} = \frac{K + (1 - r)}{2}$$
, where

$R_{об}$ is the integrate parameter of empirically estimated residual stability.

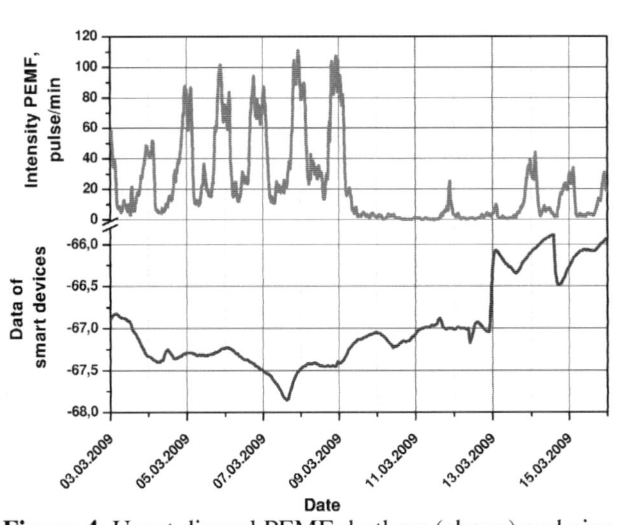


**Figure 4.** Upset diurnal PEMF rhythms (above) and pipe strain gage data (below).

An excess stability coefficient holding above 150% for three days or more indicates sliding hazard dangerous for the pipeline operation.

The suggested approach to ENPEMF data acquisition and processing provides precise and reliable assessment of slope stability (sliding activity, extension and compression, and stress directions) in real time.

Monitoring with our ENPEMF system in the course of several years has shown that different parts of the river side are dynamically developing structures and the slope stability within the landslide area can change locally even over 24 hours. ENPEMF anomalies were recorded more than ten times over the period of operation at different sites of the landslide and appeared as upset diurnal rhythms persisting for several days. On these days, data from different slope stations differed markedly.

Once such field peaks appeared, gas transport from the hazardous line was moved to a standby line. Most often the Earth's EM noise records alerted before other monitoring devices became triggered.

The ENPEMF curve in Figure 4 is recorded near the gas line with pipe strain pickup (by a smart device). The common diurnal rhythm of EM pulses became upset on March 9, before ground motion showed up as a strain peak on March 13; the ENPEMF diurnal rhythm recovered in the end of the record.

The monitored pipeline segment ran smoothly, without any accident, for over 8 years of the ENPEMF system operation. Of course, it was largely due to preventive engineering measures undertaken in cases of alert. Currently, monitoring systems of this kind are being deployed elsewhere at sites of *Gazprom* pipelines.

**References**

[1] Vorobyov A A 1970 *Russian Geology and Geophysics* **12** 3-13

[2] Mastov Sh R, Gol'd R M, Salomatin V N and Yavorovich L V 1984 *Engineering Geology* **1**

[3] Malyshkov Yu P and Malyshkov S Yu 2009 *Russian Geology and Geophysics* **50** 115-130

[4] Malyshkov Yu P and Malyshkov S Yu 2012 The Earth's Core: Structure, Properties and Dynamics *Eccentric Motion of the Earth's Core and Lithosphere: Origin of Deformation Waves and their Practical Application* ed Jon M Phillips (New York: Nova Science Publishers) pp 115-212

[5] Malyshkov Yu P *et al* 2011 The method of geophysical prospecting *Patent RF 2414726* BI **8**

[6] Malyshkov Yu P *et al* 2013 Electromagnetic Fields: Principles, Engineering Applications and Biophysical Effects *Earth's Natural Electromagnetic Noises in a Very-Low Frequency Band* ed Myung-Hee Kwang and Sang-Ook Yoon (New York: Nova Science Publishers) pp 1-86

[7] Kabanov M M, Kapustin S N, Koltun P N and Milovantsev P B 2011 *Proc. Int. Conf. VII International Congress GEO-SIBIR'* (Novosibirsk) Book 1 Part 1 pp 138-142

# Actinometrical features of basic cloud types

**Sergey V Zuev, Nikolay P Krasnenko, and Elena S Kartashova**

Institute of Monitoring of Climatic and Ecological Systems SB RAS, 10/3, Akademichesky pr., Tomsk 634055, Russia

E-mail: zuev@imces.ru

**Abstract**. In this paper, the authors propose a method for determining the main forms of clouds using an analysis of the diurnal variation of direct and diffuse solar radiation. This approach allows operating the process of determining the sky state automatically, which in most cases requires the presence of an observer at the point of measurement. For this purpose, the authors use the direct radiation transmission ratio and diffuse radiation changes with respect to cloudless sky, which have specific values for various forms and types of clouds. The method can be especially effective in cases where measurement of the basic characteristics of radiation is carried out with automatized actinometrical complexes.

## 1. Introduction

Particular attention by the climate study of areas is required to pay to the mechanisms of interaction of cloud fields and solar radiation, because climate has a strong sensitivity to them [1]. Along with the direct objective of atmospheric optics, which allows counting the radiation balance values using given parameter of the atmosphere and the underlying surface, the solving of inverse problem of atmospheric optics is of great importance, it allows recovering the atmosphere parameter through instrumental measurements of radiation characteristics results.

Studies of the relationship of solar radiation and various forms of clouds begin from the middle of the last century. One of the pioneers of such research is the S.I. Sivkov [2]. Currently, researches of mutual influence cloudiness and solar radiation in different ranges (integral, infrared, visible, and ultraviolet) are conducted mainly within the creation and development of radiative transfer models belonging to any atmospheric general circulation model. In particular, such work is carried out in the A.M. Obukhov Institute of Atmospheric Physics RAS [3], in the V.E. Zuev Institute of Atmospheric Optics SB RAS [4], and in the NASA (USA) [5].

The processes of clouds formation are influenced by many factors, each of which generates its own specific, peculiar, typical only for the particular form of clouds, microphysical, genetic and morphological features. The phase composition of the clouds (water, crystalline or mixed) determines their microphysical peculiarities. Genetic features depend on the conditions occurrence, which are attributable to stratiform clouds, waveform or cumuliform clouds. The appearance and quantity, the cloud base height, the vertical and horizontal dimensions determine the morphological characteristics of clouds. All the typical and usual features of the clouds on their own and in combination with each other have different effects on the value of coming to the surface direct $S$ and diffuse $D$ radiation, by virtue of the fact that each form and type of clouds has its own specific, peculiar actinometrical features, which can be divided into amplitude and temporal. Amplitude features depend on the microstructure and morphology of clouds and demonstrate their ability to influence the direct and diffuse radiation values. Temporal features depend on the genetics and morphology of clouds and affect the change rate of min and max values of direct and diffuse radiation, through which we can judge the speed of movement of the clouds across the sky. Thus, knowing the value of direct and diffuse radiation, it can be determined what types of clouds are most likely presented in the sky at a given time.

## 2. The influence of the basic cloud types on the direct and diffuse radiation

Figures 1 and 2 show the examples of the diurnal variation of direct and diffuse radiation for June 10, 2014, when during the day the in sky predominantly cumulus clouds were observed, according to which it can be estimated that it has a significant impact on the value of *S′* and *D* compared with their values $S'_0$ and $D_0$ in the clear sky.

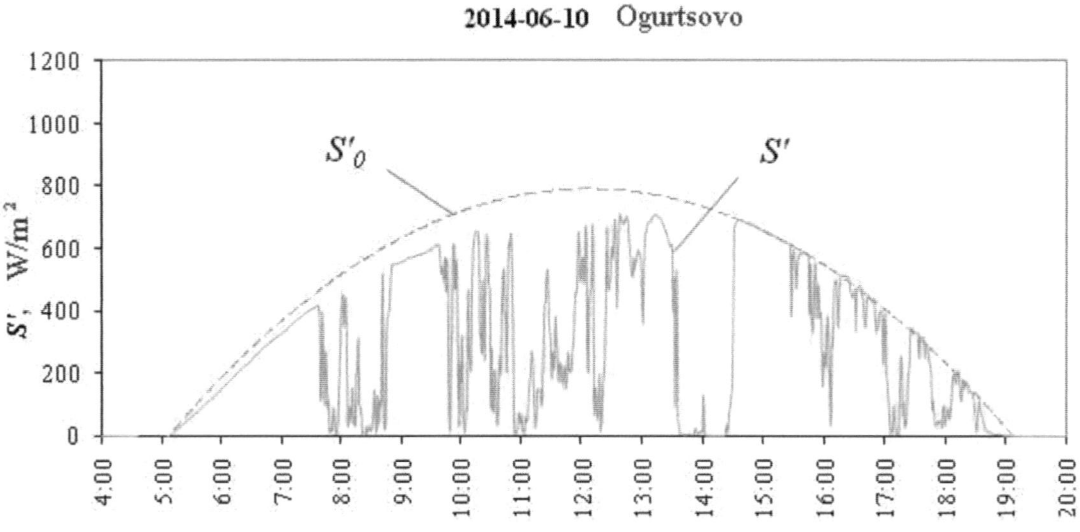


**Figure 1**. The influence of the cumuliform clouds on direct radiation.

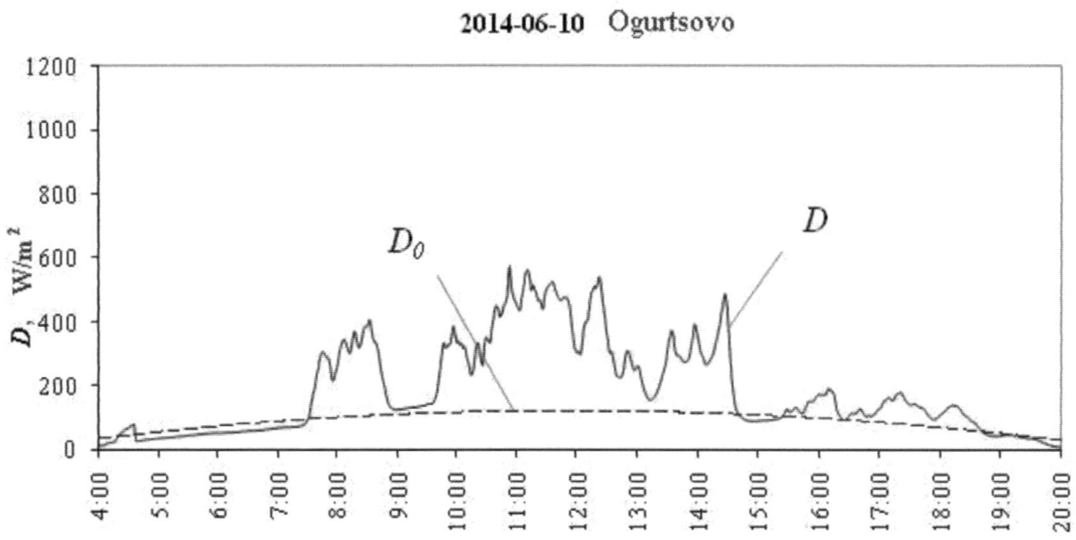


**Figure 2.** The influence of the cumuliform clouds on diffuse radiation.

Table 1 shows the typical and usual features of the basic cloud types, according to [6] in terms of their impact on the value of direct $S$ and diffused $D$ radiation (amplitude actinometrical features).

**Table 1.** The typical and usual features of the basic cloud types.

| Clouds form | Decrease of direct radiation $S$ | | | Change of diffuse radiation $D$ | | | |
|---|---|---|---|---|---|---|---|
| | | | | increase | | | |
| | weak | substantial | strong | weak | substantial | strong | decrease |
| *Ci* | x | | | x | | | |
| *Cc* | x | | | x | | | |
| *Cs* | x | | | x | | | |
| *Ac* | | x | | | x | | |
| *As* | | | x | | x | x | |
| *Ns* | | | x | | | | x |
| *Sc* | | | x | | x | | |
| *St* | | x | x | | x | x | |
| *Cu* | | | x | | x | x | |
| *Cb* | | | x | x | | | x |

It can be observed that more than in a half of cases, the basic cloud types have an unambiguous influence on the decrease of direct radiation $S$ and change of diffuse radiation $D$ (*Ci*, *Cc*, *Cs*, *Ac*, *Ns*, *Sc*). It gives an opportunity to parameterize them using amplitude actinometrical features simpler than *As*, *St*, *Cu* and *Cb,* for which it is necessary to introduce additional features, such as cloud amount and others.

## 3. Classification of clouds based on the transmission ratios of direct radiation and change of diffuse radiation

For analytical description of amplitude actinometrical features, the transmission ratio of direct radiation $C_S = S / S_0$ and change of diffuse radiation $C_D = D / D_0$ are used. Here $S$ and $D$ – measured values, and $S_0$ and $D_0$ – direct and diffuse radiation values by cloudless atmosphere at the same height of the Sun above the horizon.

**Table 2.** Ranges of values $C_S$ и $C_D$ for basic form, species and varieties of clouds.

| | $0 \leq C_D < 0.5$ | $0.5 \leq C_D < 1.6$ | $1.6 \leq C_D < 2.5$ | $2.5 \leq C_D \leq 5$ |
|---|---|---|---|---|
| $0 \leq C_S < 0.5$ | *Cb*$_{med}$<br>*Ns* | *As neb., neb. pr.*<br>*Sc op.*<br>*St fr.*<br>***Cu***$_{low}$<br>***Cb***$_{hi}$ | *Ac*<br>*As neb. op., und op.*<br>*St*<br>***Cu***$_{hi}$<br>*Cb*$_{low}$ | ***Sc***<br>***Cu***$_{med}$ |
| $0.5 \leq C_S < 0.8$ | | ***Ci sp.***$_{low}$<br>***Ci fib.***$_{hi}$<br>*Ci ing.* | *Cs fib.*<br>*Ac und., trans.,*<br>***Ac floc.***<br>***Ci sp.***$_{hi}$ | *Cs neb.*<br>*As* |
| $0.8 \leq C_S \leq 1.2$ | | *Ci*<br>***Ci fib.***$_{low}$ | | |

*Cc*

Table 2 shows the approximate ranges of values $C_S$ and $C_D$ for basic form, species and varieties of clouds (total 56 according to [6]). Indices *hi*, *med* and *low* indicate high (8-10), medium (4-7) and low (1-3) cloud amount respectively. Statistical data for pointed items are given below.

## 4. Results

Table 3 shows the statistical characteristics of ratios of $C_S$ and $C_D$ for some types of clouds, calculated according to the one-minute values of direct *S* and diffuse *D* radiation during July-August 2015 at the Sun's altitude $h \geq 30°$, obtained with the help of an automated actinometric complex of weather station "Ogurtsovo" (Novosibirsk) (WMO ID 29638 ), which is similar to the actinometric complex BSRN [7]. As $S_0$ and $D_0$ values, long-term average values of *S* and *D* were taken, when the sky is clear, according to [8].

**Table 3.** Statistical characteristics of $C_S$ and $C_D$ ratios.

| Type and species of clouds | *N* | $C_S$ | | | | $C_D$ | | | |
|---|---|---|---|---|---|---|---|---|---|
| | | Min. | Ave. | Max. | σ | Min. | Ave. | Max. | σ |
| $Cb_{hi}$ | 245 | 0 | 0.01 | 0.46 | 0.04 | 0.41 | 1.58 | 4.49 | 0.70 |
| *Cu fr.* * | 28 | *0.72* | *0.92* | *1.06* | *0.10* | 1.16 | 1.62 | 2.43 | 0.34 |
| $Cu_{low}$ | 11 | 0.01 | 0.33 | 0.49 | 0.17 | 0.97 | 1.37 | 2.06 | 0.39 |
| $Cu_{low}$ * | 606 | *0.55* | *1.02* | *1.11* | *0.07* | 0.66 | 0.99 | 2.22 | 0.28 |
| $Cu_{med}$ | 232 | 0 | 0.12 | 0.55 | 0.17 | 1.08 | ***1.94*** | 3.06 | 0.46 |
| $Cu_{med}$ * | 409 | *0.55* | *0.97* | *1.10* | *0.13* | 1.00 | ***1.93*** | 3.55 | 0.48 |
| $Cu_{hi}$ | 31 | 0 | 0.07 | 0.46 | 0.13 | 1.47 | 2.35 | 3.71 | 0.79 |
| *Sc* | 413 | 0 | 0.06 | 0.68 | 0.12 | 0.35 | **2.02** | 4.58 | 0.80 |
| *Sc* * | 11 | *1.01* | *1.02* | *1.03* | *0.00* | 2.07 | 2.14 | 2.34 | 0.08 |
| *Ac floc.* | 21 | 0.41 | 0.77 | 1.00 | 0.16 | 1.64 | 2.49 | 3.05 | 0.51 |
| *Ci sp.*$_{low}$ * | 113 | *0.58* | *0.99* | *1.08* | *0.12* | 0.67 | 0.75 | 0.92 | 0.07 |
| *Ci sp.*$_{hi}$ | 9 | 0.18 | 0.32 | 0.51 | 0.12 | 2.34 | 2.45 | 2.61 | 0.10 |
| *Ci sp.*$_{hi}$ * | 12 | *0.55* | *0.77* | *0.99* | *0.15* | 2.35 | 2.58 | 2.66 | 0.10 |
| *Ci fib.*$_{low}$ * | 276 | *0.83* | *1.02* | *1.07* | *0.05* | 0.64 | 0.88 | 1.15 | 0.18 |
| *Ci fib.*$_{hi}$ | 69 | 0.07 | 0.40 | 0.55 | 0.11 | 1.11 | 1.86 | 2.33 | 0.28 |
| *Ci fib.*$_{hi}$ * | 359 | *0.55* | *0.81* | *1.05* | *0.13* | 1.15 | 1.60 | 2.76 | 0.34 |

* - values by open Sun $\odot^2$.
*N*- measurements amount.

Form and cloud amount are determined according to the timed visual observations on this weather station (8 times a day) and to panoramic images of the whole sky, obtained at intervals of 2 minutes using the TV meter of cloud amount "Sineva" installed in the immediate vicinity of the actinometric complex and a integrated in the TV meter of cloud characteristics [9]. This cloud characteristics meter allows also determining the cloud base height, the speed and direction of the lower clouds drift. The exterior appearance of the meter and the image processing program interface are shown in Figure 3. Determination of cloud amount in this case is carried out using the RGB color model by finding of the sky blue areas free of clouds on color panoramic images. As an example, Figure 4 shows the values ranges of $C_S$ and $C_D$ ratios for cumulus and cumulonimbus clouds.

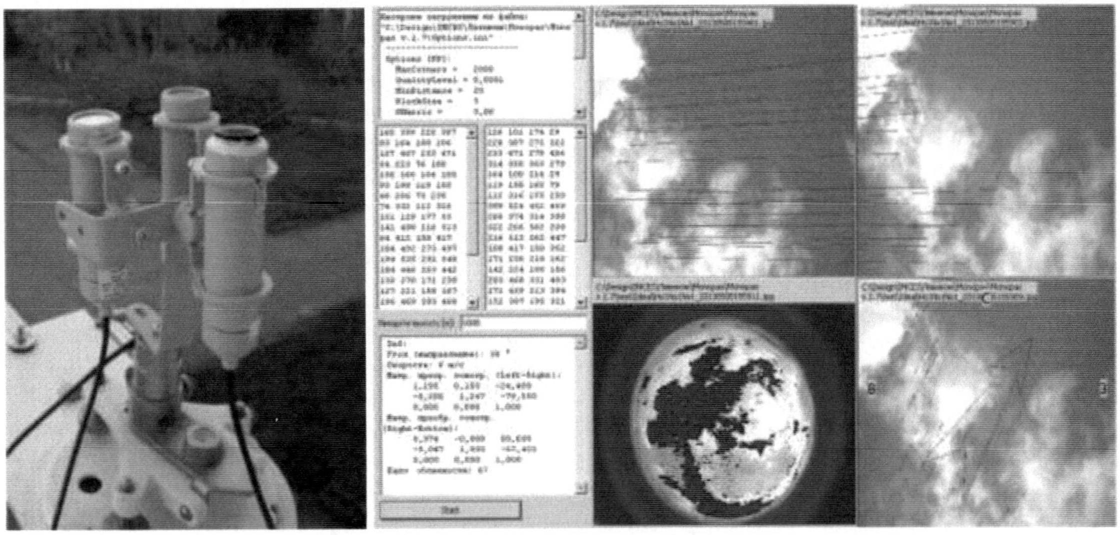

**Figure 3.** TV meter of cloud characteristics (1- ceilometer "Monopas", 2 – meter of cloud amount "Sineva") and image processing program interface.

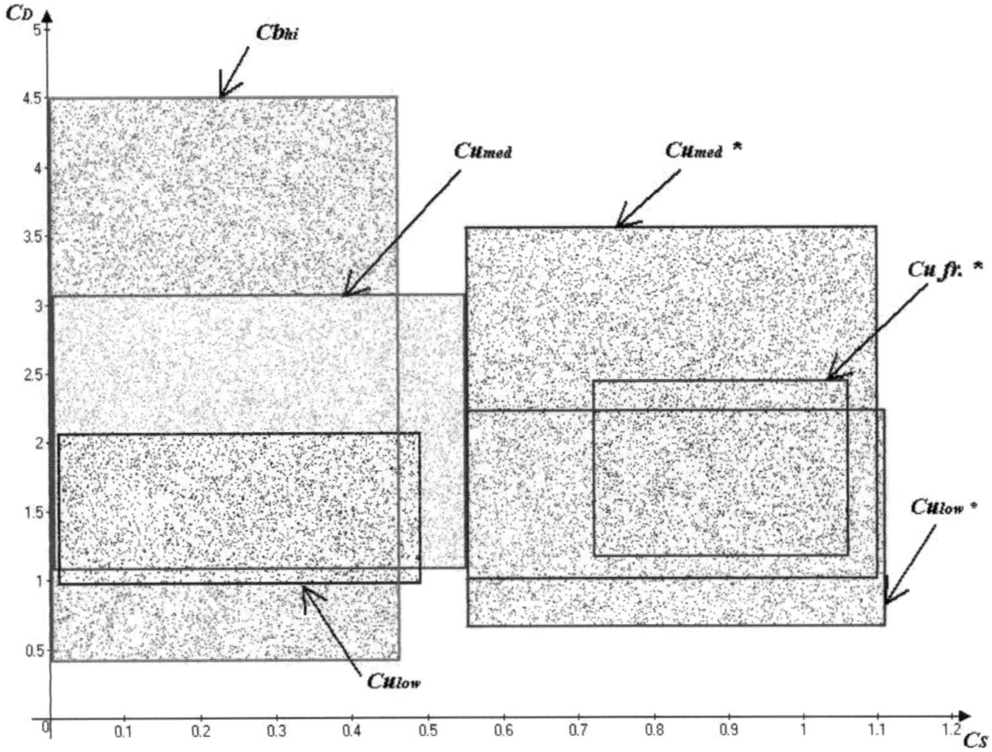


**Figure 4.** The value ranges of $C_S$ and $C_D$ ratios for cumulus and cumulonimbus clouds.

With the accumulation of data, change ranges of the $C_S$ and $C_D$ ratios for definite forms and types of clouds can be specified and corrected.

## 5. Conclusions

Different cloud types and species form their actinometrical features, which can be divided into amplitude and temporal. Cloud ability to influence the value of coming to the Earth surface direct and diffused radiation refers to amplitude features and depends on the microstructure and morphology of clouds. The change frequency of min and/or max values of direct and diffused radiation in certain short period of time in particular depends on the cloud genetics and refers to temporal features, by which we can judge the movement speed of clouds and their fragments in the sky. The use of these features allows parameterizing the basic cloud types and species and automatizing the process of estimation of sky condition for single-layered clouds.

Proposed in this study method allows defining the basic cloud types and species by their amplitude actinometrical features using analysis of the diurnal variation of the direct and diffuse solar radiation. The basic cloud types and species are divided into 3 classes according to the degree of their influence on the direct radiation $C_S$ transmission ratio, and into 4 classes according to the degree of their influence on the diffused radiation $C_D$ change ratio, taking into account the solar disk condition (open or closed by clouds) and cloud amount (high 8-10, medium 4-7, low 1-3). Such classification was conducted by the comparison of *S* and *D* values obtained with the help of an automated actinometrical complex of weather station "Ogurtsovo" (Novosibirsk) (WMO ID 29638 ) in July-August 2015, by cloud form and amount determined according to the panoramic all-sky images obtained using the TV meter of cloud amount "Sineva" installed in the immediate vicinity of the complex.

As follows from the analysis, the range of ratio values for $C_S$ и $C_D$ for cumulus clouds *Cu*, cumulonimbus clouds *Cb*, altocumulus floccus clouds *Ac floc.*, as well as for cirrus dense *Ci sp.* and cirrus fibrous clouds *Ci fib*. By different cloud amount and solar disk condition were defined.

It's worth to remark that the use of temporal actinometrical features in addition to the amplitude ones allows more precise classifying the cloud on the ground of actinometrical information analysis.

**Acknowledgments**

The research is carried out within the framework of projects VII.77.1.2 "Climatic changes in Siberia and the Arctic under conditions of aerosol loads" and VIII.80.2.2."Scientific basis for the development of optical, acoustic and electronic devices, complexes and systems for meteorological measurements and technologies of their applications for environmental monitoring issues" of the program for Basic Research of SB RAS for 2013-2016.

## References

[1] Zuev V E and Titov G A 1996 *Modern Problems of Atmospheric Optics (Vol 9. Optics of atmosphere and climate)* (Tomsk: Spektr) p 272

[2] Sivkov S I 1968 *Calculation methods of solar radiation characteristics* (Leningrad: Hydrometeorological Publishing) p 232

[3] Postylyakov O V 2004 Radiative transfer model in the spherical atmosphere with the calculation of fibered air masses and some of its applications *Math. Russian Academy of Sciences, Atmospheric and Oceanic Physics* **3 (40)** 314–329

[4] Zhuravleva T B 2008 Simulation of solar radiative transfer under different atmospheric conditions. Part I. The deterministic atmosphere *Atmos.Ocean.Opt.* **2 (21)** 81–95

[5] Gatebe Ch K , Kuznetsov A and Melnikova I 2015 Cloud optical parameters from airborne observation of diffuse solar radiation accomplished in USA and USSR in different geographical regions *International Journal of Remote Sensing* **35** 5812–29

[6] Bespalov D P *et al* 2011 *Cloud atlas* ed L K Surigina (St. Peterburg: D'ART) p 248

[7] *BSRN - World Radiation Monitoring Center* 2016 http://bsrn.awi.de

[8] *Scientific and Applied Climate Handbook. 3. Long-term data series. Parts 1-6. Issue 20. Tomsk, Novosibirsk, Kemerovo Region, Altai Territory* 1993 (St. Peterburg: Gidrometeoizdat) p 718

[9] Zuev S V, Krasnenko N P and Levikin V A 2014 TV instrument for measuring of cloud characteristics *Proceedings of TUSUR* **1 (31)** 54–5

IOP Conf. Series: Earth and Environmental Science **48** (2016) 012028 doi:10.1088/1755-1315/48/1/012028

# The algorithm for calculating the Earth's core runs on the parameters of natural pulsed electromagnetic field of the Earth

**Yury P. Malyshkov**, **Sergey Yu. Malyshkov**

Institute of Monitoring of Climatic and Ecological Systems, Siberian Branch of the Russian Academy of Science (IMCES SB RAS), 10/3, Akademichesky pr., Tomsk 634055, Russia

e-mail: msergey@imces.ru

**Abstract**. This paper presents the argumentation in favor of a hypothesis of the eccentric rotation of the Earth's shell and core. It covers the algorithm and computation of movement path and position of the Earth's core by the Earth's natural pulse electromagnetic field characteristics.

Earth's natural pulse electromagnetic fields (ENPEMF) are generated by Earth's crust heterogeneities, mechanically stressed structures, cracks and microcracks. Mechanical and electric transformations caused by actions of strain waves from the lower mantle, tidal forces, microseismic vibrations, wind and man-caused load generate pulse electromagnetic fields that create a natural electromagnetic background of a lithospheric origin [1]. Many-year observations in various regions of the Northern hemisphere showed that the ENPEMF variations demonstrate a stable diurnal and annual pattern. Spectra of such variations contain all frequencies related to the Earth's rotation around its axis, Earth's revolution about the Sun and changes of Earth's axis tilt. But diurnal and annual rotation of the Earth is not enough for such patterns to occur. Considering that an electromagnetic emission of rocks occurs only if the rocks are exposed to some disturbances, we set a goal to find the source of such disturbances. Among such actions that impact the Earth's surface there are the heat effect of the Sun on the Earth's surface or irradiation of the Earth's surface with an electrojet, but such actions cannot cause significant modulations of mechanical and electromagnetic transformations in rocks and, consequently, cannot cause stable diurnal and annual variation patterns of ENPEMF. Diurnal variations of temperature on the Earth's surface cannot actuate the EM field sources in a certain period of day. It is not average diurnal temperatures but average yearly temperatures that remain unchanged at a depth of few meters. And all the temperature processes affect only loose surface layers and can cause neither deep-seated temperature variations nor deep-seated temperature deformations. Lunar and solar tides that create considerable mechanical stresses in upper layers of the Earth's crust cannot explain a strict diurnal and annual pattern either. The most intensive tides are caused by lunar action. And the period of Moon's revolution around the Earth is not equal to 24 hours. Known frequencies of lunar tidal forces are not seen in ENPEMF spectra. Periodic patterns are also observed in variations of the atmospheric pressure affecting the Earth's crust. However the atmospheric pressure does not have a strict periodicity pattern that is typical of ENPEMF.

Due to the fact that we failed to find sources of one-way impact on the Earth's crust outside it, we assumed a hypothesis of the Earth's core shift relative to the geometric center of the Earth and eccentric rotation of the Earth's core and crust. Eccentric diurnal rotation of the Earth's crust around the Earth's shifted core causes pressure from the core and the melt surrounding it towards the mantle which bulges the shell from outside. In other areas of the planet, on the contrary, forces that draw the Earth's shell inside towards the core will occur. In case if the proposed hypothesis is true, the Earth rotation will inevitably cause the stress from the Earth's core towards the Earth's shell. Due to diurnal rotation of the Earth, the zone of pressure starts moving and, together with it, points on the Earth's crust will also start moving at the moment when disturbance caused by the shifted core is passing below (Figure 1). This leads to occurrence of diurnal rhythms of the Earth's crust. The yearly shift of the core will cause annual rhythms. We considered such a mechanism of occurrence of diurnal and annual variations of ENPEMF and seismicity in our previous publications [1].

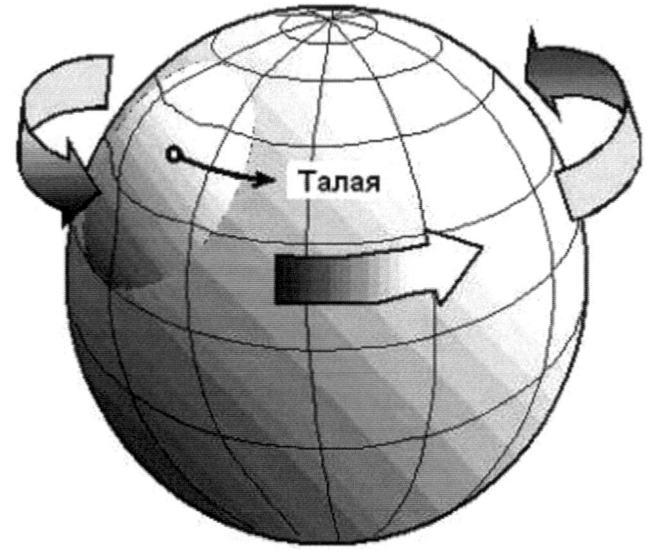


**Figure 1.** Perturbation produced by eccentric motion of core and lithosphere.

Recently a lot of indirect evidences of the fact, that the metallic core of the Earth is shifted relative to the Earth's geometric center by gravitational fields of the Sun, the Space and the Moon or by other non gravitational mechanisms, have been found [2]. Among methods of depth sounding of the Earth's structure one knows techniques based on three main physical principles, i.e. recording of seismic waves (seismic methods), recording of the gravity (gravimetric methods) and recording of a geomagnetic field of the Earth (magnetometer methods) [3]. However, these techniques have a number of disadvantages. For instance, a seismic technique of studying the Earth's subsurface structure is limited by a fast attenuation of a seismic signal when it propagating inside the Earth. Therefore, seismic methods can be used only when there are available seismic waves generated by extremely great (disastrous) earthquakes or by violent explosions similar to a nuclear explosion by yield.

Considering the fact that currently the moratorium on nuclear weapon testing is in force, it is seen impossible to apply the artificial actuation of the Earth's crust to implement our task. Disastrous earthquakes occur on the Earth no more than one per ten years. Therefore, application of a seismic method to the periodic or continuous monitoring of the Earth's core movement is impossible.

Theoretically gravimetric techniques of monitoring allow recording the gravity variations caused by the Earth's core movement because a specific density of the inner solid core is higher than that of the liquid outer core. However in practice the sensitivity of gravimetric methods is limited by gravity variations caused tidal strains of the earth's surface and by relative strains related to variations of the atmospheric pressure at the observation point (about $10^{-8}$). This level of sensitivity is not enough to record processes caused by the core movement.

Due to lack of an exact and proven theory of the Earth's geomagnetic field origin, geomagnetic fields cannot be used to identify the Earth's core position either. In addition to it, there is not any quantitative estimation of the core shift. Theoretical estimations are inconsistent and vary from almost impalpable ones up to values of few hundred meters [2].

IOP Conf. Series: Earth and Environmental Science **48** (2016) 012028 doi:10.1088/1755-1315/48/1/012028

A high sensitivity of the method based on recording the ENPEMF to processes occurring deep inside the Earth is achieved by using natural processes of mechanical and electric transformations of energy inside rock massifs in the rock natural environment. Obviously, the more strains in the Earth's crust, the more intensive the processes of energy transformations are, and the higher the energy and amplitude of an electromagnetic pulses flux coming from the lithosphere are. Thus, a unit that records a natural pulse EM field and is placed at a certain geographical point on the Earth's surface can be considered as an Earth's crust strain gauge at the point of this recording unit location.

A vast rock massif, where the natural pulse EM field of the Earth is recorded, ensures stability of such a natural energy transformer and its high sensitivity to global processes occurring deep inside the Earth. The sensitivity of the ENPEMF method to relative strains of the Earth's crust can reach up to $10^{-11}$ that is three orders higher than the sensitivity of state-of-art gravitational and strain monitoring methods.

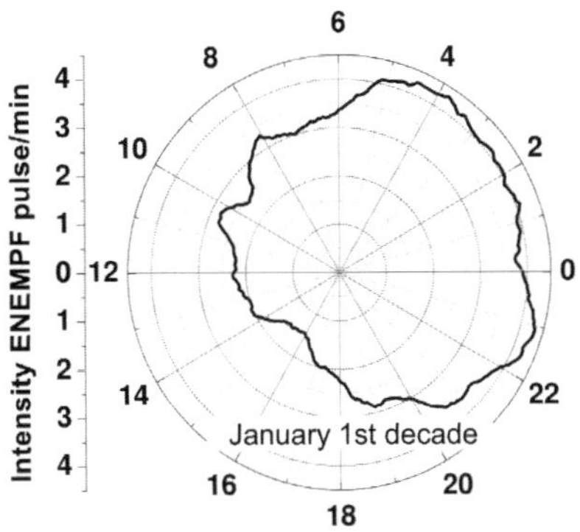


**Figure 2.** Intensity of ENEMPF in polar coordinates.

In view of the fact that a lithospheric component of ENPEMF is modulated by long-period waves from the lower mantle, units recording such fields can be used as instruments to evaluate the strain waves and calculate the path of the Earth's core which has generated such strain waves. A fixed observation station placed in middle latitudes of the globe will make a full turn above the Earth's core which is close to the Earth's center within 24 hours. Consequently diurnal ENPEMF variations plotted in polar coordinates can be considered as a certain pattern of strains (in relative units) generated by the core at the latitude of the observation station. Because it takes a year for the core to make a full turn, the strain pattern will be altered from month to month. Basing on this assumption the location of the core center inside the Earth was computed. Figure 2 illustrates diurnal ENPEMF variations smoothed by 60 minutes with a sliding window method for the first decade of several-year Januaries plotted in polar coordinates. Values of solar local time and EM field amplitudes at this local time are plotted circumferentially and radially, respectively. A "disturbing" source generating strains in the Earth's core is thought to be located at a point that can be calculated as a center of masses of a planar figure bordered by an obtained curve; the center of the globe is located in the zero point with a zero angle and vector. The vector to the center of masses shows a value (expressed in relative units) and the direction of shift of the disturbance source, i.e. the core, relative to the geometric center of the planet. Then, one should take into consideration that during a year the core moves closer to the latitude of the observation station one day but the other day it moves away from it towards the Southern hemisphere. This is evidenced by increase of diurnal means of ENPEMF intensity in the Northern hemisphere in summer time and their considerable decrease in winter time. Correction for latitudinal shift of the core can be made by normalizing a diurnal pattern for each period and making it equal for each season. After having normalized the diurnal pattern, let us determine the location of the masses center for 36 decades of the year like we did it for the first decade of January. To do this, let us plot the obtained values in polar coordinates (Figure 3). In view of the fact that computations are made as per only one observation station located in the Northern hemisphere, Figure 3 shows a planar figure of a yearly movement of the core as seen from the pole.

It is shown that the inner core is never in the geometric center of the planet but makes oscillating movements next to it along a close orbit. The plane of the core orbit is perpendicular to the plane of equator and is inclined at about $45^0$ to the direction to the Sun and to the orbital motion of the Earth around the Sun. An angular velocity of the Earth's core rotation calculated according to frequency characteristics can be an argument in favor of the proposed hypothesis. An average angular velocity of the core rotation is 1.1 degree per year higher than the angular velocity of the Earth rotation. The obtained results of studying the core movement are in line with well-known facts of unstable rotation of the Earth. Estimated values presented in references vary from 0.3 up to 2.8 degrees per year [3, 4].

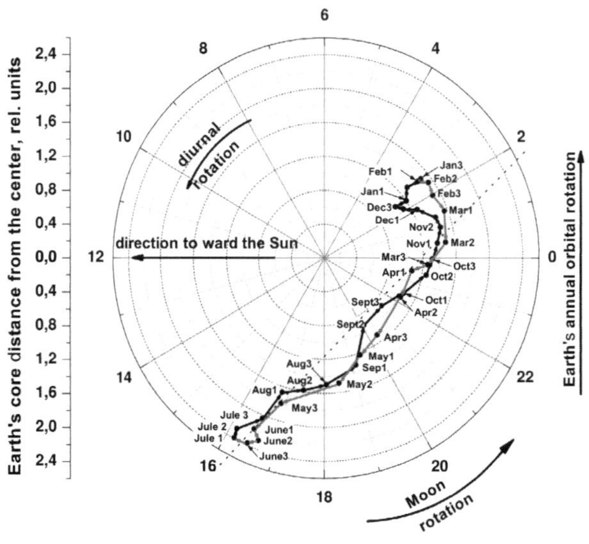


**Figure 3.** Yearly trace of Earth's core.

The work is done with the support from the program of the Presidium of the Russian Academy of Sciences №I.33П "Fundamental issues of mathematic simulation. Fundamental issues of factorization methods in various applications. Algorithms and softwares for computer-based systems of ultra-high capacity."

## References

[1] Malyshkov Yu P and Malyshkov S Yu 2009 *Russian Geology and Geophysics* **50** 115-130
[2] Antonov V A and Kondrat'ev B P 2004 *Bulletin of the Russian Academy of Sciences. Physics* **4** 63-66
[3] Avsuk Yu N, Adushkin V V and Ovchinnikov V M 2001 *Bulletin of the Russian Academy of Sciences. Physics* **8** 64-75
[4] Zhang J, Song X, Li Y, Richards P. G, Sun X and Waldhauser F 2005 *Science* **309** 1357-60

# Monitoring of carbon isotope composition of snow cover for Tomsk region

**P A Akulov[1], Y V Volkov[1,3], D A Kalashnikova[1,3], A N Markelova[1], V N Melkov[1], G V Simonova[1,2] and V A Tartakovskiy[1]**

[1]Institute of Monitoring of Climatic and Ecological Systems SB PAS, Russia, 634055, Tomsk, Akademicheskiy av. 10/3,
[2]National Research Tomsk State University, Russia, 634050, Tomsk, Lenina av. 36
[3]National Research Tomsk Polytechnic University, Russia, 634050, Tomsk, Lenina av. 30

E-mail: galina_simonova@inbox.ru

**Abstract.** This article shows the potential of using $\delta^{13}C$ values of pollutants in snow pack to study the human impact on the environment of Tomsk and its surroundings. We believe that it is possible to use a relation between the isotope compositions of a fuel and black carbon for establishing the origin of the latter. The main object of our investigation was dust accumulated by the snow pack in the winter of 2015-2016. The study of dust samples included the following steps: determination of the total carbon content in snow pack samples of Tomsk and its surroundings, extraction of black carbon from the dust, as well as the determination of $\delta^{13}C$ values of the total and black carbon accumulated in the snow pack. A snow survey was carried out on the 26th of January and on the 18th of March. The relative carbon content in the dust samples was determined by using an EA Flash 2000 element analyzer. It varied from 3 to 24%. The maximum carbon content was in the dust samples from areas of cottage building with individual heating systems. The $\delta^{13}C$ values of the total and black carbon were determined by using a DELTA V Advantage isotope mass spectrometer (TomTsKP SB RAS). The isotope composition of black carbon corresponded to that of the original fuel. This fact allowed identifying the origin of black carbon in some areas of Tomsk.

## 1. Introduction

The snow cover is a unique environment which deposits contaminants emitted into the atmosphere and then fed into the soil and surface water. The study of the chemical composition of the snow cover allows for a rapid geo-ecological assessment of urbanized areas. Observations of the snow cover are made by regional centers for monitoring environmental pollution of the Russian Hydro-Meteorological Centre, as well as by units of stationary networks for monitoring precipitation and background biosphere reserves. Control is performed mainly over 32 impurities [1]. This list does not include «black» carbon which is the main component of soot, i.e. a product of incomplete combustion of hydrocarbon fuels and biomasses. According to numerous recent studies [2], [3], it is «black» carbon that has a significant impact on public health and the climate through the process of changing the albedo of snow. «Black» carbon is released into the atmosphere as a result of forest fires, vehicle operation, activities of industrial enterprises and housing and communal services. One of the problems associated with the assessment of «black» carbon emissions is to determine the sources of its origin prevailing over a particular territory. We believe that, in order to solve this problem, it is possible to use the relation between the isotopic composition of a fuel and that of «black» carbon.

## 2. Materials and methods

The selection of snow samples was carried out on January 26 and in the period of the maximum snow accumulation on March 18, 2016. The samples were taken at 17 points (Fig. 1), six of them being background and located outside the city, and the rest being located in the areas with varying degrees of the anthropogenic load.

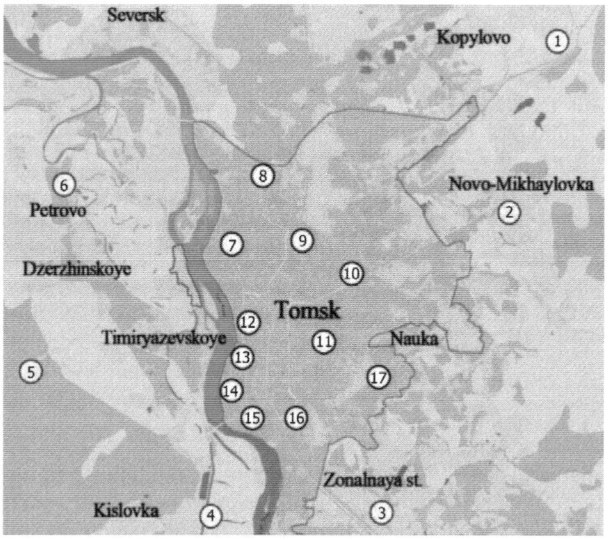


**Figure 1.** Location of points for selection of snow cover samples. ⊗ - background areas; O - areas with different degrees of anthropogenic load.

The process of sampling was carried out according to the procedure presented in [1] within areas of the size 5x5m using the envelope method. Snow cores were selected over the whole snow depth with the help of a plastic pipe with a diameter of 10cm and then packed in double sealed plastic bags. Melting of snow occurred under laboratory conditions in the packaged form. During the sampling, at each point, the average height of the snow cover was determined. The resulting melted water was filtered through the paper filter "blue ribbon". The weight of the dust deposited on the filters was determined as the difference in the weights of the clean pre-dried filter and the filter with the settling.

The dust load $P_n$ on the snow cover was calculated by the formula [4]:

$$P_n = \frac{P_0}{S \cdot t} \quad (1)$$

where $P_o$ is the dust mass in the snow sample (mg); $S$ is the area of the cross section of the pit or the sampler which was used for sampling (m$^2$); $t$ is the starting time point of the snow cover formation prior to the day of sampling. Calculations were performed for two periods: from the date of completed snow cover formation to 01.26.2016 and from 01.27.2016 to 03.18.2016. The difference between the mass of dust samples (accumulated since the formation of snow cover until 18.3.2016 and 26.1.2016), was used as the mass of dust for estimating dust load on the snow cover of the second period.

In the analysis and interpretation of the research results we used the average daily temperature and the daily precipitation calculated according to the monitoring data of the Hydro-Meteorological

Observatory of the Institute of Monitoring of Climatic and Ecological Systems SB RAS. The starting date of the completed snow cover formation was accepted as November 1, 2015.

In the dust samples we determined the relative carbon content (%) and the ratio of its stable isotopes $^{13}C/^{12}C$ in the form:

$$\delta^{13}C = [(R_s - R_{st}) / R_{st}] \cdot 1000\,ppm \tag{2}$$

where $R_s$. is the ratio of $^{13}C/^{12}C$ in the sample and $R_{st}$. is the ratio of $^{13}C/^{12}C$ in the standard, i.e. carbon of the *Belemnitella americana* fossil clam (referred to as PDB) [5].

It appeared possible to separate pyrogenic «black» carbon from some of the dust samples. For the purpose of this, we used the method of samples preparation described in [6]. The dust samples were rectified of carbonates, silicon compounds, organic substances and kerogens by a sequential treatment with three-mole hydrochloric HCl acid (3M), a mixture of hydrofluoric acid and hydrochloric acid HF (10M)/HCl(1M) and oxidizing solution containing 0,1M $K_2Cr_2O_7$ solution in 2M $H_2SO_4$ at 55°C during 60 hours. For the samples containing 500 micrograms of "black" carbon or more, we determined the ratio $\delta^{13}C$.

The relative content and the isotopic composition of carbon in the dust samples were determined using the elemental analyzer EA Flash 2000 and the isotopic mass spectrometer DELTA V Advantage, respectively (Tomsk Centre for Collective Use SB RAS).

## 3. Results and discussion

According to the observations of the Hydro-Meteorological Observatory of the Institute of Monitoring of Climatic and Ecological Systems SB RAS (HMO IMCES SB RAS), the permanent snow cover was formed by November 1, 2015. The greatest part of winter precipitation equal to 106.4 mm occurred in November and December. In these months, the increase in thickness of the snow cover showed the maximum rate (Fig. 2a, b). In January, the amount of precipitation was minimal for the current winter (only 5.8mm), precipitation being observed during 5 days. The increase in thickness of the snow cover was minimal. At the time of the first snow survey the average snow depth was 39.2cm. Within the study area, this value changed in the range of 27.6 – 47.4cm (Table 1). In February and March there appeared about 42mm of precipitation and the snow depth increased by an average of 5.6cm. At the time of the snow surveys on 03.18.16 the average height value was 44.9cm, the territory of the snow depth ranging from 31.6 to 63.6cm.

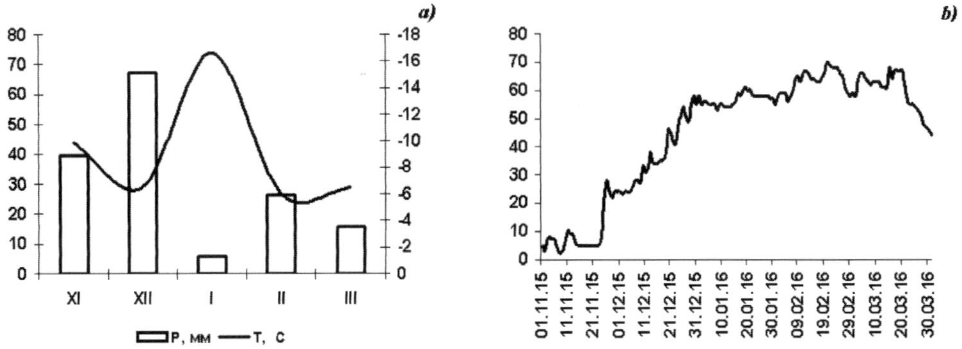


**Figure 2**. Monthly average temperatures and precipitation from November 2015 to March 2016 (a) and daily values of snow depth (b) according to HMO IMCES SB RAS.

The dust load on the snow cover corresponds to the presumed distribution of the anthropogenic load within the study area. Within the city, the dust load on the snow cover is significantly higher. In the text bellow the values of dust load are given for two investigated periods accordingly. The minimum values were observed within the background areas. There are 1.3 – 12.6 mg/(m$^2$•d) and 4.3 – 19.2 mg/(m$^2$•d). The maximum values correspond to the areas located along the roads with heavy

IOP Conf. Series: Earth and Environmental Science **48** (2016) 012029 doi:10.1088/1755-1315/48/1/012029

traffic, including cargo transport, and in the areas of industrial facilities concentration (P8, P9, P10, P14): 114 – 1055 mg/($m^2$•d) and 40.9 – 2374 mg/($m^2$•d). An abnormally high value of the dust load in the P4 area in the second period is, apparently, due to the occurrence of temporary point source of pollution. The dust load on the areas located along busy roads mainly for cars and passenger vehicles (P11, P13, P15, P16) was 24.5 – 68.3 mg/($m^2$•d) and 6.7 – 36 mg/($m^2$•d) and in the areas of cottage building with individual heating systems (P7, P12, P17) – 14.7 – 30.4 mg/($m^2$•d) and 13.5 – 137 mg/($m^2$•d)

**Table 1.** Measured properties of snow cover.

| Point of sampling | Height of snow cover (cm) | | Dust load ($mg•m^{-2}•d^{-1}$) | | Total carbon content (%) | | $\delta^{13}C_{tot}$ [a] (‰) | | $\delta^{13}C_{black\ carbon}$ [a] (‰) | |
|---|---|---|---|---|---|---|---|---|---|---|
| | 26.01.16 | 18.03.16 | 26.01.16 | 18.03.16 | 26.01.16 | 18.03.16 | 26.01.16 | 18.03.16 | 26.01.16 | 18.03.16 |
| P1 | 40.6 | 51.4 | 12.6 | 19.2 | 12 | 13 | -28.60 | -27.29 | | |
| P2 | 33.6 | 36.2 | 40.5 | 15.48 | | 6 | | -28.22 | -27.78 | |
| P3 | 37.0 | 51.0 | 9.03 | 18.8 | 7 | 9 | -26.87 | -27.26 | | |
| P4 | 27.6 | 31.6 | 2.6 | 91.8 | | 11 | | -29.13 | | |
| P5 | 53.0 | 63.6 | 1.3 | 15.9 | | 35 | | -28.32 | | |
| P6 | 31.8 | 32.4 | 11 | 4.33 | 4 | 16 | -26.67 | -17.47 | | |
| P7 | 31.2 | 32.4 | 30.4 | 13.5 | 24 | 22 | -26.30 | -27.88 | -24.69 | |
| P8 | 45.2 | 38.2 | 1055 | 2374 | 11 | 2.4 | -21.90 | -21.25 | -30.27 | -29.77 |
| P9 | 38.2 | 48.2 | 114 | 40.9 | 5 | 7.9 | -26.73 | -28.80 | -27.83 | -28.36 |
| P10 | 38.3 | 39.8 | 232 | 344 | 3 | 2.8 | -26.90 | -27.22 | -28.29 | -29.52 |
| P11 | 45.8 | 56.8 | 24.5 | 22.6 | 15 | 14 | -25.40 | -26.14 | | |
| P12 | 41.2 | 49.6 | 19.9 | 49 | 21 | 26 | -24.40 | -28.79 | | |
| P13 | 39.4 | 36.8 | 39.2 | 36 | 16 | 8.5 | -25.50 | -28.59 | | |
| P14 | 39.4 | 36.6 | 51.3 | 109 | 12 | 4 | -26.00 | -29.33 | | |
| P15 | 39.4 | 49.8 | 68.3 | - | 13 | 16 | -27.10 | -17.46 | | |
| P16 | 38.0 | 47.7 | 42.1 | 6.73 | 11 | 10 | -28.80 | -27.26 | | |
| P17 | 47.7 | 61.0 | 14.7 | 134.6 | 18 | 5 | -26.80 | -26.09 | | |

[a] The measurement error of three replicates did not exceed ± 0.15 ‰

Pollution of snow cover occurs as a result of capture and deposition of aerosols and fine dust particles by precipitation and direct deposition of pollutants from the atmosphere by gravity [1]. The ratio of the contributions made by these two mechanisms in the general contamination depends on many factors such as: the duration of the winter period, the frequency and intensity of precipitation, the size of aerosol or dust particles, the distance from the examined areas to local pollution sources and other. In the first period, the dust washout with precipitation was minimal, which was, apparently, reflected in the calculation of the dust load for two periods: 01.11.15 - 26.01.16 and 27.01.16 - 18.03.16. Within the second period the load was in most cases several times higher (Table 1).

Considering the carbon content in the dust samples obtained from the snow, we can roughly assess the contribution made by carbonaceous combustion products in the total pollution of the snow cover. In the obtained samples the relative content of carbon is small, varying between 3-24% in the samples from 26.01.16 and 6–86% in the samples from 18.03.16. The highest values were observed in the areas of cottage building for the first period 01.26.16 - 18-24% and for the second period 18.03.16 - 26-86%. In the dust samples carbon is present in the carbonates, organic substances (humic acids and plant materials) and in the form of soot. Under laboratory conditions, we determined a connection between carbon isotope relationships of different fuels and their combustion products (Table. 2) [7].

**Table 2.** Isotopic composition of $\delta^{13}$C wood, peat, coal, kerosene and soot produced after combustion.

| | $\delta^{13}$C (‰) | | | |
|---|---|---|---|---|
| Sample | *Pinus Sylvestris* | *Betula* | Coal | Kerosene |
| Original sample | -27.33±0.1 | -26.62±0.05 | -24.05±0.08 | -31.74±0.07 |
| Charcoal | -28.65±0.09 | -28.72±0.12 | – | – |
| Soot | -29.27±0.06 | -25.12±0.1 | -24.20±0.05 | -30.02±0.06 |

Considering the isotopic composition of the total carbon contained in the dust samples obtained from the snow, it is difficult to make a conclusion about a particular source of carbon origin (Tables 1 and 2). However, the isotopic composition of «black» carbon can be used to assess the original substance - fuel. Particularly, if within the investigated area «black» carbon is formed primarily as a result of combustion of one certain fuel. For example, the value $\delta^{13}$C=–24.7‰ of «black» carbon in the sample from the site located within a cottage area P7 corresponds to the value of coal soot. The isotopic composition of the P2 and P9 samples is the closest to the isotopic composition of soot resulting from burning wood. In case of furnace heating, the most commonly used wood is birch, pine being used less frequently. The isotopic composition of «black» carbon from the sample P8 (a road with heavy traffic of cars and trucks, located in the vicinity of a tank farm) corresponds to the isotopic composition of soot resulting from combustion of petroleum products.

## 4. Conclusions

The dust samples obtained from the snow cover formed in the winter 2015-2016 were investigated. As it was expected, the dust load was significantly higher in the city than outside. The maximum values of it were observed within the areas of industrial facilities concentration and in the areas located along the roads with heavy traffic, including cargo transport. However, the total carbon content was the highest in the dust samples from the areas of cottage building with individual heating systems. It was able to separate the sufficient amount of black carbon only from several dust samples. It turned out that in some areas of the city the carbon isotopic composition corresponds to supposed source of its origin. In the areas of cottage buildings with individual heating systems, the isotopic composition of black carbon corresponds to the coal and wood. In the vicinity of rural settlements the isotopic composition of black carbon corresponds to the wood. In the north of the city along the highway with heavy truck traffic near the oil depot the isotopic composition of black carbon corresponds to the fuel, obtained in the processing of petroleum products.

## 5. Acknowledgements

The research was supported by the Russian Foundation for Basic Research (research project no.16-45-700941).

## References

[1] Zarina L M and Gildin S M 2011 *Geo-ecological workshop: Study guide* (St. Petersburg: Herzen State Pedagogical University of Russia) p 60
[2] Mouteva G O, Czimczik C I and Fahrni O E 2015 *Global Biogeochemical Cycles* **29** 1977–2000
[3] Saiz G, Wynn J G, Wurster C M, Goodrick I, Nelson P N and Bird M I 2015 *Biogeosciences* **12** 1849–63
[4] Letuvninkas A I 2002 *Anthropogenic geochemical anomalies and natural environment* (Tomsk: Science and Research Publishing) p 290
[5] Lebedev A T 2013 *Mass spectrometry for analyzing environmental objects* (Moscow: «Tekhnosfera») p 632
[6] Liu L, Song Y, Cui L and Hao Z 2013 *Palaeogeogr. Palaeoclim. Palaeoecol* **388** 109–14

[7] Kalashnikova D A, Markelova A N, Melkov V N and Simonova G V 2016 *Chemistry for Sustainable Development* **24** 467-71

# Development of climate data storage and processing model

**I G Okladnikov[1,2,3], E P Gordov[1,2,3,4] and A G Titov[1,2,3]**

[1]Instutite of Monitoring of Climatic and Ecological Systems of the Siberian Branch of the Russian Academy of Sciences, 10/3, Akademichesky pr., Tomsk 634055, Russia
[2]Institute of Computational Technologies of the Siberian Branch of the Russian Academy of Sciences, Tomsk Branch, 10/4, Akademichesky pr., Tomsk, Russia
[3]Instutite of Atmospheric Optics of the Siberian Branch of the Russian Academy of Sciences, 1, V.E. Zuev sq., Tomsk 634055, Russia
[4]Tomsk State University, 36, Lenina str., Tomsk 634050, Russia

**Abstract.** We present a storage and processing model for climate datasets elaborated in the framework of a virtual research environment (VRE) for climate and environmental monitoring and analysis of the impact of climate change on the socio-economic processes on local and regional scales. The model is based on a «shared nothing» distributed computing architecture and assumes using a computing network where each computing node is independent and self-sufficient. Each node holds a dedicated software for the processing and visualization of geospatial data providing programming interfaces to communicate with the other nodes. The nodes are interconnected by a local network or the Internet and exchange data and control instructions via SSH connections and web services. Geospatial data is represented by collections of netCDF files stored in a hierarchy of directories in the framework of a file system. To speed up data reading and processing, three approaches are proposed: a pre-calculation of intermediate products, a distribution of data across multiple storage systems (with or without redundancy), and caching and reuse of the previously obtained products. For a fast search and retrieval of the required data, according to the data storage and processing model, a metadata database is developed. It contains descriptions of the space-time features of the datasets available for processing, their locations, as well as descriptions and run options of the software components for data analysis and visualization. The model and the metadata database together will provide a reliable technological basis for development of a high-performance virtual research environment for climatic and environmental monitoring.

## Introduction

The present state of the art of modern environment monitoring tools and Earth system models is characterized by a continuous growth of the produced geospatial data volume. These data are used for monitoring and analysis of the climate change impact on the environment [1]. The data produced by various scientific and commercial institutions within the framework of hundreds of scientific projects usually have significant differences in the sets of physical variables, file formats, as well as different syntax and semantics, hampering the use of a common terminology. Due to the increasing volume of modeling and observational (especially, satellite) data, reaching several petabytes for a single dataset, there arise substantial computational limitations on their processing. Researchers regularly encounter those while working with large geospatial datasets in such scientific fields as meteorology, physics [2], and in biological and ecological studies [3]. The significant increase in the variety, volume, and speed of generation of meteorological and climatic data makes it impossible to collect and process them on a single computational node utilizing the traditional approaches [4]. A solution of the problems outlined is using a dedicated hardware and software infrastructure based on some new approaches to storing, retrieving, and processing of climatic geospatial data.

We present a model of storing and processing large climatic datasets, elaborated in the framework of development of a virtual research environment for the climate and environmental monitoring and analysis of the impact of climate change on the socio-economic processes on local and regional scales. This environment integrates known and new sets of climatic data, software implementations of classic and new methods of statistical analysis of large data sets. It provides an opportunity to scientists and decision-makers to use different geographically distributed spatially referenced data, processing resources, and services through a web browser by integrating distributed systems that store, process and provide information via a Geoportal.

## 1. Climate datasets storage and processing model

### *1.1. General concept*

Our approach to storing and processing large spatial data is based on a distributed computing architecture called "shared nothing" [5]. In this approach, each computing (or working) node is independent and self-sufficient, without shared memory and storage space. The advantages of this architecture are: high reliability of computer network (a node failure does not affect the whole network functioning), high data processing speed (multi-node parallel data processing), and simple network scalability (the new nodes are easily integrated into the existing infrastructure). Each working node holds a dedicated software for processing and visualizing spatial data ("computing core"), which provides an application programming interface (API) for the interconnecting working nodes. The nodes are linked by data networks. They exchange data and control instructions via a secure connection or web processing service (WPS) (see Figure 1). The data processing tasks ("jobs") are distributed among the computing cores by a special task manager ("Job Manager"), located on the master node. The common functions of the task manager are:

- to get data processing and visualization tasks from clients;
- to take into account the locations of datasets to be processed ("data awareness") and to divide the jobs into individual tasks for the computing cores;
- to pass individual tasks to the working nodes corresponding to the datasets to be processed;
- to collect the individual results of processing from the working nodes; and
- to aggregate the individual results into a final result and provide it to the clients.

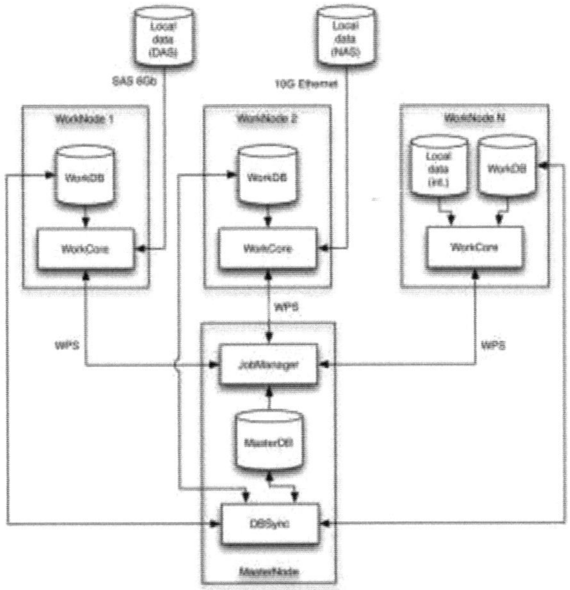


**Figure 1.** Computing network architecture.

### *1.2. Data*

Currently, two major approaches to storing geospatial data are used: geospatial databases and file collections. The database approach utilizes relational and non-relational, local or distributed spatial databases such as Apache HBase (https://hbase.apache.org/), Esri Geodatabase (http://www.esri.com/products/arcgis-capabilities/data-management), Paradigm4 SciDB (http://www.paradigm4.com/), etc. This approach requires inserting all data into a spatial database before their actual use, which is a quite disk space- and time-consuming operation. The file collections approach relies on storing data as a collection of data files in file system directories. Usually, self-describing formats are used for storing geospatial data. It was shown [6] that the retrieval of data chunks larger than 40 Mb from a spatial database is less effective than from a simple collection of data files. On the other hand, in most cases the file-based approach provides fast data extraction without redundant preprocessing. However, this approach requires the development of an additional software layer to provide API adapters for storing and processing distributed file collections. In our case, the simplicity and flexibility of the file-based data storage approach prevailed over the other one.

Network Common Data Form (netCDF) was chosen as the major file format for most geospatial data in our data archive. This format is formally acknowledged by scientific institutions (including UCAR - University Corporation for Atmospheric Research) and OGC (Open Geospatial Consortium) as a standards' candidate for storing geospatial data and stimulating data exchange in binary form. It is perfectly suited for storing geospatial data and supported by a wide range of commercial and open source software.

The data are stored on data storage systems as collections of netCDF files and arranged in a strict hierarchy of directories:

- /<data root directory>/
- <data collection name>/
- <spatial domain resolution>/
- <time domain resolution>/
- <files and directories with data>

Here, *<data root directory>* is the root location of data collections, *<data collection name>* is the name of a directory containing a single data collection, *<spatial domain resolution>* is the name of a directory containing data with the same horizontal resolution, *<time domain resolution>* is the name of a directory containing data with the same time step. All data files (sometimes grouped in subdirectories) are located deeper in the hierarchy. The names of files and subdirectories are not regulated and determined by the specifics of a particular data set. Each data file contains one or more multi-dimensional arrays of meteorological parameters.

The data are stored on working nodes or on connected data storage systems (Local data) according to a “data awareness” principle. Data awareness is some knowledge about all of the data available within each working node and where those data are located [7]. According to this information, the datasets are processed locally, on the nodes containing them. The metadata describing geospatial datasets are stored in a distributed database on the master node (MasterDB) and on working nodes (WorkDB). The metadata include a description of the spatio-temporal characteristics of each dataset, a list of meteorological parameters and physical locations of datasets on data storage systems. The metadata regularly or on demand are synchronized (DBSync) among all nodes in order to maintain up-to-date information about datasets available for processing and their locations.

To speedup the data reading and processing, three approaches are used: a preliminary calculation and storage of the frequently used intermediate products, a distribution of the data across multiple storage systems (with or without redundancy), and caching and reuse of the previously obtained results. The first approach involves storing not only the geospatial datasets, but also the pre-calculated intermediate products obtained on their basis. First of all, these products include the results of spatial and/or temporal averaging: data with coarser spatial resolution, and averages per day, month, and year. In addition, some independent preliminary intermediate products, whose calculation takes a considerable time, can be obtained for the future use. The intermediate datasets will also be available as netCDF files, and their metadata will be stored in the database together with the information on the original datasets. The second approach involves a distribution of the datasets among the working nodes (possibly with redundancy), thereby providing increased performance due to parallel data reading and processing on multiple nodes. The third approach lies in a short-term storage of the intermediate and a long-term storage of the final results obtained by the users with automatic inclusion of their metadata into the database. When a task for the calculation of a cached dataset is passed to a working node, this result is not calculated again but extracted from the cache.

### *1.3. Metadata database*

To describe the geospatial datasets and their processing routines and provide effective VRE functioning, a dedicated metadata database (MDDB) is required. Currently, there is a lack of such database in the Earth sciences. We suggest a comprehensive description of the climatic geospatial datasets and processing routines in a single database, which is characterized by the following features (Figure 2). The database contains structured spatial and temporal characteristics of the available geospatial datasets, their locations, and run options of the software components for data analysis. The structure contains "Dataset" and “Dataset collection” levels. Dataset is a set of data which is a) given on a single temporal and spatial grid, b) covers the same time range and c) obtained under the same simulation or observation conditions (if applicable). Spatial datasets according to the chosen data storage model are mostly represented by collections of netCDF files grouped by the spatio-temporal features and placed in the hierarchy of directories on the data storage systems. Each netCDF file includes one or more variables containing values of meteorological parameters given on a spatio-temporal domain. The file names in the dataset are unified. It is necessary to distinguish the term "parameter" (meteorological parameter) and "variable". A meteorological parameter is the name of some meteorological characteristic: temperature, pressure, humidity, etc. A variable is the unique name of a multidimensional array in a netCDF file. The names of meteorological parameters are standardized. In contrast, the names of variables in different datasets may be different and usually depend on the preferences of the institution that produced them. Along with data, the netCDF files contain horizontal, vertical, and time domain grids. A dataset collection is a collection of datasets created by an organization within a specific project, but specified on different spatial and/or temporal grids, or for different scenarios. A collection may consist of one dataset. MDDB contains two major parts: 1) a description of climate datasets; and 2) a description of data processing software components (computing modules).

*1.3.1. Description of climate datasets.* The climate datasets available for processing are described by metadata stored in several interconnected tables. Each dataset is represented by a record in the table dataset which links together a unique set of dataset features. The dataset features are given in the following tables: collection, dataset_root_directory, scenario, resolution, time_step, as well as time_span, filetype. The table collection contains short names of data collections (e.g., "ERA40", "MERRA") and URLs of the corresponding internet sites with a detailed description. The related table collection_tr contains multi-lingual information about the collections: long names (e.g., "ERA-40 Reanalysis", "MERRA Reanalysis"), short descriptions (e.g.: "ECMWF 40 Year Re-analysis", " NASA/GMAO Modern-Era Retrospective Analysis for Research and Applications"), and names of organizations of origin (e.g., "ECMWF, UK", "NASA/GMAO, USA"). Table dataset_root_directory contains root paths to datasets. Usually these paths point to dedicated directories on data storage systems relative to the file system root directory on a computing node. For example, "/mnt/storage1/data" or "/nfs/megaraid1/data". The table scenario contains names of data acquisition conditions or modeling scenarios and the first part of a path to the data files relative to the dataset root directory. For example, climate projection modeling scenarios, such as "RCP45" or "RCP85". The table resolution contains a list of horizontal spatial resolutions and the second part of the path to data files. The resolutions can be given directly, in some units: "0.67x9.67 deg.", "30x30 m.", or by using topical notation: "T62", "T106". The table time_step contains a list of short notations of temporal grid steps and the third part of the path to data files. Typical notations of temporal grid steps for meteorological data are: "3h" (3 hours), "6h" (6 hours), "1d" (1 day), "1m" (1 month), "1y" (1 year). Related table time_step_tr contains multi-lingual full names of temporal grid steps (e.g., "6 hours"). Being concatenated, all the path fragments corresponding to the selected collection, scenario, horizontal resolution and temporal grid resolution form a full path to data files: / <dataset_root_directory.rootpath> / <scenario.subpath0> / <resolution.subpath1> / <time_step.subpath2> /. The table time_span contains short and long names of a temporal span covered by one NetCDF file, i.e., length of a time grid in one file. For example, "6h" and "6-hours span", "1d" and "1-day span", "1m" and "1-month span". In the latter case it means that one data file holds data obtained for a month-long period. Thus, data for one year will take 12 files, and for 50 years 600 files. Usually, a time-stamp is included in the data file names. The table filetype contains types of data files in the datasets. For example, such well-known types as "netcdf", "grid", or "shapefile".

Each meteorological parameter in a dataset given on the whole spatio-temporal domain of this dataset and one vertical level, and stored in a single variable with the same name in all files of the dataset is treated as a "distributed multidimensional data array". Each data array is represented by a record in the table *data* which links dataset (*data*.dataset_id), variable (*data*.variable_id), vertical level (*data*.levels_id), as well as data file name template (*data*.file_id) and a variable containing a list of vertical levels (*data*.level_variable_id). The table *variable* contains names of variables storing various meteorological parameters (e.g., "air"). It is linked with a table *parameter* containing IDs and a table *parameter_tr* containing multi-lingual standard names of meteoparameters (e.g., "Air temperature"). Also it is linked with a table *units* containing IDs and a table *units_tr* containing multi-lingual names of meteoparameter units (e.g., "K" – Kelvin degrees). The table *levels* contains lists of vertical levels at which meteoparameters are given (e.g., ":300:400:500:600:700:850:925:1000:" or ":msl (Mean sea level):"). Related table *levels_tr* contains multi-lingual names of corresponding vertical levels (e.g., "Mean sea level"). Also, the table *levels* is linked with a table *level_type* containing types of vertical levels. The table *level_type* is linked with a table *level_type_tr* containing corresponding multi-lingual names of types of vertical levels (e.g., "Isobaric level") and tables *units*/*units_tr* containing multi-lingual names of units (e.g., "hPa"). The table *file* contains data file name templates and time ranges covered by a corresponding collection of data files. For example, a template "%year%/macc_tmp_%year%%mm%.nc" uses two placeholders: "%year%" corresponding to a four-digit year and "%mm%" corresponding to a two-digit month number (e.g., 1990/macc_tmp_199005.nc). The data file name template can include many different placeholders or not include them at all. On the basis of the time range required and a file name template the dedicated data-access software modules prepare a list of names of data files. Then required variables are read from these files. The table *level_variable* contains names of auxiliary variables storing lists of vertical levels in NetCDF-files (e.g., "levels") and is used by service software modules for searching and reading data arrays from files.

The first problem which is solved with the help of the MDDB is the determination of the full path to data files and NetCDF file names containing the required meteorological parameter from a selected climate dataset and given on a selected spatio-temporal domain. Besides, there are interrelated tables in the MDDB containing multi-lingual descriptions of collections (*collection_tr*), resolutions of horizontal (*resolution_tr*) and temporal (*time_step_tr*) grids, vertical levels (*levels_tr*, *level_type_tr*), meteoparameters (*parameter_tr*) and units (*units_tr*). Together they describe all datasets in a human-readable form and are used in drop-down menu of the graphical user interface. The second problem which is solved with the help of the MDDB is providing a graphical user interface with relevant interrelated information describing the spatial datasets available for analysis. The advantage of using the MDDB is a rapid actualization and delivery to the user the information concerning available datasets just after their integration into the VRE.

*1.3.2. Description of data processing software components (software modules) and their run options.* The dedicated software procedures ("computing modules" or "processing modules") are used for climate datasets processing. "Module" is a functionally finished fragment of a computing core implementing a number of methods and designed to solve one elementary task. Each module has at least one input dataset (argument) and at least one output dataset (result). To control the module functionality, some run options can be passed to it as a set of different constants. Any module's result can be an argument for another one. When several modules are run consequently and results are passed from the output of one module to the input of another one a "computing pipeline" is formed. Run and execution control of the computing pipeline is performed by the computing core of the VRE. Some module options can be set or selected from a list of valid values using a graphical user interface (GUI). The following option types are valid: 1) Boolean; 2) string; 3) number. Each option type has a corresponding GUI element: 1) Boolean – checkbox; 2) string – dropdown list; 3) number – textbox. A set of certain options determines the only way of data processing by a particular computing pipeline. That way of processing is called hereinafter "a process". The processes based on the same computing pipeline differing only in run options are combined in a "class of processes". A process is described in an XML "task-file". Only one task-file corresponds to each process. If a task-file describes a processing pipeline, intermediate datasets and particular run options, while the descriptions of datasets to be processed and spatio-temporal domain are omitted it becomes a "task-template". The omitted descriptions and options are inserted into a task-template after the user made the corresponding selections in the GUI. The prepared task-file is passed to the computing core for further processing.

Each class of processes is associated with a class type which is used to establish a correspondence between the data arrays and the applicable processes. The table *class_type* contains the names and descriptions of class types. It is linked to the table *handled_data* where correspondence between data arrays and class types are set. The table *process_class* contains IDs of classes of processes, and *process_class_tr* , their multi-lingual names. A class of processes combines the processes which differ only in run options. The table *option* contains the IDs of all available options of all processes represented in the MDDB, and related table *option_tr* , their multi-lingual names. The options of one type related to the same process are combined into groups. The table *options_group* contains the IDs of groups of options, and a related table *options_group_tr*, their multi-language names which can be used as captions of GUI elements. Another related table, *group_type*, contains types of groups of options. The type of a group defines the type of the GUI element used for displaying the options in the group. For example, "checkbox" corresponds to a check box element. The options and groups are linked in the table *grouped_options*. The groups of options and classes of processes are linked in the table *class_options*. Thus, all possible run options linked to a class of processes define all possible processes of this class. Selection of a class of processes and a certain set of run options by the user defines a specific process. All valid combinations of run options for each class that the user can select are specified in the table *selected_options*. This table is linked to the table *options_set* containing IDs of valid combinations. Each process is represented by a record in the table *process* linking a process class (*process*.process_class_id) and a combination of run options (*process*.options_set_id). The related table *task_template*_files contains the file names of template-tasks corresponding to each process. These template-tasks are used by VRE's middleware to prepare file-tasks for the computing core.

The third problem which is solved with the help of the MDDB is the determination of file names of template-tasks using processing options selected by the user in the graphical user interface. Besides, there are interrelated tables in the MDDB containing multi-lingual names of classes of processes (*process_class_tr*), names of options of processes (*option_tr*), and captions of the GUI elements used for displaying these options (*options_group_tr*). Together they describe spatial data processing options in a human-readable form and are used as data for filling the elements of the graphical user interface. The fourth problem which is solved with the help of the MDDB is providing a graphical user interface with relevant interrelated information describing the available modules for spatial data processing. The advantage of using the MDDB is rapid actualization and delivery to the user of the information concerning the available data processing modules just after their integration into the VRE.

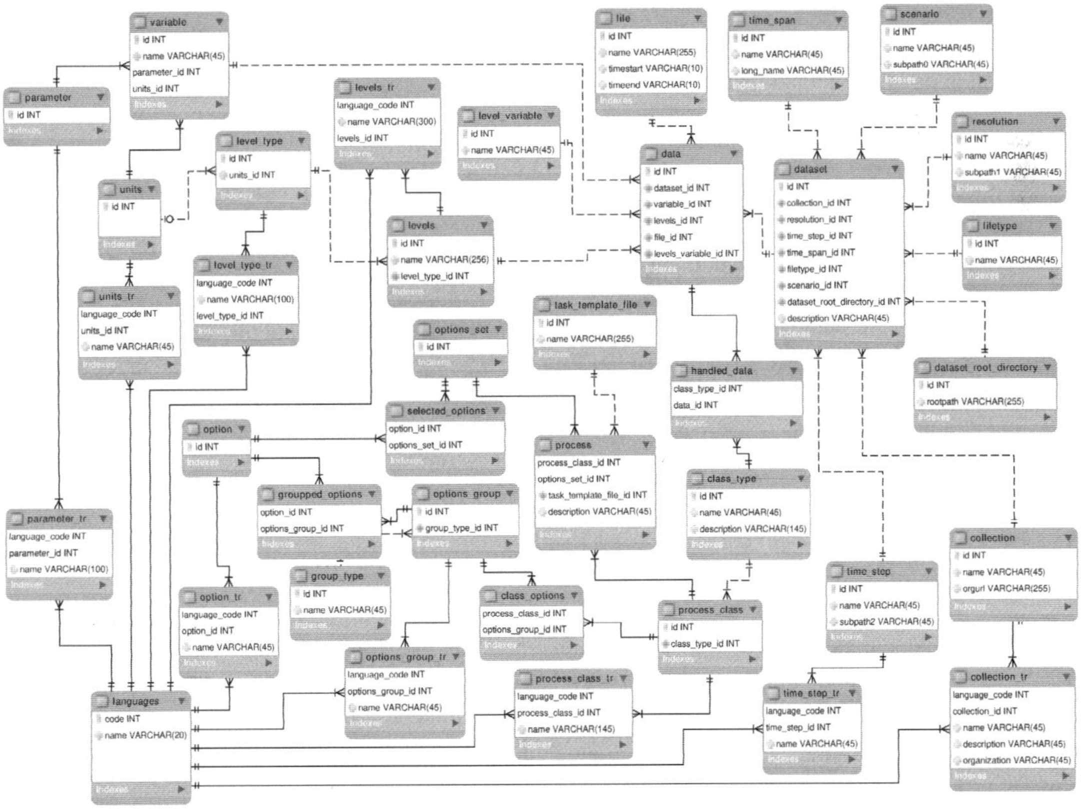

**Figure 2.** ER-diagram of metadata database describing features of available spatial datasets, data processing computing modules and their inter-relations.

## 2. Conclusions

The above-presented climate data storage and processing model is designed for a virtual research environment of climatic and environmental monitoring and analysis of the impact of climate change on the socio-economic processes on local and regional scales. It will provide a computing network with reliability and scalability, as well as high computational performance, due to the "shared nothing" basis, netCDF file collections approach, and the use of a metadata database. To date, there is no formalized description of the metadata database for large sets of geospatial meteorological and climatic data in the literature. The database presented in this study is the first attempt to address this crucial data intensive domain problem. The developed metadata database has several advantages. It provides data awareness for the working nodes by minimizing the inter-node data transfer and increasing the processing performance. Also, it facilitates an easy and fast expansion of the existing and new datasets and data processing computing modules into the VRE. This reduces the time required for extending the VRE's functionality. Finally, it unifies the data processing routines resulting in a stable and solid program code.

The first application of the metadata database demonstrated high efficiency of its use [8]. It provided easy and fast addition of the new data processing computing modules into the information and computation platform "Climate" [9]. It shows that the climate data storage and processing model and the metadata database together will provide a reliable technological basis for the development of a high-performance virtual research environment for climatic and environmental monitoring and analysis of the impact of climate change on the socio-economic processes on local and regional scales.

**Acknowledgements**
The authors thank the Russian Science Foundation for support of this work under grant №16-19-10257.

**References**

[1] Lykosov V N, Glazunov A V, Kulyamin D V, Mortikov E V and Stepanenko V M 2012 *Supercomputer Simulations in the Climate System Physics* (Moscow: Moscow State Universoty publishing) p 295

[2] Sandia sees data management challenges spiral *Scientific computing world* 4 August 2009 http://www.scientific-computing.com/news/news_story.php?news_id=922.

[3] Reichman O J, Jones M B, Schildhauer M P Challenges and Opportunities of Open Data in Ecology 2011 *Science* **331** 6018 703–5

[4] Gordov E P, Kabanov M V, Lykosov V N 2006 Information-computational technologies for environmental science: young scientists training *J. Computational Technologies* (Special issue 1) **11** 3–15

[5] Valduriez P 2009 Shared-Nothing Architecture *Encyclopedia of Database Systems* ed L Liu and M Tamer Özsu pp 2638–9

[6] Santokhee A, Blower J, Haines K 2005 Storing and Manipulating Gridded Data In Spatial Databases (Reading: Reading E-science Center, University of Reading) http://go-essp.gfdl.noaa.gov/presentations/06_06_05/Santokhee/Adit_Sank.ppt%20[Read-Only].pdf

[7] Brackett M 2012 *Data Resource Integration: Understanding and Resolving a Disparate Data Resource* (Technics Publications, LLC publishing) p 540

[8] Ryazanova A A, Voropay N N, Okladnikov I G 2016 Application of information and computing web system «Climate» for estimation of aridity of South Siberia *Proc. of Int. Conf. and Early Career Scientists School on Environmental Observations, Modeling and Information Systems ENVIROMIS-2016* (July 11-16, 2016, Tomsk, Russia) pp 358–62

[9] Okladnikov I G, Gordov E P, Titov A G and Shulgina T M 2015 *Proc. of the XVII Int. Conf. on Data Analytics and Management in Data Intensive Domains* (*October 13–16, 2015, Obninsk, Russia)* (*Electronic Materials*) ed L Kalinichenko and S Starkov pp 76–80

# Methods and algorithms for statistical processing of instantaneous meteorological parameters from ultrasonic measurements

**D S Rohmistrov[1], A Ya Bogushevich[2] and I A Botygin[3]**

[1] Undergraduate, National Research Tomsk Polytechnic University, Tomsk, Russia
[2] Senior Researcher, Institute of Monitoring of Climatic and Ecological Systems SB RAS, Tomsk, Russia
[3] Assistant Professor, National Research Tomsk Polytechnic University, Tomsk, Russia

E-mail: obozhe12@gmail.com, bay@imces.ru, bia@tpu.ru

**Abstract**. This paper describes a software system designed to support atmospheric studies with ultrasonic thermo-anemometer data processing. The system is capable of processing files containing sets of immediate values of temperature, three orthogonal wind velocity components, humidity, and pressure. The paper presents a technological scheme for selecting the necessary meteorological parameters depending on the observation time, the averaging interval, and the period between the immediate values. The data processing consists of three stages. At the initial stage, a query for the necessary meteorological parameters is executed. At the second stage, the system calculates the standard statistical characteristics of the meteorological fields, such as mean values, dispersion, standard deviation, asymmetric coefficients, kurtosis, correlation, etc. The third stage prepares to compute the atmospheric turbulence parameters. The system creates new arrays of data to process and calculate the second order statistical moments that are important for solving problems of atmospheric surface layer physics, predicting the pollutant dispersion in the atmosphere, etc. The calculation results are visualized and stored on a hard disk.

## 1. Introduction

The meteorological measurements are an essential part of scientific and applied knowledge on the atmosphere properties. The measurement data allows researchers to learn the specifics of physical and chemical processes in the atmosphere, to determine its composition and structure, to find out its thermal conditions and moisture exchange, research the formation conditions for a variety of optical and acoustical phenomena and, most importantly, modeling weather and climate forecasts.

Weather data is obtained by a complex of various equipment for meteorological measurements [1]. Ultrasonic anemometer is one of the most up-to-date devices used at weather stations [2-12]. The idea of using ultrasonic methods for atmosphere meteorological parameters estimation has appeared in the late 1940s. Ultrasonic methods are largely free of such drawbacks as inertia, radiation error etc. It allows researchers to accurately estimate the turbulent characteristics that describe the small-scale high-frequency spectral structure of meteorological observations.

This work presents a software that extends the functionality of the meteorological complex AMK-03, including its modifications (developed by Tomsk Scientific Center SB RAS). The principle of ultrasonic anemometer operation is based on measuring the sound speed, which varies depending on air flow (wind direction) vector orientation relative to sound propagation path. A standard implementation of an automatic ultrasonic weather station allows to measure such parameters as air temperature (T), various components of the wind velocity ($v_s$, $v_e$, w), pressure (P), humidity (r). The immediate values of the meteorological properties are calculated during the preparation stage for processing raw measurement data obtained by ultrasonic thermo-anemometer and stored at the local PC, directly connected to the device. The preparation stage includes:

- reading the raw measurement data stream from the device;
- testing the raw measurement data on operation and transmit failures, quality accessing;
- real-time calculation of the immediate meteorological values, such as temperature, horizontal wind velocity, horizontal wind direction, vertical wind velocity, relative air humidity, atmospheric pressure;
- automatically storing the meteorological value calculation results on the PC hard drive as files of various types, indexed by the measurement date.

Those files are stored in a special archive and they allow to calculate various meteorological parameters for any past period of time, including measurement means and statistic functions for atmosphere temperature as well as wind fields. Traditionally, they are used for analysis of the thermo dynamical condition of the ground layer of atmosphere and numerical estimations of the conventional atmospheric turbulence characteristics.

## 2. Stages of processing

The presented work describes a mathematical processing of meteorological parameters obtained from standard ultrasonic weather station. The data is stored in files and structured as ten minute observation periods with 12.5 millisecond measurement intervals. The meteorological values undergo a standard statistical processing, based upon existing formulae. Authors of this work are responsible for developing the necessary algorithms for analysis described below and implementing them in Java programming language. The statistical processing is done for all of the measurements, filtered by the registration time according to observation time and data statistical analysis (averaging) interval. The goal of the processing is to create informative atmosphere characteristic timelines that reflect statistically stable value trends for longer time periods compared to the initial one (day, month, year). The developed software allows to select data from any observation time and for any averaging interval, as well as to set the period between immediate values of meteorological parameters.

The application form screenshots that illustrate query processing are shown in Figures 1 and 2.

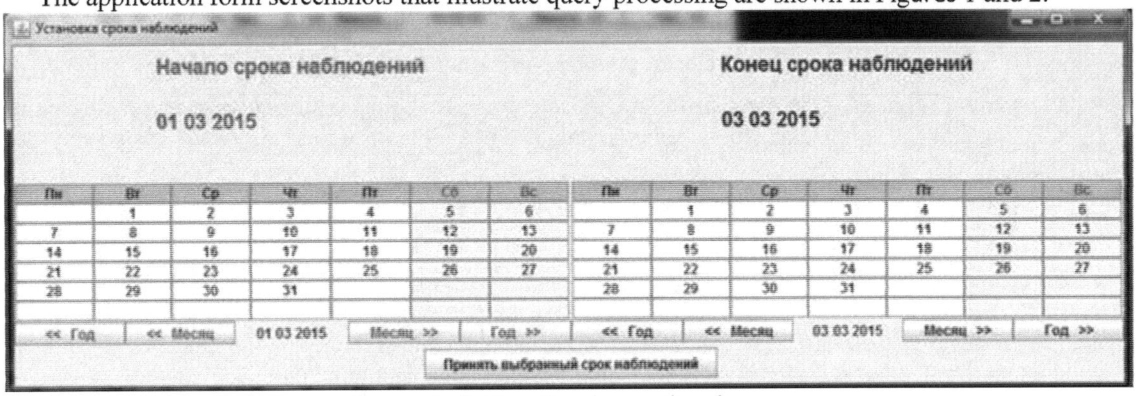


**Figure 1.** Setting the observation time.

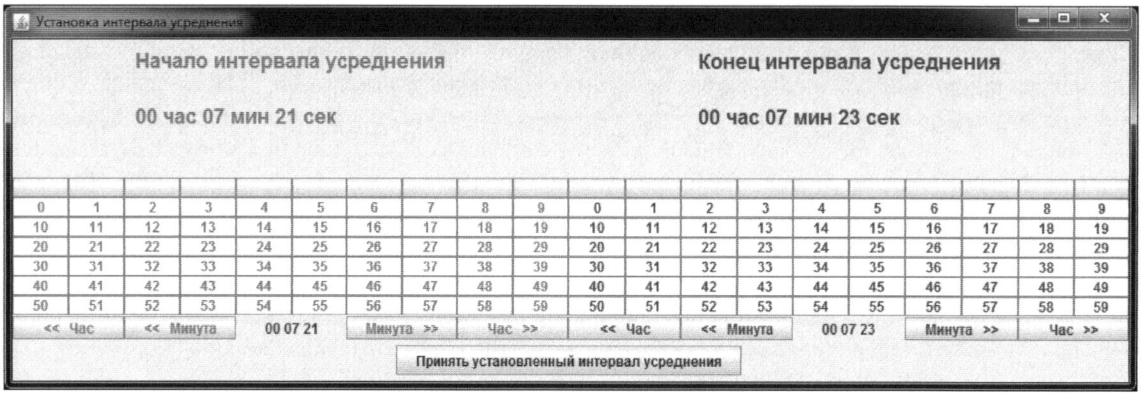


**Figure 2.** Setting the averaging interval.

The initial data for processing is measured from $t_{per} - t_{avg}$ to $t_{per}$ points of time according statistical data analysis interval selected according to some criteria.

Stage 1. Calculating averages for meteorological values and other physical parameters describing the quasi-equilibrium atmosphere state at measurement location for the given time period.

The initial immediate values (T, $v_s$, $v_e$, w, r and P) are obtained from binary files containing ultrasonic thermo-anemometer observation results. They are used to calculate the averages of meteorological values for the given observation time $t_{per}$ and analysis interval $t_{avg}$ (while accounting for requirements set by World meteorological organization and Roshydromet). The calculation results are also used to evaluate other atmosphere air physical parameters that are necessary for the numerical analysis of atmospheric turbulence and related tasks, but usually ignored at standard weather stations.

Stage 2. Preparing to calculate the atmosphere turbulence parameters, creating new data arrays for processing and calculating the statistical moments of second order (or higher).

Stage 3. Calculating the atmosphere turbulence parameters.

The basic mathematical equations used for the initial data processing and repeated for every selected $t_н$ time interval are listed below.

The averages and dispersion for $\xi(t)$ random process are calculated for $T$, $v_s$, $v_e$, w, r and P meteorological values describing the quasi-equilibrium (for a specific $t_{avg}$ time interval) state at the measurement location for the given time period.

In the case of ultrasonic weather station, for each meteorological $\xi_i$ value ($i$ stands for value type) we only have a set of its discrete samples $\xi_{i,k}$ (where k is the ordinal number registered from the $t_{per} - t_{avg}$ point of time). Averages are calculated with equation $\langle \xi_i \rangle = \frac{1}{N}\sum_{k=0}^{N-1}\xi_{i,k}$, where $N \approx F_d \cdot t_{avg}$ (whole number) – number of the existing $\xi_{i,k}$ discrete samples in the analyzed time interval ranging from $t_{per} - t_{avg}$ to $t_н$, $F_d = 80$ Hz – the weather station hardware measurement frequency (data sampling rate). The $D_i$ dispersion of $\xi_{i,k}$ value can be calculated as follows: $D_i = \frac{1}{N-1}\sum_{k=0}^{N-1}\left(\xi_{i,k} - \langle \xi_i \rangle\right)^2$.

## 3. Calculated parameters

The averages of orthogonal wind velocity components ($\langle v_s \rangle$ – south component of wind velocity horizontal vector, $\langle v_e \rangle$ – eastern component of horizontal wind velocity vector and $\langle w \rangle$ – vertical component of full wind velocity vector) are used to evaluate the following parameters:

The absolute value of wind velocity mean vector – $\langle V \rangle = \sqrt{\langle v_s \rangle^2 + \langle v_e \rangle^2 + \langle w \rangle^2}$.

Horizontal wind average speed – $\langle V_h \rangle = \sqrt{\langle v_s \rangle^2 + \langle v_e \rangle^2}$.

The inclination angle to the horizontal of wind velocity mean vector –$\psi = \arccos(\frac{\langle V_h \rangle}{\langle V \rangle})$.

The direction of horizontal wind average velocity $\langle D \rangle$:

$$\langle D\rangle = \begin{cases} \pi + f, \langle v_s\rangle > 0{,}01 \\ f, \langle v_s\rangle < -0{,}01; \langle v_e\rangle \geq 0 \\ 2\pi + f, \langle v_s\rangle < -0{,}01; \langle v_e\rangle < 0 \\ \frac{\pi}{2}, |\langle v_s\rangle| \leq 0{,}01; \langle v_e\rangle \geq 0 \\ \frac{3\pi}{2}, |\langle v_s\rangle| \leq 0{,}01; \langle v_e\rangle < 0 \end{cases},$$

where $f = \mathrm{arctg}(-\frac{\langle v_e\rangle}{\langle v_s\rangle})$.

Saturated vapor pressure – $e_w = \left(6.112 * e^{\frac{17.62T}{243.12+T}}\right) * (1.0016 + 3.15 * 10^{-6}P - 0.074P^{-1})$, where T stands for temperature, P stands for atmospheric pressure.

Average pressure of atmosphere water vapor (e, GPa) – $e = \frac{e_w * r}{100\%}$, where r is relative air humidity.

Moisture deficit ($E_d$, GPa) – $E_d = e_w - e$.

Absolute air humidity (q, g/m$^3$) – $q = \frac{r*P*M}{R*T*100\%}$, where M is the molar mass of dry air, R is the universal gas constant.

Air density (ρ, g/m$^3$) – $\rho = \frac{P*M}{R*T}$.

The preparation for atmospheric turbulence calculation includes creating new arrays of data to process and evaluating the statistical moments of second order (or higher). The first part of this stage is the computing process and use the immediate values of ultrasonic measurements T, $v_s$, $v_e$ and w to form new time series that describe the turbulent fluctuations:

For temperature – $T' = T - \langle T\rangle$.

For vertical wind velocity component – $w' = w - \langle w\rangle$.

For longitudinal wind velocity component – $u' = (\langle v_s\rangle * (v_s - \langle v_s\rangle) + \langle v_e\rangle * (v_e - \langle v_e\rangle)/\langle V_h\rangle$.

For transverse wind velocity component $v' = (-\langle v_s\rangle * (v_e - \langle v_e\rangle) + \langle v_e\rangle * (v_s - \langle v_s\rangle)/\langle V_h\rangle$.

To analyze turbulence for the ground atmosphere layer, it is important to evaluate the second order correlational statistical moments:

Heat flux momentum – $\langle T' * w'\rangle$.

Moment-of-momentum flux – $\langle u' * w'\rangle$.

Atmospheric turbulence parameters (average from $t_н$ - $t_{уср}$ to $t_н$):

The total energy of turbulent motions – $E_v = (\sigma_u^2 + \sigma_v^2 + \sigma_w^2)/2$, where $\sigma_u^2, \sigma_v^2, \sigma_w^2$ are dispersions of turbulent fluctuations for three wind velocity components u', v', w'.

The relative intensity of wind velocity fluctuations – $I_v = E_v/{V_m}^2$.

The energy of temperature fluctuations – $E_t = \sigma_t^2/2$, where $\sigma_t^2$ is the dispersion of turbulent fluctuations of T' temperature.

Vertical momentum flux – $\tau = -\rho\langle u' * w'\rangle$, where $\rho$ – air density.

Vertical heat flux – $H = c_p * \rho * \langle T' * w'\rangle$, where $c_p$ is air specific heat capacity at the constant pressure.

Wind friction velocity – $v^* = \sqrt{-\langle u' * w'\rangle}$.

Temperature scale – $T^* = -\langle T' * w'\rangle/v^*$.

Monin-Obukhov length – $L^* = \langle T\rangle * (v^*)^2/\chi * g * T^*$, where $\chi = 0{,}4$ and $g = 9{,}81$ m/s$^2$.

Flow resistance coefficient – $C_d = (\frac{v^*}{\langle V\rangle})^2$.

The structure constant of temperature fluctuations $C_T^2 = \langle[T'(t + \Delta t - T'(t)]^2\rangle * (\langle V\rangle\Delta t)^{-2/3}$, where V is the absolute value of mean wind velocity vector; $\Delta t$ – the time between immediate meteorological parameters measurements.

The structure constant of wind fluctuations $C_V^2 = \langle[u'(t + \Delta t - u'(t)]^2\rangle * (\langle V\rangle\Delta t)^{-2/3}$.

The structure constant of acoustic refraction index fluctuations $C_{na}^2 = \frac{C_T^2}{4\langle T\rangle^2} + \frac{C_V^2}{\langle c\rangle^2}$, where $T_k$ is air temperature measured in kelvins, c is sound speed.

The structure constant of optical refractive index fluctuations $C_{no}^2 = \{8 * 10^{-5} * \langle P\rangle/\langle T\rangle^2\}^2 * C_T^2$, where P is atmospheric pressure measured in hPa.

The Monin-Obukhov similarity theory [13, 14] is used to predict values of a parameter group. The calculations use $\varphi_V$ and $\varphi_T$ – the universal similarity functions of the dimensionless height $\xi = z/L^*$ for wind and thermal stratification, respectively. The universal similarity functions use $\varphi_V$ and $\varphi_T$ of the dimensionless height parameter $\xi = z/L^*$. Those parameters are semi-empirical, since their general form is determined by the theoretical analysis of a ground turbulent flux and can be found using the following formulas [15, 16]:

$$\varphi_V(\xi) = \begin{cases} 1 + 4{,}7\xi, .. \xi > 0 \\ (1 - 15\xi)^{-1/4}, .. \xi < 0 \end{cases}$$

$$\varphi_T(\xi) = \begin{cases} 0{,}74 + 4{,}7\xi, .. \xi > 0 \\ 0{,}74 * (1 - 9\xi)^{-1/2}, .. \xi < 0 \end{cases}$$

Local gradient of wind velocity at $z_m$ height:

$$\frac{du}{dz} = \frac{v^*}{\chi}\varphi_V\left(\frac{z_m}{L^*}\right)$$

Local gradient of potential temperature at $z_m$ height:

$$\frac{d\theta}{dz} = \frac{T^*}{\chi}\varphi_T\left(\frac{z_m}{L^*}\right)$$

Local gradient of temperature at $z_m$ height:

$$\frac{dT}{dz} = \frac{d\theta}{dz} - 0{,}0098$$

The turbulent exchange coefficient for momentum:

$$K_m = v^{*2}/\frac{du}{dz}$$

The turbulent heat exchange coefficient:

$$K_h = v^* * T^*/\frac{d\theta}{dz}$$

Turbulence outer scale:

$$L_0 = \left(K_m/\left|\frac{du}{dz}\right|\right)^{1/2}$$

Energy dissipation rate for wind fluctuations:

$$\varepsilon = K_m\left(\frac{du}{dz}\right)^2 - \left(\frac{g}{\langle T\rangle}\right)K_h\frac{d\theta}{dz}$$

Energy dissipation rate for heat fluctuations:

$$N = K_h\left(\frac{d\theta}{dz}\right)^2$$

Richardson gradient number:

$$R_i = \left(\frac{g}{\langle T\rangle}\right)\left(\frac{d\theta}{dz}\right)/\left(\frac{du}{dz}\right)$$

## 4. Screens interface

The developed means of visualization are shown in figures 3 – 8, displaying the meteorological data calculated means, describing the quasi-equilibrium state (for a given averaging interval) at the measurement location in a specific point of time (or for a given observation period).

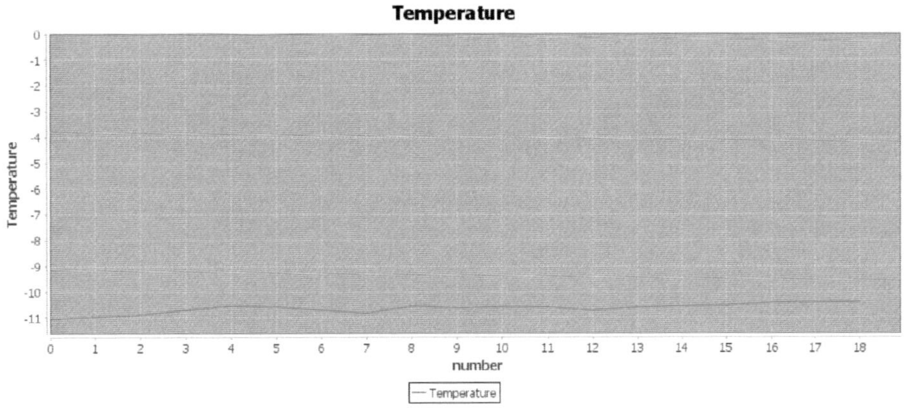


**Figure 3.** Average temperature chart.

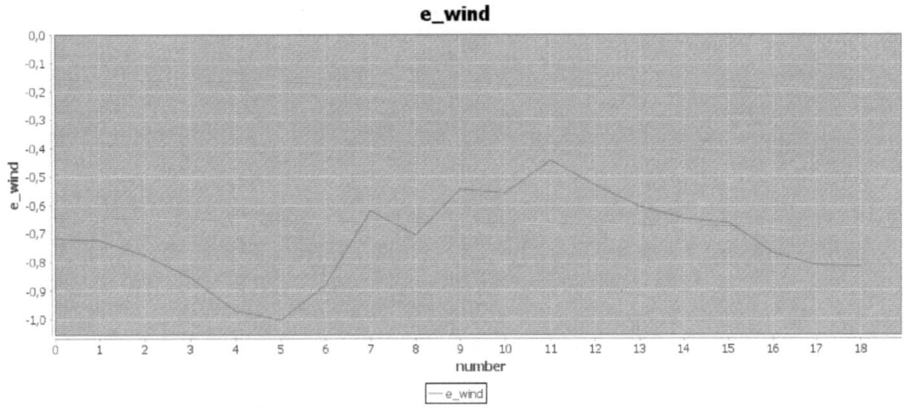


**Figure 4.** Eastern wind component mean vector chart.

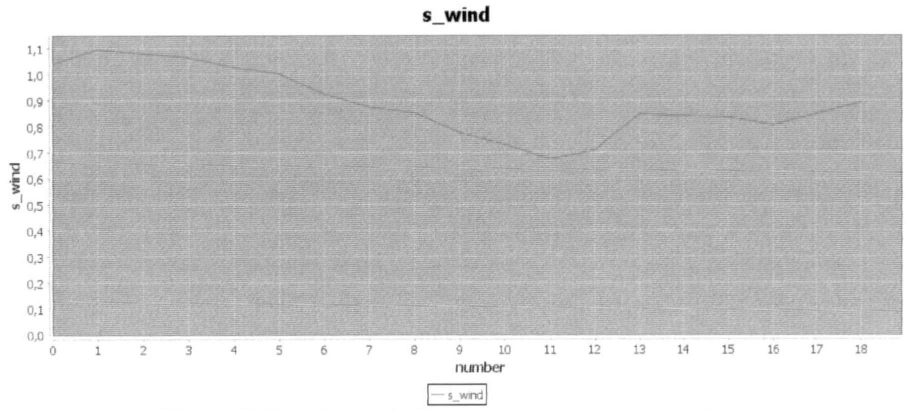


**Figure 5.** Southern wind component mean vector chart.

IOP Conf. Series: Earth and Environmental Science **48** (2016) 012031 doi:10.1088/1755-1315/48/1/012031

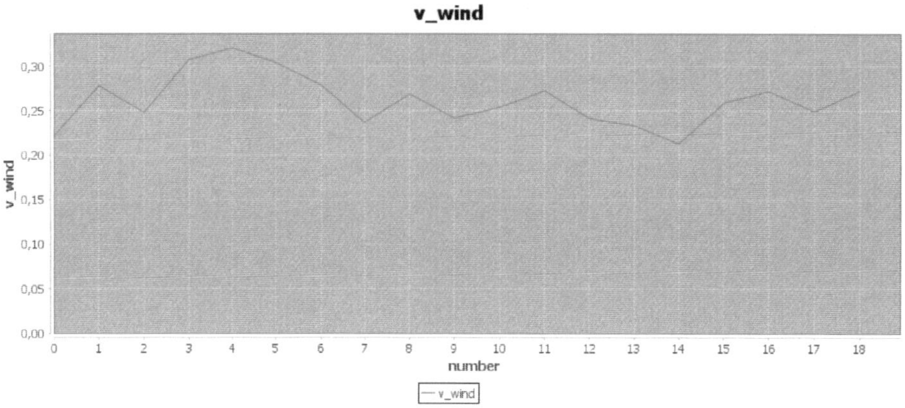


**Figure 6.** Vertical wind component mean vector chart.

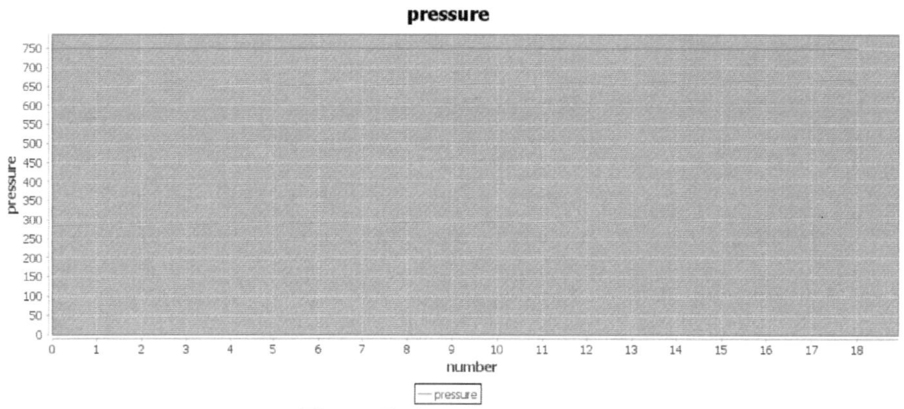


**Figure 7.** Average pressure chart.

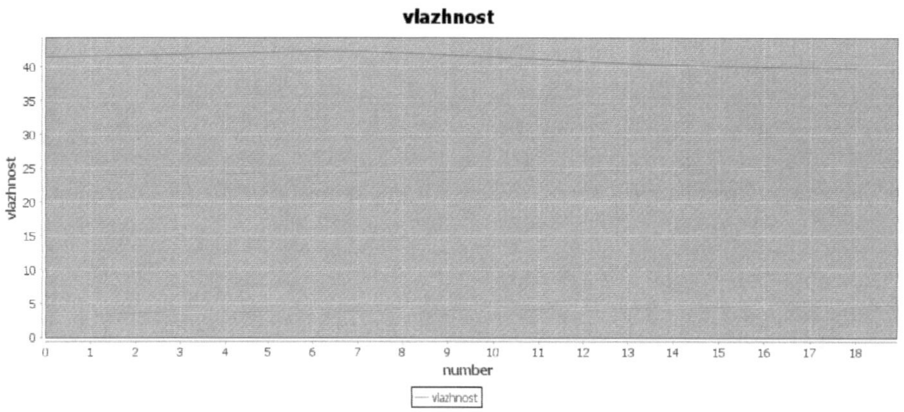


**Figure 8.** Average humidity chart.

A fragment of the web-interface window for the developed software is shown in Figure 9 below.

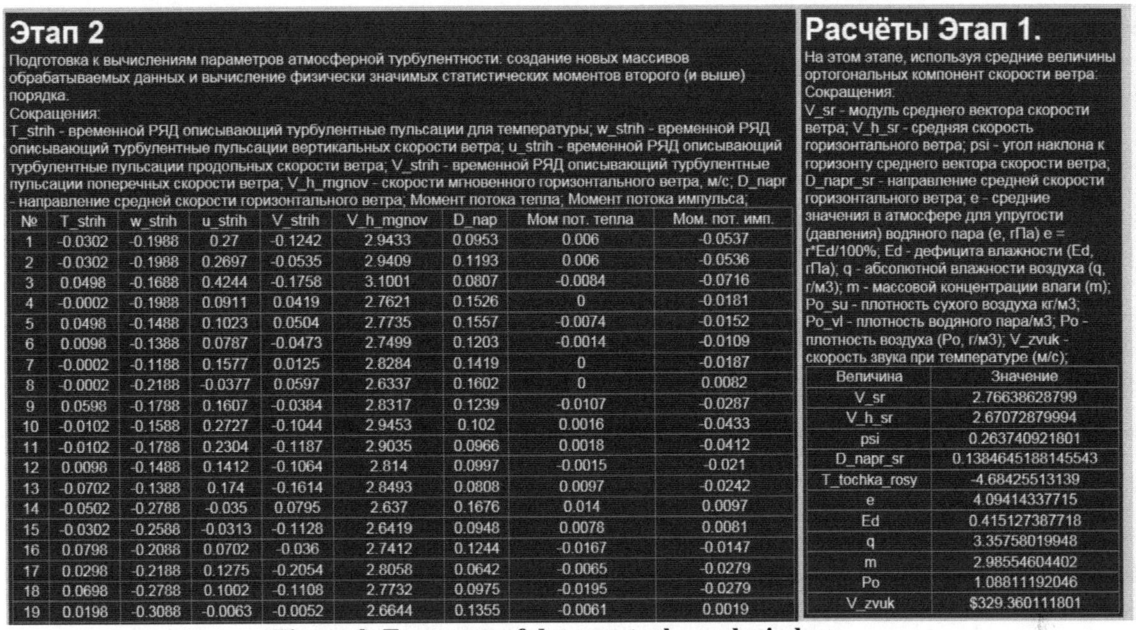

**Этап 2**

Подготовка к вычислениям параметров атмосферной турбулентности: создание новых массивов обрабатываемых данных и вычисление физически значимых статистических моментов второго (и выше) порядка.

Сокращения:

T_strih - временной РЯД описывающий турбулентные пульсации для температуры; w_strih - временной РЯД описывающий турбулентные пульсации вертикальных скорости ветра; u_strih - временной РЯД описывающий турбулентные пульсации продольных скорости ветра; V_strih - временной РЯД описывающий турбулентные пульсации поперечных скорости ветра; V_h_mgnov - скорости мгновенного горизонтального ветра, м/с; D_napr - направление средней скорости горизонтального ветра; Момент потока тепла; Момент потока импульса;

| № | T_strih | w_strih | u_strih | V_strih | V_h_mgnov | D_nap | Мом пот. тепла | Мом. пот. имп. |
|---|---|---|---|---|---|---|---|---|
| 1 | -0.0302 | -0.1988 | 0.27 | -0.1242 | 2.9433 | 0.0953 | 0.006 | -0.0537 |
| 2 | -0.0302 | -0.1988 | 0.2697 | -0.0535 | 2.9409 | 0.1193 | 0.006 | -0.0536 |
| 3 | 0.0498 | -0.1688 | 0.4244 | -0.1758 | 3.1001 | 0.0807 | -0.0084 | -0.0716 |
| 4 | -0.0002 | -0.1988 | 0.0911 | 0.0419 | 2.7621 | 0.1526 | 0 | -0.0181 |
| 5 | 0.0498 | -0.1488 | 0.1023 | 0.0504 | 2.7735 | 0.1557 | -0.0074 | -0.0152 |
| 6 | 0.0098 | -0.1388 | 0.0787 | -0.0473 | 2.7499 | 0.1203 | -0.0014 | -0.0109 |
| 7 | -0.0002 | -0.1188 | 0.1577 | 0.0125 | 2.8284 | 0.1419 | 0 | -0.0187 |
| 8 | -0.0002 | -0.2188 | -0.0377 | 0.0597 | 2.6337 | 0.1602 | 0 | 0.0082 |
| 9 | 0.0598 | -0.1788 | 0.1607 | -0.0384 | 2.8317 | 0.1239 | -0.0107 | -0.0287 |
| 10 | -0.0102 | -0.1588 | 0.2727 | -0.1044 | 2.9453 | 0.102 | 0.0016 | -0.0433 |
| 11 | -0.0102 | -0.1788 | 0.2304 | -0.1187 | 2.9035 | 0.0966 | 0.0018 | -0.0412 |
| 12 | 0.0098 | -0.1488 | 0.1412 | -0.1064 | 2.814 | 0.0997 | -0.0015 | -0.021 |
| 13 | -0.0702 | -0.1388 | 0.174 | -0.1614 | 2.8493 | 0.0808 | 0.0097 | -0.0242 |
| 14 | -0.0502 | -0.2788 | -0.035 | 0.0795 | 2.637 | 0.1676 | 0.014 | 0.0097 |
| 15 | -0.0302 | -0.2588 | -0.0313 | -0.1128 | 2.6419 | 0.0948 | 0.0078 | 0.0081 |
| 16 | 0.0798 | -0.2088 | 0.0702 | -0.036 | 2.7412 | 0.1244 | -0.0167 | -0.0147 |
| 17 | 0.0298 | -0.2188 | 0.1275 | -0.2054 | 2.8058 | 0.0642 | -0.0065 | -0.0279 |
| 18 | 0.0698 | -0.2788 | 0.1002 | -0.1108 | 2.7732 | 0.0975 | -0.0195 | -0.0279 |
| 19 | 0.0198 | -0.3088 | -0.0063 | -0.0052 | 2.6644 | 0.1355 | -0.0061 | 0.0019 |

**Расчёты Этап 1.**

На этом этапе, используя средние величины ортогональных компонент скорости ветра:

Сокращения:

V_sr - модуль среднего вектора скорости ветра; V_h_sr - средняя скорость горизонтального ветра; psi - угол наклона к горизонту среднего вектора скорости ветра; D_napr_sr - направление средней скорости горизонтального ветра; e - средние значения в атмосфере для упругости (давления) водяного пара (e, гПа) e = r*Ed/100%; Ed - дефицита влажности (Ed, гПа); q - абсолютной влажности воздуха (q, г/м3); m - массовой концентрации влаги (m); Po_su - плотность сухого воздуха кг/м3; Po_vl - плотность водяного пара/м3; Po - плотность воздуха (Po, г/м3); V_zvuk - скорость звука при температуре (м/с);

| Величина | Значение |
|---|---|
| V_sr | 2.76638628799 |
| V_h_sr | 2.67072879994 |
| psi | 0.263740921801 |
| D_napr_sr | 0.1384645188145543 |
| T_tochka_rosy | -4.68425513139 |
| e | 4.09414337715 |
| Ed | 0.415127387718 |
| q | 3.35758019948 |
| m | 2.98554604402 |
| Po | 1.08811192046 |
| V_zvuk | $329.360111801 |

**Figure 9. Fragment of data control panel window.**

## 5. Conclusions

The developed software is capable of using ultrasonic thermo-anemometer measurements data to calculate averages of meteorological values, as well as parameters and statistical functions for temperature and wind atmospheric fields, traditionally used to conduct the atmosphere ground layer thermodynamic condition analysis and numerical evaluation of its turbulence properties.

The importance of this work lies in its high practicality – it suggests the hardware and software that allow to approach automated computation of meteorological fields' secondary parameters row. The final version of the hardware-software complex will be useful to the specialists in related fields as an "instrument" for gaining new knowledge.

The results are obtained in the framework of the project "Development and research of intelligent information-analytical system for analysis and prediction of climatic processes on the basis of high-performance clusters" of RAS 1.33P fundamental research program "Fundamental problems of mathematical modeling. Fundamental problems of factorization methods in different fields. Algorithms and software for ultra-high performance computing systems".

## References

[1] Tikhomirov A A 2010 *Atmos. Ocean. Opt.* **23** 585–600

[2] Tuton J and Pethica C 2010 J. Atmos. Ocean. Technol. **27** 2031–2038

[3] Azbukin A A, Bogushevich A Ya, Illichivski V S, Korolkov V A and Tikhomirov A A 2006 *Russian Meteorology and Hydrology* 89–97

[4] Azbukin A A, Bogyshevich A Ya, Kobzev A A, Korolkov V A, Tikhomirov A A and Schelevoy V D 2012 *Sensors and Systems* 47–52

[5] Tikhomirov A A, Korolkov V A, Bogyshevich, Azbukin A A, A Ya and Schelevoy V D 2008 *Bulletin of the Academy of Military Sciences* 144–149

[6] Kaimal J C and Gaynor J E 1991 *Bound-Lay. Meteorol.* **56** 410–418

[7] Kabanov M V 1997 *Regional atmospheric monitoring. Part 1: Scientifically-methodical bases* (Tomsk: Publishing House of the SB RAS) p 211

[8] Kabanov M V 1997 *Regional atmospheric monitoring. Part 2: New tools and measurement techniques* (Tomsk: Publishing House of the SB RAS) p 295

[9] Sherstnev V S, Sherstneva A I, Botygin I A and Kustov D A 2016 *High Technology: Research*

*and Applications (Key Engineering Materials vol 685)* (Switzerland: Trans Tech Publications) pp 867–871

[10] World Community Grid *http://www.worldcommunitygrid.org/* Accessed: January 2016

[11] Botygin I A, Popov V N, Tartakovsky V A and Sherstnev V S 2015 *Proc. 21st Int. Symp. Atmospheric and Ocean Optics: Atmospheric Physics* (Tomsk: SPIE) vol. 9680 pp 96800J1–96800J4

[12] Popov V N, Botygin I A and Koshelev N V. 2016 *High Technology: Research and Applications (Key Engineering Materials vol 685)* (Switzerland: Trans Tech Publications) pp 925–929

[13] Monin A S, Obukhov A M 1954 *Proc. Geophysical Institute of the USSR Academy of Sciences* **24** 163–187

[14] Zilitinkevich S S 1970 *The dynamics of the atmospheric boundary layer* (Leningrad: Hydrometeorological Publishing House) p 292

[15] Businger J A, J.C. Wyngard J C, Isumi Y and Breadley E F 1971 *J. Atmos. Sci.* **28** 181–189

[16] Dyer A J 1974 *Boundary-Layer Meteorol.* **7** 363–372

# Development of computational module of regional aridity for web-GIS "Climate"

**A A Riazanova[1,2], N N Voropay[1,3], I G Okladnikov[1,4] and E P Gordov[1,4,5]**

[1]Institute of Monitoring of Climatic and Ecological Systems SB RAS, Tomsk, Russia
[2] V. E. Zuev Institute of Atmopsheric Optics SB RAS, Tomsk, Russia
[3] V.B. Sochava Institute of Geography SB RAS, Irkutsk, Russia
[4]Institute of Computational Technologies, Tomsk Branch, Tomsk, Russia
[5]Tomsk State University, Tomsk, Russia

**Abstract.** Modern global climate changes are characterized by a significant warming started in the late 1970s. According to the observation data, the average rate of warming in Russia for 1976-2012 was 0.43 °C/10 years. Additionally, the trend of annual precipitation for the same period in most parts of Russia is positive. To study the integral climate characteristics of a region, in particular, hydrothermal conditions, is of great interest. A new software module for the assessment of aridity was developed and added to an information and computing web system called "Climate". The well-known hydrothermal coefficient of Selyaninov (HTC) was used with ECMWF ERA Interim reanalysis data for the period from 1979 to 2010. The input data (precipitation, temperature, HTC) were validated using observation data from weather stations. It was found that the ERA Interim data substantially (by 50%) underestimated the measured precipitation in the study region. The precipitation was corrected by the results. New spatially distributed precipitation and HTC data were obtained. The web system "Climate" is based on web and GIS techniques and is part of a hardware and software complex for "cloud" analysis of climatic data. It includes several climatological and meteorological datasets, as well as dedicated software modules for searching, selection, extraction, and visualization of data.

## 1. Introduction

Modern global climate changes are characterized by a significant warming in the late XX – early XXI centuries (starting in the late 1970s). According to observational data, the trend of average global warming reached 0.166 °C per 10 years for the period of 1976-2012, and 0.075 °C per 10 years for the period 1901-2012 [1]. The observed warming rate in Russia for the period of 1976-2012 is 0.43 °C per 10 years [2]. The trend of annual precipitation for 1976-2012 is positive in most parts of Russia. To study the integral characteristics of a region, especially hydrothermal conditions, is of great interest. These are described by coefficients depending on air temperature and precipitation. In this study, the well-known Selyaninov's hydrothermal coefficient (HTC) [3] is used for the estimation of moistening and drought conditions of territories and the determination of moist and dry periods.

Traditionally, HTC is calculated in a manual or semi-automatic mode for a given territory with data from weather stations. This approach is very labour-consuming, requires significant time for multi-year observational data processing, and characterizes HTC only in regions with the weather stations. In combination with data of high sparsity of the weather stations network on the territories of Siberia and Far East, this significantly reduces the reliability of the obtained results.

In contrast to local observational data, global retrospective analysis (re-analysis) data are obtained on a regular grid and have global coverage. The recent development of the meteorological models and numerical schemes of observational data assimilation resulted in an improvement of quality and an increase in the spatial resolution of the global re-analyses. This allows using them for spatial distribution estimation of the territories' moisture conditions. However, the big volumes of data (up to several terabytes) require powerful computing complexes for their processing, and the dedicated file formats used for storing data require specific programming skills. One of the most perspective approaches to solving this problem is using "cloud" computing, i.e., storing and processing of data on powerful servers providing remote access to the analysis tools and processing results via the internet [4].

In this study we estimate the aridity of South Siberia territory with re-analysis data using an information-computing web system (ICWS) called "Climate". For this purpose, a new software module of ICWS for HTC calculation was developed. The web system "Climate" (http://climate.scert.ru) is based on a combined use of web and GIS approaches. It is part of a hardware and software complex for "cloud" analysis of climatic data with various archives of climatic data, as well as dedicated algorithms for their searching, extraction, processing, and visualization [5]. Use of this ICWS significantly facilitates and accelerates analysis of big volumes of geospatial data, allowing researchers to perform complex climate data analysis using desktop PCs with internet connection.

## 2. Methods and approaches

### *2.1. Drought estimation techniques*

Heat and moisture are the main factors of agricultural productivity. Herein the greatest productivity loss is done by droughts due to air temperature anomalies and deficit of precipitation either during the vegetation period or in the preceding autumn-winter period [6]. All historical drought observations show high vulnerability of the population and economy to them in the world. A drought occurs when precipitation, during some period of time, is observed below the average value. A detailed study of the agro-climatic conditions of a territory by their aridity degree is required to develop a land-use planning strategy in regions with high agricultural activity. In numerous papers devoted to the study of droughts, different approaches to droughts detection and determination of their characteristic types and forming conditions are considered; the most representative aridity indices are developed; and catastrophic droughts analysis and catalogization are performed [3, 7-10]. However, D.A. Ped showed [11] that there is no any recognized quantitative index of a drought. Also, studying droughts, some other dedicated indices are used. These indices are calculated using indicators of soil water balance and soil moisture deviation from the optimal condition, duration of these periods, amplitude of moisture reserves variation, monthly anomalies in soil moisture, drought periods length and average monthly values of temperature and precipitation, positive amplitudes of temperature anomalies and negative precipitation anomalies [12-15].

According to a Russian drought classification [9], the following types are distinguished: atmospheric, soil, and atmospheric-soil. The Russian classification is more oriented to the drought effects and takes into account the environment of a drought. It differs from the American one [16]. That is why the Russian classification is more useful for understanding the physical nature of droughts. For example, it was shown [17] that an atmospheric-soil drought occurs due to a prolonged atmospheric drought and is intensified by a soil drought. Also, during some years a soil drought was intensified by an atmospheric one, and an atmospheric-soil drought is characterized by a combination of soil and atmospheric droughts.

The adverse agroclimatic phenomena significantly reduce the agroclimatic potential. During an atmospheric drought we observe active heating and relative desiccation of the air processes in the absence of rainfalls and a large income of solar radiation.

Selyaninov's hydrothermal coefficient (HTC) is a well-known technique for territory moisture conditions estimation in Russia. It is assumed that the sum of temperatures is close to evaporation from an optimally moist field and can, therefore, be taken as the maximum evaporation, i.e. evaporability. This coefficient represents the ratio of precipitation for some time period to the sum of temperatures higher than 10 °C divided by 10, for the same period. Thus, the hydrothermal coefficient, as an index of territory estimation by moisture conditions, has an advantage over the precipitation, characterizing not only the income water balance (precipitation), but also the non-productive outcome of water (surface and vegetation evaporation, etc.) closely linked to the overall level of the air temperature. According to many studies [18], this index is not only applicable for comparative moisture estimation of areas with significantly different thermal conditions, but also used as a comparative moist characteristic of the individual months as well.

### *2.2. HTC computing module*

To calculate Selyaninov's hydrothermal coefficient, the ICWS "Climate" functionality was extended by a dedicated software module. The module was developed using Interactive Data Language (IDL) v. 6.0. The algorithm is based on a well-known formula for calculating the HTC [3]:

$$HTC = \frac{\sum r}{\sum t_{>10}}. \quad (1)$$

Here $\sum r$ is precipitation during a vegetation period with air temperature above 10 °C, $\sum t_{>10}$ is the sum of temperatures during a vegetation period with air temperature above 10 °C.

The vegetation period was calculated on the basis of a technique presented by Ped [19].

To estimate the accuracy of the aridity index HTC calculated using reanalysis data, the results obtained were compared to the HTC calculated using observation data from weather stations in the study area. For this, the HTC values calculated at the nodes of the re-analysis grid were interpolated into the coordinates of the stations using a linear interpolation method. With the results of the comparison, a correction of the re-analysis precipitation was performed using linear regression at all nodes of the re-analysis grid for the study area.

The software module was integrated into the software framework of ICWS "Climate" and linked to all its resources, including various datasets and modules for processing and visualization of results.

## 3. Results

Using the above-developed software module and the ECMWF (European Centre for Medium-Range Weather Forecasts) ERA Interim [20] re-analysisspatial fields of temperature and precipitation, Selyaninov's hydrothermal coefficient was calculated for the territory of Southern Siberia (50°-65° N, 60°-115° E) for the period from 1979 to 2010. For quality assessment of the results obtained, HTC was also calculated using local observation data from the weather stations. Then temperature, precipitation and calculated HTC values were interpolated from the re-analysis grid to the coordinates of the weather stations.

A reanalysis and observation data comparison showed a slight difference in the temperature values (average -0.86 °C or -6.77%) and a significant discrepancy in the precipitation and hydrothermal coefficients (average 55% and 50%, respectively). Detailed results of the calculations are presented in Table 1. Similar findings were obtained by T.M. Shulgina [21].

Next, a correction of the precipitation data was performed using linear regression. The observational data on daily precipitation were taken as the dependent variable, and the re-analysis daily precipitation interpolated to the coordinates of weather stations was taken as the independent variable. The correction was performed separately for each year and month of the vegetation period (from May to September) for the study area. After the correction the mean deviation between the re-analysis and observational daily precipitation significantly decreased. Similarly, the HTC values also changed. For the corrected daily precipitation and HTC, the mean deviations during the vegetation period are -3.38% and 3.02%, and RMSs are 10.74% and 21.48%, respectively. Detailed results of the calculations are presented in Table 2.

**Table 1.** Reanalysis and observation data comparison.

| | May | June | July | August | September |
|---|---|---|---|---|---|
| | | **Precipitation, %** | | | |
| **Average** | -51.58 | -56.99 | -55.51 | -55.76 | -54.74 |
| **RMS** | 7.65 | 4.61 | 6.00 | 4.84 | 6.04 |
| **Maximum** | -26.34 | -40.18 | -39.12 | -45.16 | -31.71 |
| **Minimum** | -64.03 | -64.32 | -66.63 | -63.89 | -63.65 |
| | | **Temperature, °C** | | | |
| **Average** | 1.04 | 1.06 | 0.93 | 0.80 | 0.48 |
| **RMS** | 1.37 | 1.85 | 2.11 | 1.92 | 1.19 |
| **Maximum** | 4.14 | 5.26 | 5.85 | 4.91 | 3.30 |
| **Minimum** | -0.62 | -1.73 | -1.38 | -1.35 | -0.88 |
| | | **HTC, %** | | | |
| **Average** | -46.18 | -51.60 | -55.44 | -51.90 | -48.80 |
| **RMS** | 17.28 | 10.22 | 10.40 | 8.83 | 12.64 |
| **Maximum** | -14.74 | -22.64 | -27.28 | -31.01 | -8.92 |
| **Minimum** | -72.32 | -64.10 | -83.00 | -75.40 | -68.63 |

**Table 2.** Corrected reanalysis and observation data comparison.

| | May | June | July | August | September |
|---|---|---|---|---|---|
| | | **Corrected precipitation, %** | | | |
| **Average** | -3.41 | -3.73 | -3.58 | -2.87 | -3.31 |
| **RMS** | 11.54 | 8.80 | 10.52 | 11.91 | 10.94 |
| **Maximum** | 25.94 | 10.50 | 22.47 | 20.84 | 25.62 |
| **Minimum** | -22.46 | -21.07 | -29.54 | -26.23 | -22.43 |
| | | **Corrected HTC, %** | | | |
| **Average** | 4.61 | 5.64 | -5.02 | 2.93 | 6.93 |
| **RMS** | 29.12 | 18.78 | 18.71 | 16.55 | 24.24 |
| **Maximum** | 116.70 | 47.37 | 24.08 | 39.42 | 71.75 |
| **Minimum** | -48.02 | -16.67 | -63.41 | -24.25 | -34.94 |

The differences between the long-term HTC values before and after the daily precipitation correction are shown in Figures 1 and 2.

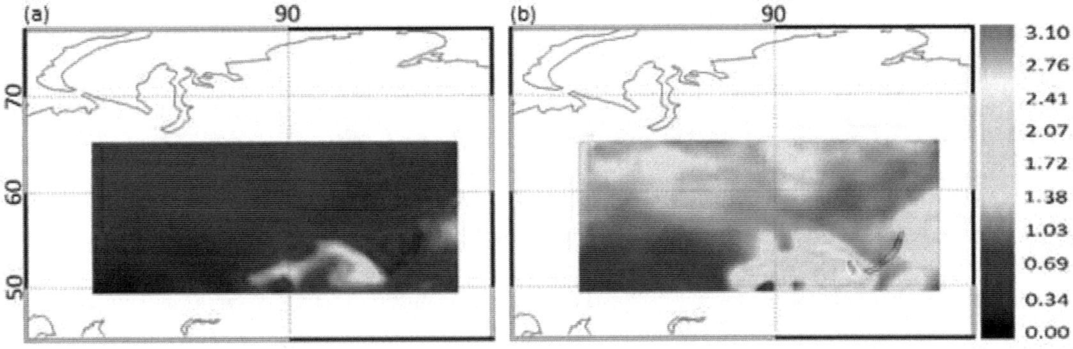


**Figure 1.** Long-term HTC values for Southern Siberia for the vegetation season calculated using ECMWF ERA Interim data for 1979-2010: a – uncorrected daily precipitation, b – corrected daily precipitation.

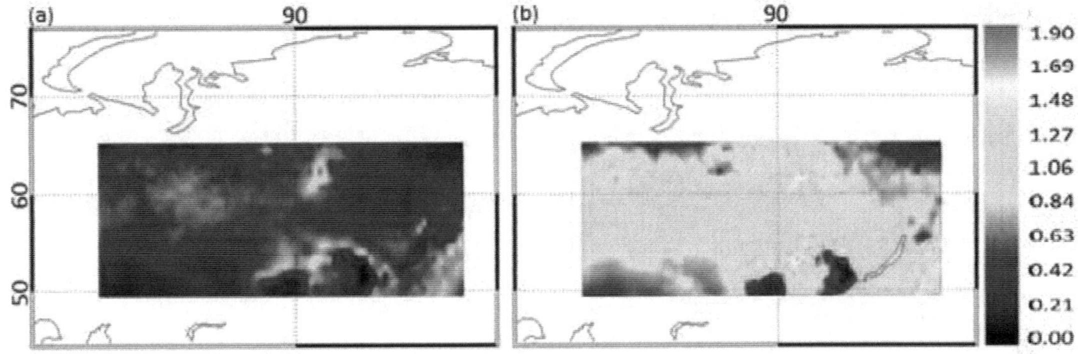


**Figure 2.** Long-term HTC values for Southern Siberia in August calculated using ECMWF ERA Interim data for 1979-2010: a – uncorrected daily precipitation, b – corrected daily precipitation.

## 4. Conclusions

The above-developed software module and the ICWS "Climate" significantly facilitate and speed up the calculations of aridity indices using observation and re-analysis data. In this work, the ICWS "Climate" provided functionality to calculate and visualize HTC for the territory of Southern Siberia from 1979 to 2010 (the user can change the boundary conditions). It allowed assessing the geographical distribution of hydrothermal conditions for the region. The results are represented by net CDF files, georeferenced maps and layers available as downloadable files or web services [22]. They can be easily transferred to online and desktop GISs for further cross-analysis with other spatial products, e.g. for estimation of the drought impact on vegetation productivity in the region. Since the hydrothermal indices are integral characteristics of the heat-moisture conditions of the area, special attention should be paid to the quality of the initial data. It was found that the Era Interium data substantially (by 50%) underestimate the measured precipitation in the study region. Adjustment of the re-analysis daily precipitation using the linear regression coefficients allows us to correctly estimate the spatial distribution of the HTC over the area, thus increasing the quality of droughts detection and estimation of extra moisture conditions in the study area.

**Acknowledgements**
The authors thank the Russian Science Foundation for support of this work under grant № 16-19-10257.

**References**

[1] Stocker T F, Qin D, Plattner G-K, Tignor M, Allen S K, Boschung J, Nauels A, Xia Y, Bex V and Midgley P M 2013 *Climate Change 2013: The Physical Science Basis. Contribution of Working Group I to the Fifth Assessment Report of the Intergovernmental Panel on Climate Change* (Cambridge: Cambridge University Press and NY, USA) p 1535
[2] Kattsov V M and et al 2014 *Second Roshydromet assessment report on climate change and its consequences in Russian Federation* (Moscow: Federal service for hydrometeorology and environmental monitoring (Roshydromet)) p 56
[3] Selyanoinov G T 1928 About agricultural estimate of climate *Proc. of Agricultural meteorology* (Leningrad: Hydrometeo Pub.) **20** pp 165–77
[4] Schäffer B, Baranski B and Foerster T 2010 Towards spatial data infrastructures in the Clouds *Geospatial thinking* ed M Painho and H Pundt (Berlin: Springer Verlag) pp 399–418.
[5] Gordov E P, Okladnikov I G, Titov A G, Bogomolov V Yu, Shulgina T M and Genina E Yu 2012 Geo-information system for investigation of regional climatic changes and first results obtained *Atmos. Ocean. Opt.* **25** 137–43
[6] Hydrometeorological disasters 2001 ed Golicin G S and Vasilev A A (Moscow: KRUK Publisher) p 296
[7] Meschcerskaay A V, Blazhevich V G, Golod M P and Belyankina I G 1986 Long-term fluctuations of aridity indexes in warm period of year in main agricultural zone of USSR *Proc. of the Voeikov Main Geophysical Observatory* (Leningrad: Hydrometeo Pub) **505** pp 120–29
[8] Meschcerskaay A V, Blazhevich V G and Zhitorchuk U V 1978 Hydrothermal coefficient and connection with yields of agricultural crops *Proc. of the Voeikov Main Geophysical Observatory* (Leningrad: Hydrometeo Pub) **400** pp 134–49
[9] Drozdov O A 1980 *Droughts and dynamic of wetting* (Leningrad: Hydrometeo Pub) p 96
[10] Zolotokrylin A N, Cherenkova E A 2013 Moistening tendencies in grain-growth zone of Russia in beginning of 21st century*Problems of Ecological Monitoring and Ecosystem Modelling* **25** (Moscow: IGCE Roshydromet and RAS) pp 251–64
[11] Ped D A 1975 About index of drought and excessive wetting *Proc. of the Hydrometcentre of USSR* **156** pp 19–38
[12] Wiiliam A 1984 The Palmer Drought Severity Index: Limitations and Assumptions *Journal of Climate and Applied Meteorology* **23** 1100–09
[13] Dai A and et al 2004 A Global Dataset of Palmer Drought Severity Index for 1870–2002: Relationship with Soil Moisture and Effects of Surface Warming *Journal of Hydrometeorology* **5-6** 1117–30
[14] Guttman N B 1998 Comparing the Palmer Drought Index and the Standardized Precipitation Index *Journal of the American Water Resources Association* **34-1** 113–21
[15] Shafer B A, Dezman L E 1982 Development of a Surface Water Supply Index (SWSI) to assess the severity of drought conditions in snowpack runoff areas *Western Snow Conference* (Reno, Colorado State University, Fort Collins) pp 164–75
[16] Wilhite D A 2004 *DroughtInternational Perspectives on Natural Disasters: Occurrence, Mitigation, and Consequences* ed J P Stolman and et al (Netherlands: Kluwer Academic Publishers) pp 147–62
[17] Sadokov V P, Kozelceva V F and Kuznecova N N 2002 Emergence of atmospheric and soil drought with considering atmospheric and soil aridity *Proc. of the Hydrometcentre of Russia* **337** pp 48–56
[18] Gordeev A V, Kleschenko A D, Chernyakov B A and Sirotenko O 2006 Bio-climatic potential

of Russia: theory and practice ed A V Gordeev (Moscow: M.: Association of Scientific Publications KMK) p 512.
[19] Ped D A 1951 About definition of dates when temperatures consistently rose above definite values *Meteorology and Hydrology* **10** 38–9
[20] Dee D P and et al 2011 The ERA-Interim reanalysis: configuration and performance of the data assimilation system *Quarterly Journal of the Royal Meteorological Society* **137** 553–97
[21] Shulgina T M 2012 Development and application of software complex for regional climate changes analysis using modelling data *Synopsis of the Ph.D. thesis* (Tomsk) p 15
[22] Okladnikov I G, Gordov E P, Titov A G and Shulgina T M 2015 *Proc. of the XVII Int. Conf. on Data Analytics and Management in Data Intensive Domains* (*October 13–16, 2015, Obninsk, Russia)* (*Electronic Materials*) ed L Kalinichenko and S Starkov pp 76–80

# Development of Distributed Research Center for analysis of regional climatic and environmental changes

**E Gordov[1,2,3], A Shiklomanov[4], I Okladnikov[1,3], A Prusevich[4] and A Titov[1,3]**

[1]Institute of Monitoring of Climatic and Ecological Systems of the Siberian Branch of the Russian Academy of Sciences, 10/3, Akademichesky pr., Tomsk 634055, Russia
[2]Tomsk State University, 36, Lenina str.,Tomsk 634050, Russia
[3]Institute of Computational Technologies of the Siberian Branch of the Russian Academy of Sciences, Tomsk Branch, 10/4 Akademichesky pr.,Tomsk 634055, Russia
[4]Earth Systems Research Center, Morse Hall, 8 College Road, Durham, NH, USA

**Abstract.** We present an approach and first results of a collaborative project being carried out by a joint team of researchers from the Institute of Monitoring of Climatic and Ecological Systems, Russia and Earth Systems Research Center UNH, USA. Its main objective is development of a hardware and software platform prototype of a Distributed Research Center (DRC) for monitoring and projecting of regional climatic and environmental changes in the Northern extratropical areas. The DRC should provide the specialists working in climate related sciences and decision-makers with accurate and detailed climatic characteristics for the selected area and reliable and affordable tools for their in-depth statistical analysis and studies of the effects of climate change. Within the framework of the project, new approaches to cloud processing and analysis of large geospatial datasets (big geospatial data) inherent to climate change studies are developed and deployed on technical platforms of both institutions. We discuss here the state of the art in this domain, describe web based information-computational systems developed by the partners, justify the methods chosen to reach the project goal, and briefly list the results obtained so far.

## Introduction

The significant increase in the environmental data volumes (up to a petabyte level) and the necessity to store, search, retrieve, process, visualize, and analyze geospatial datasets made the traditional approaches questionable. It required new distributed data access and processing technologies to be created and turned this IT area into a new scientific field for the approaches and tools developed in the recently appeared data intensive application domain [1-5]. The geospatial datasets used in climate and environmental research require the application of a spatial data infrastructure (SDI) approach [6]. This approach, in particular, includes the development of a web portal (geoportal) [7-8] considered as a single access point to SDI resources [9-10]. It provides a functionality of searching the geographic information resources, data sampling and processing, as well as cartographical visualization services along with the corresponding client applications [11]. Such data processing infrastructure should also employ the Web 2.0 standards providing multidisciplinary geographically dispersed research teams with functional capabilities to perform joint investigations [12].

Currently, it is generally accepted that the development of client applications as integrated elements of such infrastructure should be based on the use of modern web and GIS technologies [13-15].

According to the general requirements of the INSPIRE Directive to geospatial data visualization [16], it is necessary to provide such features as data overview, image navigation, scrolling, zooming and graphical overlay, as well as displaying map legends and related meta information. That is, the well-known and widely used basic functionality of a standard GIS should be provided for the potential users. The benefits of the Web-based GIS technologies are quite obvious: theoretical independence of web browsers and operating systems, automatic software location and upgrade, combined use of geographically distributed data sources, and joint use of centralized data warehouses. To perform a reliable analysis of the responses of natural ecosystems and society to the ongoing climate changes, some professional skills and expertise are needed to deal with complex numerical models, modern statistical analysis methods, and programming techniques. Such skills are not very typical for the specialists engaged in related scientific studies. Application of Web-GIS technologies provides environmental scientists and decision makers with reliable comprehensible tools for investigating the climate change impacts using a well-known graphical user interface.

There are several computational and/or information web-systems working with geophysical data linked to environmental science. One of them deals with the analysis of real-time meteorological observational data [17]. A very sophisticated and functional system for analysis and visualization of satellite data was developed at NASA: GES-DISC Interactive Online Visualization ANd aNalysis Infrastructure (GIOVANNI), (http://daac.gsfc.nasa.gov/techlab/giovanni/) [18]. Among those RIMS (http://RIMS.unh.edu/) and CLIMATE (http://climate.climate.scert.ru/) systems are developed by the project partners.

A fully integrated advanced system created at the University of New Hampshire for online analysis of heterogeneous climate, hydrology, and remote sensing data, RIMS was a successful attempt to realize a multi-functional GIS as a web application using MapServer (http://mapserver.org/). RIMS is widely used for various research projects [19-20] including NEESPI (Northern Eurasian Earth Science Partnership Initiative, http://NEESPI.sr.unh.edu/). The main feature of RIMS is that it deals directly with geospatial datasets instead of the corresponding cartographical images, and allows access to the displayed data at each pixel of the map, as well as search and access to data aggregated according to some space and/or time criteria.

A Web-GIS platform called CLIMATE developed at the Institute of Monitoring of Climatic and Ecological Systems SB RAS is intended for processing and visualization of meteorological and climate data [21], thus providing an integration of the results of modeling and observations with the functionality of a retrospective analysis. The platform is accessed via a geoportal which serves as an integration point providing web applications, web mapping services, and metadata management. The platform includes: structured geospatial datasets along with relevant metadata; a computational core (an independent set of modules written in GDL–GNU Data Language, http://gnudatalanguage.sourceforge.net) and Python (http://www.python.org)); spatial data services, geoportal and Web-GIS client. It also provides access to complex numerical models and to the calculation results obtained. In addition, the platform provides researchers with the thematic communication environment and professional education support required for virtual training of young scientists. The developed prototype of a hardware and software platform supports integrated regional climate change studies in Siberia within the framework of the NEESPI Program [22]. It combines Web 2.0 capabilities, tools for running climate models, processing and visualization of large geospatial datasets, and joint application development by dispersed research teams. It also supports virtual training courses for Tomsk State University students and postgraduates [23].

To develop specialized DRC supporting interactive processing and analysis of large geospatial datasets using comprehensive computational and visualization tools the above-mentioned CLIMATE and RIMS systems have been chosen as a basis. Below we describe these in detail and present an approach elaborated to reach the project ultimate objective. It will provide the specialists working in climate and related sciences focused on climate change impact assessment, adaptation strategies, and other climate related activities with accurate and detailed climatic characteristics and reliable tools for an in-depth statistical analysis.

## 1. Web based information-computational systems developed by the partners

### *1.1. Web-GIS platform CLIMATE*

The platform CLIMATE (http://climate.climate.scert.ru/) is aimed to support regional climatic and ecological changes research [24-26]. In particular, it allows the user to perform calculations of statistical characteristics of meteorological parameters, analysis of periodic and non-periodic processes, and estimation of the derivative climatic indices on the basis of observational and modeling datasets. Also, the platform allows running the mathematical models integrated into it and can be used for supporting collaborative scientific research projects and education of high-grade and post-graduate students of the corresponding departments. It incorporates software components elaborated within the framework of such international projects as ATMOS (http://atmos.iao.ru/ and http://atmos.scert.ru/ ); RISKS (http://climate.risks.scert.ru/) and ENVIROMIS (http://enviromis.scert.ru/en/) aimed at the development of informational-computational systems. Detailed description of those systems is given in [8, 27-31].

Software of the platform was developed on the base of the special software framework [32]. This framework allows development of thematic information-computation web mapping systems according to specifics of application tasks by connecting the new software elements to its basis. These new software elements might include:

- geospatial datasets represented by modeling and observational data grouped by features and placed on remote storage systems;
- server-side external software modules of computational core;
- client-side GUI components;
- the integrated climate model Planet Simulator [33] and the mesoscale meteorological model WRF [34];
- integrated Web 2.0 tools supporting interaction of distributed groups of researchers and software developers (forums, wiki, blogs, and control system versions).

At present the software of the platform is hosted at IMCES SB RAS (Institute of Monitoring of Climatic and Ecological Systems, Siberian Branch of Russian Academy of Sciences, Tomsk, Russia) on a hardware that includes: high performance server HP Proliant DL585 G7 (four 12-core processors AMD Opteron 6172 and 32 Gb RAM) as the computational core; dedicated server (Intel Pentium 4 and 2 Gb RAM) for the geoportal; and dedicated server (two Intel Xeon 5130 and 4 Gb RAM) for Geoserver software. All servers are interconnected using 1 Gb Ethernet. Collections of spatial datasets are prepared for using and stored on three storage systems with a total capacity of 252 Tb. These datasets include several reanalyses and some results of climatic modeling. Also, modeling results of "Planet simulator" and "WRF" models, archives of remote sensing data for LandSat 4-7, Global Land Survey (GLS) and MODIS (http://glovis.tsc.ru/), as well as observational data from meteorological stations located on the Russian Federation territory are included in the platform. All datasets were validated, and their quality was approved by the data origin institution. Each dataset was reformatted and placed in the storage system in a unified manner. The core modules provide functions for data search, retrieval and pre-processing. Additional software modules were developed to provide the platform with the required thematic area data processing and visualization functionality. They were connected to the computational core of the platform using a dedicated API. Graphical user interface of the platform was designed to meet the requirements of the application tasks. It provides the user with intuitive browser-based instrument similar to such widespread desktop GIS applications as uDIG, QuantumGIS, etc. (see Figure 1).

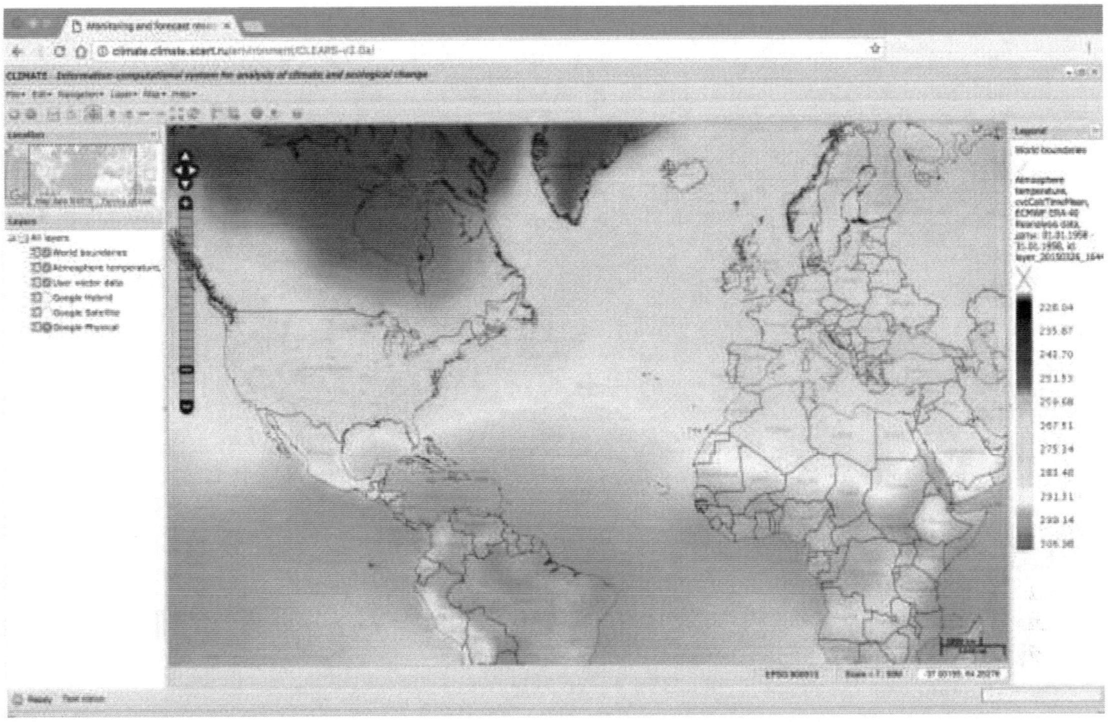

**Figure 1.** Web-application graphical interface.

The web interface main function is to formalize the requests of a user solving a particular computational task in the field of climatic and ecological research, to compile a task technical description for the computational core, and to present the computational results in digital and graphical formats using modern web GIS technologies in a way similar to the conventional desktop GIS. The environment aimed at organizing the interactions of distributed research groups represents a complex solution for the platform information support and includes capabilities for creating user groups with access rights, public and limited (to particular users and groups) blogs, links to publications (Internet links, presentations, papers, analysis results), and collaborative software and documentation development (version control system).

At present, the platform developed is used for studying the ongoing climatic changes and their consequences in Siberia. The obtained results of an analysis of the surface temperature and precipitation changes are in a good agreement with the results obtained by other researchers, thus approving the reliability and effectiveness of the platform for climate and ecosystems change analysis [35]. The educational components of the platform give an opportunity to use it for education of high-grade and postgraduate students [36]. Since 2012 students of the Meteorology and Climatology Department of Tomsk State University are using the platform to study the basics of climate change analysis and to perform on-line computational labs supporting courses on "Regional climate change analysis" and "Future climate analysis".

### *1.2. RIMS: Rapid Integrated Mapping and analysis System*

RIMS (Rapid Integrated Mapping and analysis System) is developed to test the usability of Web interfaces and services with a wide range of geospatial Earth system and socio-economic information [37]. It is predominantly a data oriented system interfacing the digital content of the datasets rather than their image representation such as data viewing Web tools, e.g. Google Maps/Earth applications. In addition to visualization/mapping, the major Web services of RIMS are specifically designed for interfacing the following digital data content:

- dataset search (Catalog) services. Presently there are over 10000 raster and vector time series and single layer datasets (in 100+ file formats readable by the GDAL library) in our data holdings;
- Data Cube views of time series and/or spatially aggregated data (prototype of Web Aggregation Service: WAS). This is the ability to view different temporal aggregation subsets of the same time series data (e.g. daily, monthly, yearly, daily climatology or long term averages, etc.) as well as their spatial aggregations (e.g. per country, per watershed, etc. integral averages);
- instant access to raw data values for each pixel of client-side displayed maps (WFS/WCS). This is a raw data layer on the client side displayed on the top of the map imagery;
- on-the-fly data production and manipulation tools (a Data Calculator that does algebra, calculus, and built-in server side modeling (e.g. water balance model) over arbitrary sets of data) that utilize concepts of Web Production Service (WPS) extensions;
- tools for easy mounting and management of vector, raster and point/station data and metadata.

An illustration of the Web-client application of RIMS is given in Figure 2.

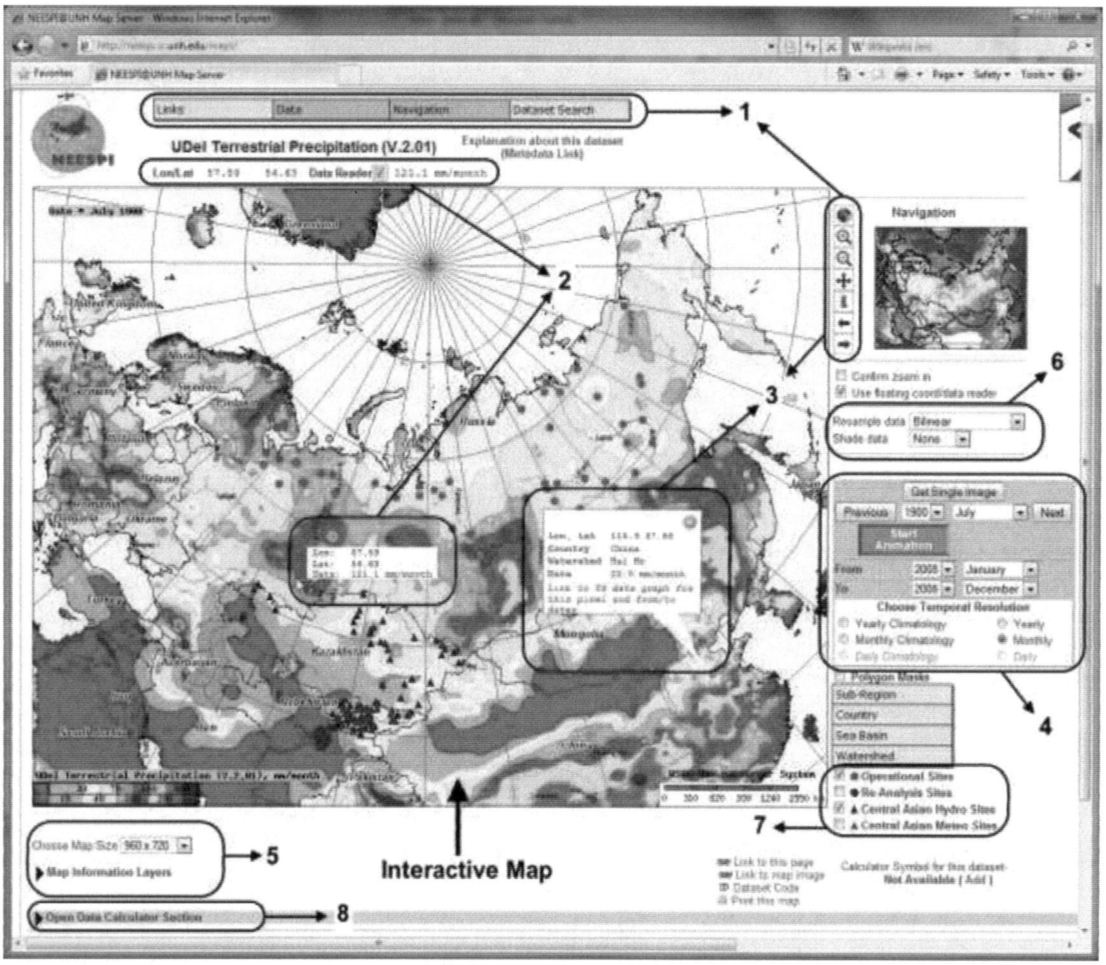


**Figure 2.** Web-client application of an adaptive projection (North always up) customized for the NEESPI site with adaptive dynamic re-projection (always North up). Keys: (1) basic data and map navigation tools include drop-down menu for data search/selection, sidebar with zoom/pan, back/forward map history tools for spatial/data navigation, metadata link, etc.; (2) interactive

coordinate and map data value reader for the mouse over locations (loads raw map digital data to the browser); (3) pixel query tool (i-tool) gets coordinates, country, watershed, and map data values in a callout frame that has a link to time-series graphs and spreadsheet data for this pixel as well as pre-built polygon aggregations for this pixel location; (4) time series navigation tool allows the user to choose and display maps/animations for a given date(s) and switch temporal resolutions for the selected dataset (see DataCube concept section); (5) map size and base layer choices including a DEM river network option; (6) data interpolation and shading tools allow choices for smoothed/interpolated visual interpretations; (7) point/station data list: selected datasets (stations) are displayed on the map with clickable symbols that open station pages in a separate browser window; (8) fold-out section to run the Data Calculator application to perform mathematical and logical functions over gridded or vector datasets to build custom maps, graphs and spatial queries (see Data Calculator section).

The idea of the DataCube web aggregation service (WAS) is to combine data mapping and query in three virtual dimensions: (a) scaled time aggregation, (b) non-scaled or climatological time aggregation, and (c) polygon aggregation. Only one dataset in the DataCube represents the original core data, while the others are pre-built (predetermined) aggregation derivatives, as illustrated in Figure 3. The DataCube concept allows the user to explore the same dataset in many different ways without the need to run their own aggregation software. For example, a functional Time-Series Navigation tool in RIMS is built on the WAS concept (Figure 2, item #4).

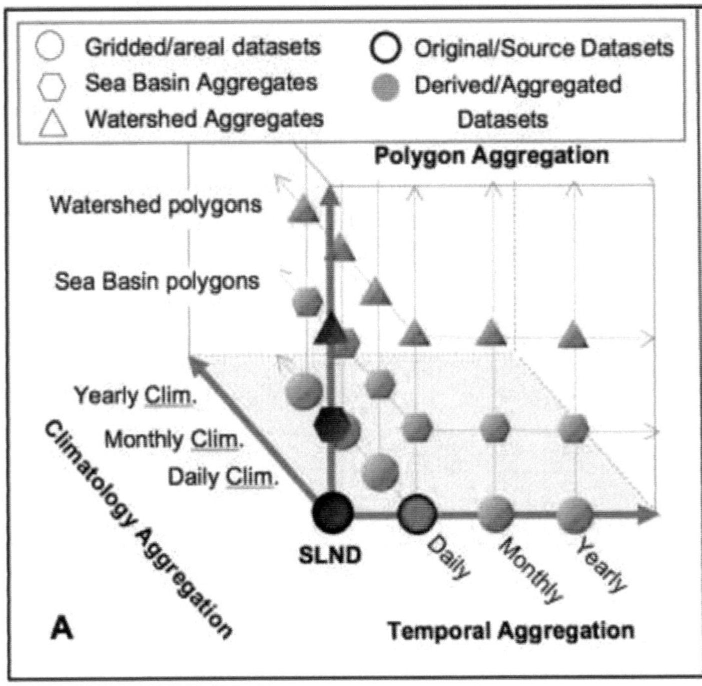


| Aggregation Type | Data Example |
|---|---|
| 1. Mean | Temperature |
| 2. Sum | Population |
| 3. Maximum Frequency | Land Cover |
| 4. Resultant Vector | Wind speed and direction |

B

**Figure 3.** Concept of data aggregation scheme used in RIMS. Original Daily dataset (e.g. NCEP daily temperature at 2 m) can be aggregated along the temporal scale to monthly and yearly derivative datasets, and along the climatology scale to daily, monthly and yearly climatology (long-term averages) derivative sub-datasets. In turn, each of these can be aggregated by any number of polygon sets (on the polygon aggregation scale) to polygon averages or cumulative (e.g. average temperature per country). Single layer non-dated datasets (e.g. elevation) can be aggregated only along the polygon aggregation scale (e.g. average elevation of a watershed). Aggregation method can be one of the following types: (1) average, e.g. temperature; (2) cumulative, e.g. population; (3) max frequency, e.g. land cover; (4) vector average, e.g. wind.

The Data Calculator concept for the Web Processing Service (WPS) extension originated from the idea to allow the user to perform interactive on-the-fly arbitrary mathematical, statistical, custom functions, and/or logical statements/queries on a pixel or at the polygon (area integral) level. In the existing prototype it works with any number of datasets and their time series data layers. This service will be provided by a dedicated Data Production Server, which first loads data hyperslabs from the WCS or directly from the primitive data DB and then validates and evaluates the submitted equations and/or logical expressions over the data. The WPS returns the calculation results in any combination of the following formats: (1) maps; (2) data value frequency histograms; (3) georeferenced data files (e.g. NetCDF, AAIGrid or other GDAL formats); (4) polygon time series data (spreadsheet format); (5) graph of polygon data; and (6) statistical moments of the resulting data.

## 2. The approach

The project includes several major directions of research (Tasks) listed and described below. Existing geospatial datasets used in climate and environmental research are very heterogeneous in storage structure and file formats; they cover different spatial and temporal scales, consist of different sets of climatic and environmental characteristics, and often have a huge size that significantly complicates their analysis. Thus, the first step is pre-processing of these geospatial datasets unifying their storage formats and access tools that allow searching and the retrieval of multidimensional subsets for specific research needs of various users. We shall develop an approach that will unify access to distributed archives of environmental data hosted at ESRC and IMCES and extract (generate) principal metadata for these datasets.

Network Common Data Form (netCDF) was chosen as the major file format for most geospatial data in our data archive. This format is formally acknowledged by scientific institutions (including UCAR - University Corporation for Atmospheric Research, http://www2.ucar.edu/) and OGC (Open Geospatial Consortium, http://www.opengeospatial.org/) as a standards' candidate for storing geospatial data and stimulating data exchange in binary form. It is perfectly suited for storing geospatial data and supported by a wide range of commercial and open source software.

The data are stored on data storage systems as collections of netCDF files and arranged in a strict hierarchy of directories:

```
/<data root directory>/
        <data collection name>/
                <spatial domain resolution>/
                        <time domain resolution>/
                                <files and directories with data>
```

Here, *<data root directory>* is the root location of data collections, *<data collection name>* is the name of a directory containing a single data collection, *<spatial domain resolution>* is the name of a directory containing data with the same horizontal resolution, *<time domain resolution>* is the name of a directory containing data with the same time step. All data files (sometimes grouped in subdirectories) are located deeper in the hierarchy. The names of files and subdirectories are not regulated and determined by the specifics of a particular data set. Each data file contains one or more multi-dimensional arrays of meteorological parameters. The data are stored on working nodes or on connected data storage systems (Local data) according to a "data awareness" principle. Data awareness is some knowledge about all of the data available within each working node and where those data are located [38]. According to this information, the datasets are processed locally, on the nodes containing them. A detailed description of the data storage and processing model is represented in [39].

Due to the huge volume of geospatial datasets, additional measures are taken to increase the speed of data searching, viewing, and sampling. Within the framework of the project, a dedicated database of metadata describing the available at ESRC and IMCES multidimensional arrays of climatic, environmental, demographic, and socio-economic characteristics will be developed on the base of the

DBMS MySQL. This database will facilitate and accelerate searching of required information on storage systems.

The rapidly increasing volumes of information during the past decade significantly decreased the effectiveness of single-server Web GISs, like a thematic Web GIS for cloud analysis of multidimensional climatic characteristics recently developed at IMCES SB RAS. This project seeks to develop an online Distributed Research Center with information and computing software package for distributed processing of large datasets. A paradigm of "share nothing" provides the best approach to solving this problem. This technology, when the data are residing where they are and only upfront software handles the information needs of the customers, will be used for the development of a new computing software complex, i.e., the Center itself. All efforts to minimize the typical shortcomings of this paradigm will be made within the framework of the project.

More than 90% of queries for climatic and environmental data services are quite simple. The users request spatial pattern of one or more variables, their temporal dynamics and a suite of their basic statistics. They may also ask for specific analyses (e.g., "what if") made on-line. Scientists prefer to do their own assessments and require original (but preprocessed) data and comprehensive information about their specifics and quality (metadata). Only in rare situations an in-depth analysis for specific issues is required from the Data Services. The proposed DRC will respond to all three types of these queries. The first two types will be handled on-line, and the third type of requests will be processed in-house by several dedicated software packages on an order-response basis. The DRC staff will collect (or project) typical requests for the third, most demanding type of queries and develop a suite of software packages to allow expediently addressing them.

Summarizing the approach we chose the following features for the DRC under development:

- geoportal with user-friendly and intuitive clear GUI as a single access point to distributed archives of spatial data and tools for their processing;
- cloud data analysis and visualization;
- flexible modular computational engine with verified data processing routines;
- export of processing results and metadata as image (PNG, JPEG, GIF, GeoTIFF), as NetCDF files as well as access via WMS and WFS.

To meet the above requirements, the following tasks are carried out now:

- development of an architecture and defining the major hardware and software components of the DRC for monitoring and projecting the regional environmental changes;
- development of an information database and computing software suite for distributed processing and analysis of large geospatial data hosted at ESRC and IMCES;
- development of geoportal, thematic web client and web services providing international research teams with an access to cloud computing resources at DRC; two options will be executed: access through a basic graphical web browser and using geographic information systems.

Using an output of these tasks, compilation of the DRC prototype, its validation, and testing the DRC feasibility for analyses of the recent regional environmental changes over Northern Eurasia and North America and public services is planned to be done this fall.

## 3. First results

### *3.1. Metadata database*

To describe geospatial datasets and their processing routines and provide effective DRC functioning, a dedicated metadata database (MDDB) was developed. This database contains spatial and temporal characteristics of available geospatial datasets, their locations, and run options of software components for data analysis. The tables in MDDB are divided into "technical" and "interface" ones. The technical tables contain data intended for computing software components. The interface tables hold textual multilingual content for a graphical user interface.

There are two major parts of MDDB providing description of climate datasets and description of data processing software components (computing modules).

Each climate dataset in MDDB is uniquely identified by its four major characteristics: name of the data collection, resolution of the horizontal grid, resolution of the time grid and name of the modeling scenario (if applicable). And each dataset includes one or several data arrays containing values of various meteorological parameters given on spatial and temporal grids. The information about all datasets available for analysis is stored in the first part of MDDB. It is used to locate the data files and to provide the metadata on request.

Geospatial data processing is performed by a set of dedicated computing modules. These modules are run in accordance to a pipelined call sequence. This sequence is prepared by the web portal on the basis of user interactions with the graphical user interface. Second part of MDDB contains a description of various call sequences and their options. Since some data analysis routines are designed to process only specific meteorological parameters, the connections between computing modules and data arrays are set in MDDB.

### *3.2. Web-GIS client*

The boundless/OpenGeo architecture was used as the basis for a Web-GIS client development. The cartographical web application (Web-GIS client) for working with archive of geospatial NetCDF datasets contains 3 basic tiers [24]:

- tier of NetCDF metadata in JSON format;
- middleware tier of JavaScript objects implementing methods to work with:
  - NetCDF metadata;
  - XML file of selected calculations configuration (XML task);
  - WMS/WFS cartographical services;
- graphical user interface tier representing JavaScript objects realizing general application business logic.

*3.2.1. NetCDF metadata tier.* Web-GIS client metadata tier represents a set of interconnected JSON objects, created on the base of MySQL metadata relations, and presenting NetCDF datasets information (spatial and temporal resolutions, meteorological parameters available, acceptable processing procedures, etc.). Generally, there are two kinds of objects:

- objects with the structure conforming to the corresponding metadata database relations, for instance, object of measurement units;
- objects based on complex SQL queries to metadata relation sets that allow fast retrieving of necessary information using MySQL indices as associative array keys.

The structure of JSON objects was chosen according to the following criteria:

- efficiency of filling out graphical user interface interactive forms;
- optimization of process of creating and editing of XML file of selected calculations configuration (XML task).

It might be concluded that by virtue of the approach chosen above the processes of interaction between user and metadata database via Web-GIS graphical interface are optimized.

*3.2.2. Middleware tier of JavaScript objects.* This tier implements methods to work with NetCDF metadata, XML task file and WMS/WFS cartographical services, and appears to be a middleware, which connects JSON metadata and graphical user interface tiers. The methods include such procedures as:

- loading and updating of metadata JSON objects using AJAX technology;
- creating, editing, serialization of XML calculation task object;
- launching and tracking the task execution process located on the remote calculation node;

- working with WMS/WFS cartographical services: obtaining a list of available layers, presenting layers on a map, export layers into various formats according to a user request, obtaining and presenting a layer legend with the selected SLD style applied.

*3.2.3. Graphical user interface.* The tier is based on a conjunction of JavaScript libraries, such as OpenLayers, GeoExt and ExtJS, and represents a set of software components either standalone (information panels, buttons, list of layers, etc.), or implementing general application business logic (menu, toolbars, wizards, mouse and keyboard event handlers, and so on). The graphical interface performs two main functions: providing functional capabilities for editing XML task file, and a visual presentation of cartographical information for the end user. It is similar to the interfaces of such popular classic GIS applications as uDig, QuantumGIS, etc. The basic elements of the graphical user interface include (figure 1):

- panel displaying user cartographical layers on the map. Google maps are used as the base layer by default, but there's a possibility to set an arbitrary base layer including newly created by the user;
- layer tree allowing to toggle layer display;
- layer legends display panel;
- map information panel (scaling, cartographical projections, cursor geographical coordinates)
- application general status panel;
- overview map panel;
- general application menu;
- toolbar (adding/removing layer, saving NetCDF data, panning, map refresh, obtaining of information related to given geographical point, etc.);
- application context menu;
- wizard creating cartographical layers based on results of computational processing of geospatial datasets available to the system.

The toolbar, application, and context menus contain mouse and keyboard event handlers which uniquely define Web-GIS behavior depending on the user actions with the execution context applied. Web-GIS client complies with the general INSPIRE standard requirements and provides computational processing services launching to support solving tasks in the area of environmental monitoring, as well as presenting the calculation results in the form of WMS/WFS cartographical layers in raster (PNG, JPG, GeoTIFF), vector (KML, GML, Shape), and binary (NetCDF) formats.

It should be noted that the geospatial data cartographical services based on Geoserver software can be used in the Web-GIS client considered as well as in the standard desktop GIS applications.

## 4. Specific results

To date, there is no formalized description of the metadata database for large sets of geospatial meteorological and climatic data. The architecture presented above is the first attempt to address this crucial for data intensive domain problem. The designed metadata database solves three main tasks: 1) provides a content for the graphical user interface; 2) provides the geoportal with the information needed to generate a correct task file for the computational core; 3) contains information on the structure and arrangement of data sets allowing the computational core to read and process them efficiently. The use of this database organizes the information on the available data sets, facilitates an automatic retrieval of the data files, and improves the scalability and flexibility of the computations.

The developed GIS Web client is based on the architecture Boundless / OpenGeo. The first version of the GUI uses interrelated JavaScript libraries, OpenLayers, GeoExt and ExtJS and represents a set of software components including independent components (dashboards, buttons, layer lists), and those implementing a general logic of the software application (menus, toolbars, wizards, mouse and keyboard event handlers, etc.).

The first application of the above-developed metadata database and user interface showed that their combined usage facilitates the expanding of the set of data archives available for analysis and adding new statistical processing procedures [40].

The results obtained show that the thus developed VRE and tools will be useful for the decision makers and specialists working in the related sciences, with the focus of their work as socio-economic impact assessment, ecological impact assessment, adaptation strategies, science policy administration, and other climate related activities. On this basis they will get the reliable climate related characteristics required for studies of the economic, political, and social consequences of global climate change at a regional level.

## 5. Conclusions

The obtained and anticipated project results will create a pathway for the development and deployment of thematic international virtual research centers focused on interdisciplinary studies by international research teams.

The software developed in this project will be utilized for processing the partners datasets and other datasets including observations and reanalysis products in order to perform a space-time analysis of the recent and possible future dynamics of climate and all of its components with a special emphasis on the extreme climatic phenomena in the Northern extratropics.

The following main goals will be achieved upon the completion of the project:

- data analysis products made available to researchers and stakeholders in the form of digital atlases and data arrays in electronic form through the project website and other recognized media;
- interactive web tools developed for a comprehensive analysis of climate variables and their derivatives provided by the thematic geoportal;
- dialog with stakeholders established for further improvement of the developed technology.

The DRC will provide an opportunity for scientists, professionals, and decision-makers to use various geographically distributed and georeferenced resources and processing services via a web browser by integrating the geoportal distributed storage, processing, and delivery of information. In particular, it will allow carrying out a simultaneous analysis of several georeferenced thematic data sets using modern statistics and thus identify the responses of ecological processes to climate change.

**Acknowledgment**

This work is funded by RF Ministry of Education and Science Project 14.613.21.0037 RFMEFI61315X0037 and by the Iola Hubbard Climate Change Endowment.

**References**

[1] Big Data Definition *MIKE 2.0* http://mike2.openmethodology.org/wiki/Big_Data_Definition
[2] Kusnetzky D 16 February 2010 What is "Big Data?" *ZDNet* http://www.zdnet.com/article/what-is-big-data/
[3] Vance A 22 April 2010 Start-Up Goes After Big Data With Hadoop Helper *New York Times Blog* http://bits.blogs.nytimes.com/2010/04/22/start-up-goes-after-big-data-with-hadoop-helper/?dbk
[4] Kalinichenko L A, Volnova A A, Gordov E P, Kiseleva N N, Kovaleva D A, Malkov O Yu, Okladnikov I G, Podkolodny N L, Pozanenko A S, Ponomareva N V, *et al* 2016 Problems of access to data in Data Intensive Domain in Russia *Informatics and Applications* **10** 3–23
[5] Hilbert M 2016 Big Data for Development: A Review of Promises and Challenges. *Development Policy Review* **34** (1) 135–74 doi: http://doi.org/10.1111/dpr.12142
[6] Steiniger S and Hunter A J S 2012 Free and open source GIS software for building a spatial data infrastructure *Geospatial Free and Open Source Software in the 21st Century* ed Bocher E, Neteler M (Heidelberg: LNGC, Springer) pp 247–61
[7] Gordov E P and Fazliev A Z 2004 *J. Computational Technologies (Spec. Iss.)* **9** 123–26

[8] Gordov E P, Lykosov V N and Fazliev A Z 2006 *Advances in Geosciences* **8** 33–38
[9] Koshkarev A V, Ryakhovskii A V and Serebryakov V A 2010 Infrastructure of distributed environment of storage, search and transformation of geospatial data *Open Education* **5** 61–73
[10] Krasnopeev S M 2001 Experience of deployment of key elements of special data infrastructure on the base of web-services *Proc. XIV All-Russia Conf. on Internet and Modern Society* (November 20-23, 2001, St. Petersburg, Russia) pp 92–9
[11] Koshkarev A V 2008 Geoportal as a tool to control spatial data and services *Spatial data* **2** 6–14
[12] Allan R J 2009 *Virtual Research Environments: from Portals to Science Gateways* (Chandos: Oxford Publishing)
[13] Yakubailik O E 2007 Geoformation geoportal *J. Computational Technologies (Special Issue 3)* **12** 116–25
[14] Dragicevic S, Balram S and Lewis J 2000 The role of Web GIS tools in the environmental modeling and decision-making process *4th Int. Conf. on Integrating GIS and Environmental Modeling (GIS/EM4): Problems, Prospects and Research Needs* (September 2-8, 2000, Banff, Alberta, Canada) http://www.srcosmos.gr/srcosmos/showpub.aspx?aa=5823
[15] Frans J M van der Wel 2005 Spatial data infrastructure for meteorological and climatic data *Meteorol. Appl.* **12** 7–8
[16] Vatsavai R R, Thomas E B, Wilson B T and Shekhar S 2000 A Web-based browsing and spatial analysis system for regional natural resource analysis and mapping *Proc. of the 8th ACM Int. symp. on Advances in geographic information systems* (Noveber 6-11, 2000, Washington, D.C., US) pp 95–101
[17] Samet R and Tural S 2010 Web based real-time meteorological data analysis and mapping information system *WSEAS Transactions on Information Science and Applications* **7**(9) 1115–25
[18] Berrick S W, Leptoukh G, Farley J D and Rui H 2009 GIOVANNI: A Web service workflow-based data visualization and analysis system *IEEE Trans. Geosci. Remote Sens.* **47**(1) 106–13.
[19] Welp L R, Randerson J T, Finlay J C, Davydov S P, Zimova G M, Davydova A I and Zimov S A 2005 A high-resolution time series of oxygen isotopes from the Kolyma River: Implications for the seasonal dynamics of discharge and basin-scale water use *Geophys. Res. Lett.* **32** L14401 doi: 10.1029/2005GL022857.
[20] Shiklomanov A I 2012 Regional Integrated Monitoring System for the Pan-Arctic (RIMS) *Proc. of the Northern Eurasian Earth Science Partnership Initiative (NEESPI) Regional Science Team Meeting devoted to the High Latitudes* (June 2-6, 2012, Helsinki, Finland) pp 128–30
[21] Gordov E P, Lykosov V N, Krupchatnikov V N, *et al* 2013 *Computational-Information Technologies for Monitoring and Modeling of Climate Change and its Consequences* (Novosibirsk: Nauka)
[22] Gordov E, Bryant K, Bulygina O, Csiszar I, Eberle J, Fritz S, Gerasimov I, Gerlach R, Hese S and Kraxner F et al 2013 Development of Information-Computational Infrastructure for Environmental research in Siberia as a baseline component of the Northern Eurasia Earth Science Partnership Initiative (NEESPI) Studies *Regional Environmental Changes in Siberia and Their Global Consequences (Series: Springer Environmental Science and Engineering)* ed P Ya Groisman and G Gutman **XII** pp 19–55
[23] Gordova Yu E, Genina E Yu, Gorbatenko V P, Gordov E P, Kuzhevskaya I V, Martynova Yu V, Okladnikov I G, Titov A G, Shulgina T M and Barashkova N K 2013 Support of the educational process in the field of modern climatology based on Web GIS "Climate" platform *Open Distance Education* **1**(49) 14–9
[24] Okladnikov I G, Gordov E P and Titov A G 2012 Hardware-software platform «Climate» as a basis for local spatial data infrastructure geoportal *Vestnik NSU (Series: Information Technologies)* **10** (4) 105–11

[25] Okladnikov I G, Gordov E P, Titov A G, Bogomolov V Yu, Shulgina T M and Genina E Yu 2012 Geoinformation system for investigation of regional climatic changes and first results obtained *Atmos.Ocean. Opt.* **25** (2) 137–43

[26] Gordov E P, Okladnikov I G and Titov A G 2016 Information and Computing Web-System for Interactive Analysis of Georeferenced Climatological Data Sets *Vestnik NSU (Series: Information Technologies)* **14** (1) 13–22

[27] Baklanov A and Gordov E P 2006 Man-induced environmental risks: monitoring, management and remediation of man-made changes in Siberia *J. Computational technologies (Special issue)* **11** 162–71

[28] Gordov E P and Lykossov V N 2007 Development of information-computational infrastructure for integrated study of Siberia environment *J. Computational technologies (Special issue)* **12** 19–30

[29] Gordov E P, Okladnikov I G and Titov A G 2007 Development of elements of a web-based information-computational system for studies of regional environment processes *J. Computational technologies (Special issue 3)* **12** 20–8

[30] Gordov E P, Okladnikov I G and Titov A. G 2011 Application of web mapping technologies for development of information-computational systems for georeferenced data analysis *Vestnik NSU (Series: Information Technologies)* **9** (4) 94–102

[31] Shulgina T M, Gordov E P, Okladnikov I G, Titov A G, Genina E Y, Gorbatenko V P, Kuzhevskaya I V and Akhmetshina A S Computational module for regional climate change analysis *Vestnik NSU (Series: Information Technologies)* **11** (1) 124–31

[32] Titov A G and Okladnikov I G 2014 Architecture of Web-GIS for climate monitoring based on geospatial data services *Vestnik NSU (Series: Information Technologies)* **12** (1) 79–88

[33] Fraedrich K, Jansen H, Kirk E *et al* 2005 The Planet Simulator: Towards a user friendly model *Meteorologische Zeitschrift* **14** 299–304

[34] Skamarock W C, Klemp J B, Dudhia J, Gill D O, Barker D M, Duda M G, Huang X Y, Wang W and Powers J G 2008 *A Description of the Advanced Research WRF Version 3* (Technical note NCAR/TN-475+STR, June 2008) http://www2.mmm.ucar.edu/wrf/users/docs/arw_v3.pdf

[35] Shulgina T M, Genina E Yu and Gordov E P 2011 *Environmental Research Letters* **6** 045210

[36] Gordova Yu E, Martynova Yu V and Shulgina T M 2014 *Bulletin of Irkutsk State Univ. (Series: Earth's Sciences)* **9** 55–68

[37] Prusevich A A, Shiklomanov A I and Lammers R B 2012 RIMS: An Integrated Mapping and Analysis System with Application to Siberia *Regional Environmental Changes in Siberia and Their Global Consequences* ed P Ya Groisman and G Gutman (New York: Elsevier) pp 46–50

[38] Brackett M 2012 *Data Resource Integration: Understanding and Resolving a Disparate Data Resource* (Technics Publications, LLC publishing) p 540

[39] Okladnikov I G, Gordov E P and Titov A G 2016 Development of data storage and processing model *IOP Conf. Ser.:Earth and Environmental Science* In press

[40] Ryazanova A A, Voropay N N and Okladnikov I G 2016 Application of information and computing web system «Climate» for estimation of aridity of South Siberia *Proc. of Int. Conf. and Early Career Scientists School on Environmental Observations, Modeling and Information Systems ENVIROMIS-2016 (*July 11-16, 2016, Tomsk, Russia) pp 358–62

# Environmental science applications with Rapid Integrated Mapping and analysis System (RIMS)

**A Shiklomanov [1, 2, 3], A Prusevich [1], E Gordov [4,5,6], I Okladnikov [4,6] and A Titov [4,6]**

[1] Earth Systems Research Center, UNH, Durham (NH), USA
[2] P.P.Shirshov Institute of Oceanology, Russian Academy of Sciences, Moscow, Russia
[3] Arctic and Antarctic Research Institute, St. Petersburg, Russia
[4] Institute of Monitoring of Climatic and Ecological Systems of the Siberian Branch of the Russian Academy of Sciences, Tomsk, Russia
[5] Tomsk State University, Tomsk, Russia
[6] Institute of Computational Technologies of the Siberian Branch of the Russian Academy of Sciences, Tomsk Branch, Tomsk, Russia

**Abstract.** The Rapid Integrated Mapping and analysis System (RIMS) has been developed at the University of New Hampshire as an online instrument for multidisciplinary data visualization, analysis and manipulation with a focus on hydrological applications. Recently it was enriched with data and tools to allow more sophisticated analysis of interdisciplinary data. Three different examples of specific scientific applications with RIMS are demonstrated and discussed. Analysis of historical changes in major components of the Eurasian pan-Arctic water budget is based on historical discharge data, gridded observational meteorological fields, and remote sensing data for sea ice area. Express analysis of the extremely hot and dry summer of 2010 across European Russia is performed using a combination of near-real time and historical data to evaluate the intensity and spatial distribution of this event and its socioeconomic impacts. Integrative analysis of hydrological, water management, and population data for Central Asia over the last 30 years provides an assessment of regional water security due to changes in climate, water use and demography. The presented case studies demonstrate the capabilities of RIMS as a powerful instrument for hydrological and coupled human-natural systems research.

## 1. Introduction

The Earth's environment is undergoing rapid transformations and there is alarming evidence that important tipping points, leading to irreversible changes in major ecosystems and the planetary climate system, may already have been reached or passed [1]. Wide recognition of the climate change problem by the scientific community has stimulated significant efforts in developing new monitoring [2,3], modeling, and analysis tools [4,5] to improve our understanding of ongoing environmental changes. However, in spite of a growing number of spatially and temporally coherent environmental data sets, there is a lack of user friendly systems with the capability to discover, integrate, and analyze multiple geophysical and socioeconomic data sets [6]. Because of the significant, accelerating changes, it is imperative that researchers [7], decision makers, and students [8] are equipped with appropriate data and interactive tools to identify, analyze, and communicate these changes in a rapid manner using applications.

The integration of web-based services with Earth science observational and modeling frameworks [6,9] is not new and has been previously implemented for specific projects in the fields of climatology [10], hydrology [11], oceanography [12], and global ecosystems [13], where GIS technologies are an important part of the scientific research. The introduction of Open Geospatial Consortium-OGC standards for web-based mapping and data querying and processing [14] as well as the maturity of other web-based technologies (e.g. Open Layers) lead to further popularity of such system. However, many fundamental environmental science data needs, such as multi-layer datasets representing time series of environmental variables (e.g. temperature, precipitation, etc.), are still not implemented in the OGC standards and other related web-based technologies [15].

## 2. Online Rapid Integrated Mapping and analysis System (RIMS)

To facilitate operation and analysis of diverse data, researchers at the University of New Hampshire (UNH) developed RIMS (Rapid Integrated Mapping and analysis System), an online informatics tool for multidisciplinary data visualization, analysis, and manipulation with a focus on hydrological applications [16,17]. Presently, the RIMS data pool covers a variety of themes including climate, hydrology, land cover, demographics, economics, and others (see Table 1).

**Table 1.** Summary of RIMS data holdings.

| Earth System Science Data Category | Key Sources | Examples of Major Parameters | Current Dataset Count | |
|---|---|---|---|---|
| | | | Source | Source + aggregations |
| Hydrology | UNH, CCNY, UAF | Discharge, runoff, evapotranspiration, river networks, irrigation, water use | 650 | 4000 |
| Past and Present Climate | NASA, NOAA, UDel, Princeton University | Temperature, precipitation, evapotranspiration (ET), solar radiation, pressure, wind, humidity | 100 | 400 |
| | NCEP, MERRA reanalysis | | 62 | 210 |
| Future Climate and Hydrology | IPCC, UNH | Temperature, precipitation, ET, snow, runoff, discharge | 800 | 5000 |
| Remote Sensing | MODIS, UNH, Univ. of Oklahoma, NSIDC | Vegetation indices, soil moisture, clouds, snow, fires, temperature | 88 | 126 |
| Physical Geography | NASA, USGS, UNH | Elevation, bathymetry, Blue Marble, Lon/Lat | 28 | 32 |
| Oceanography | NOAA, NCOF | SST-sea temperature, sea ice | 5 | 8 |
| Land Cover | UM, NASA, USGS | Land cover, vegetation, permafrost, freeze/thaw | 80 | 120 |
| Sociology and Economics | CIESIN, World Bank, US CIA, UNH | Population, GDP, industry, mortality/ birth/malnutrition rates, dams | 50 | 100 |
| Agriculture | UW-Madison, Various | Crop land, crops, fertilizer loads, greenhouse emissions | 160 | 200 |
| Polygon Masks | UNH | Watershed, sea/ocean catchments, continents, countries, administrative units | 18 | 18 |
| Station Data | UNH, USGS, NCDC | Hydrology, climate, public health | 12 | 12 |
| | | Total | ~2050 | ~10200 |

Acronyms: UNH- University of New Hampshire; CCNY-City College of New York; UAF-University of Alaska at Fairbanks; NASA- National Aeronautic and Space Agency; NOAA- National Oceanic and Atmospheric Administration; UDel- University of Delaware; NCEP- National Centers for Environmental Prediction; NSIDC – National Snow and Ice Data Center; MERRA- Modern Era Retrospective-Analysis for Research and Applications; IPCC- International Panel for Climate Change; USGS- United States

Geological Survey; NCOF - National Centre for Ocean Forecasting; UM – University of Maryland; CIESIN- Center for International Earth Science Information Network; UW-Madison – University of Wisconsin, US CIA – Central Intelligence Agency; NCDC- National Climate Data Center.

It comprises over two thousand single layer (e.g. elevation) and time series (e.g. daily runoff) composite datasets as well as a number of climate and hydrology gridded and station/point network datasets. It employs a near-real-time data flow architecture where new temporal data is generated by computer clusters chained in data mining → processing → modeling → data product delivery in a similar manner described in [5]. In addition to visualization/mapping, RIMS contains web services designed for digital data content interfacing: i) A dataset search service among our data holdings, including more than 10000 raster and vector time series and single layer data (in 200+ file formats readable by the GDAL library); ii) the ability to view different temporal aggregation sub-sets of the same time series data (e.g. daily, monthly, yearly, daily climatology or long term averages, etc.) as well as their spatial aggregations (e.g. integral averages per country, per watershed, per administrative unit, etc.); iii) instant access to the raw data values for each pixel of the client-side displayed maps (WFS/WCS); iv) on-line computation and creation of new data layers via the new data calculator tool; v) tools for easy mounting and management of new vector, raster, and point/station data and metadata.

Several RIMS clones for different global, continental and regional applications (http://earthatlas.sr.unh.edu; http://NEESPI.sr.unh.edu; http://hydroepscor.sr.unh.edu) [16,17] have been developed at the Water Systems Analysis Group of the University of New Hampshire (WSAG UNH) to facilitate integrated analysis of multiple data sources including: gridded meteorological data from several re-analysis products, regional climate and global circulation models (GCM); gridded outputs from hydrological models; observational hydrometeorological in situ data, land cover information, vegetation indices from remote sensing, and census-based demographic and socio-economic data.

RIMS is an important resource for research and exploration in different fields of Earth science. This web-based system can be successfully applied to address multiple research problems using an extensive data archive and embedded tools for data analysis, visualization, modification, and distribution. The Water Systems Analysis Group (WSAG) at UNH widely uses the RIMS for various scientific applications and several examples of data analyses made with use of RIMS data and tools are discussed below. These examples show RIMS capabilities for integrated analysis of various data sets.

## 3. Research applications with RIMS

### *3.1. Analysis of historical, contemporary and future changes in climate, hydrology and water management*

The massive RIMS archive of hydro-climatological information includes observational data, climate re-analysis products as well as outputs from many climatic and hydrological models. Temporal and spatial aggregation online tools allow estimation of historical and future trends for different regions/watersheds/administrative units/stations, construction of rasterized maps of these changes and to evaluate uncertainties of the modeled data relative to observations [18-20].

River flow is an integrated characteristic reflecting numerous environmental processes and their changes aggregated over large areas. River runoff plays a significant role in the freshwater budget of the Arctic Ocean, and ocean salinity and sea ice formation are critically affected by river input [19]. Changes in the fresh water flux to the ocean can exert significant control over global ocean circulation via the North Atlantic deep water formation [21]. Eurasia contributes 75% of the total terrestrial runoff to the Arctic Ocean [22] and contains three of the four major arctic rivers. Figure 1 demonstrates combined analysis of long-term changes in observed river discharge, major water balance components, and some climatic characteristics for Eurasian pan-Arctic drainage basin.

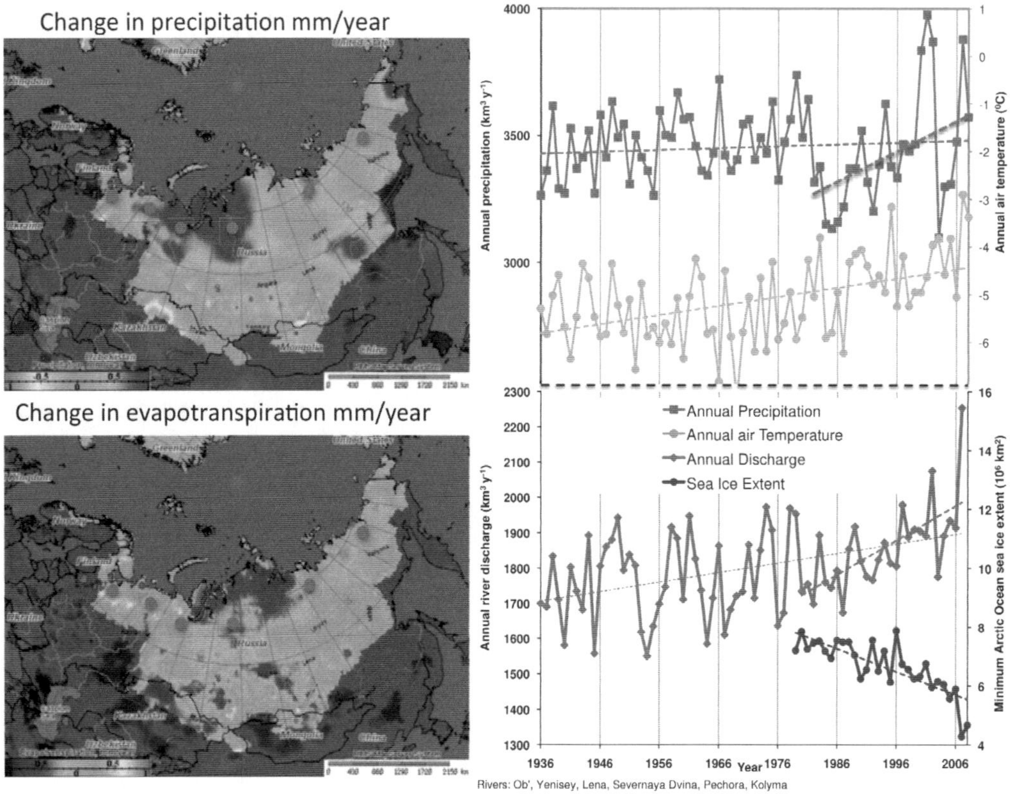


**Figure 1.** Analysis of changes in major water balance components for Eurasian pan-Arctic drainage basin. Upper map shows linear slope of the total annual precipitation over 1936-2010 based on monthly gridded data from University of Delaware (UDel) with $0.5^0$ spatial resolution. Lower map demonstrates linear slope over 1936-2010 of potential evapotranspiration computed on UNH water balance model using UDel meteorological data. Blue areas represent wetter conditions or more precipitation and less evapotranspiration. Right plot shows annual time series for the total river discharge of 6 largest Eurasian rivers flowing to the Arctic Ocean - Ob, Yenisey, Lena, Kolyma, Sev. Dvina and Pechora (red line, downstream river gauges are shown as red dots on the maps), spatially aggregated annual precipitation (green line) and air temperature (brown line) for the 6 river basins based on UDel gridded data, and the total Arctic Ocean sea ice coverage in September based on remote sensing data over 1978-2008 (blue line). All these data were derived from http://neespi.sr.unh.edu using embedded tools.

Observations from the six largest Russian north-flowing rivers have shown an increase of 7% in combined river discharge over the period 1936-1999 [23]. More recent estimates have shown this increase has continued into the 21st Century with 2007 showing a new historical maximum (figure 1, red line; [24]). RIMS contains data sets and tools allowing exploration of possible causes of the changes in Eurasian pan-Arctic river flow. The time series of annual precipitation aggregated over 6 large Eurasian river basins has a very slight (insignificant) increasing trend overall (figure 1, green line) but much higher interannual variation and a significant positive trend of 1.2 mm/year since the 1980s. Thus, although precipitation changes cannot be the primarily cause of streamflow alterations over the long-term historical period (1936-present), observed rapid increase in river discharge since 1980 can be explained by precipitation rise. The spatial changes in precipitation (P) and evapotranspiration (ET) towards wetter conditions (more P and less ET) are most significant in the northern parts of the Ob and Yenisey watersheds and upper Lena basin. Decline in precipitation is observed in the southern parts of Ob and Yenisey river basins and in the European part. However, in some regions, especially with dry climate (Kazakhstan, Mongolia), decrease in precipitation is partly compensated with lower evapotranspiration (see maps in figure 1). The obvious long-term tendency towards wetter and colder conditions in the northern part of Kara Sea basin may be related to increasing cyclonic activities in the area [25,26] and is mainly explained by significant retreat of the Arctic sea ice during last 30 years (figure 1, blue line). The long-term variation in September sea ice area and Eurasian river discharge demonstrate a good correspondence with negative correlation ($r = -0.7$; figure 1) suggesting that (a) both rivers and sea ice were responding to changes in large-scale hemispheric climate patterns; and (b) an increasingly ice-free summer in the Arctic Ocean contributed to wetter conditions on the land via atmospheric moisture transport from open sea areas [24].

The given example of multiple historical data analysis within RIMS environment can be easily extended to monthly and daily data for historical and future time frames as well as information about changes in land cover and permafrost.

*3.2. Express analysis of extreme climatic events their anomalies and impacts on socio-economic factors such as population, crops etc.*

RIMS data archive and tools allow rapid scientific analysis of various anomalous phenomena. Combining regularly updated near real time data from re-analysis products and station observations with historical data on climate, hydrology, land use, fire, and population, we can evaluate characteristics and impacts of the extreme climatic and hydrological events.

The record-breaking heat wave during summer 2010 in western Russia had a significant negative impact on people (health, mortality), agricultural yield, and forestry (wild fires, tree mortality). Applying RIMS data and tools, we present an example of express analysis for this extreme climatic event (see figure 2). The analysis was made at the end of this heat wave in August 2010 using available in RIMS regularly updated near real time data and historical archives. The spatial distribution of daily maximum air temperature anomaly (figure 2a) and the long-term variation of daily maximum air temperature for meteorological station in Moscow (figure 2c) are shown along with maps highlighting the area and population significantly impacted by the heat wave (figure 2b), defined as the area where the mean daily air temperature from July 1 to August 18, 2010 was 5°C higher than the long-term mean for this period. The number of people residing this area was about 89 million or approximately 60% of total Russia population. The estimates were made in RIMS using embedded calculator tool with daily NCEP re-analysis products [27] and gridded population density data [28]. Comparison of NCEP maximum daily air temperature data with ground observations from meteorological station located in the Moscow (figure 2c) revealed that NCEP re-analysis has a tendency to underestimate the air temperature in the large city like Moscow. This is probably due to relatively coarse spatial resolution of NCEP data that cannot take into account additional heating of urban areas. The maximum daily air temperature for period from 07/05/2010 to 08/17/05 was in average 8°C higher than long-term mean and in some days the difference exceeds 11 °C. This hot period in central European Russia was also extremely dry. The total precipitation for July in Moscow region was less than 5 mm when the norm is 80 mm. Such dry and warm conditions led to extreme fire hazard conditions in many regions of the European part of Russia. The simple dryness index suggested by Ped' [29] and based on anomalies of precipitation and air temperature was estimated within RIMS environment using NCEP data aggregated for the most impacted region (figure 3b) to define anomaly of summer 2010. The index (S) is defined as:

$$S = \frac{\Delta T}{\sigma_T} - \frac{\Delta R}{\sigma_R} \quad (1)$$

where $\Delta T, \Delta R$ are deviations from the long-term mean of monthly air temperatures and precipitation over the June-August, and $\sigma_T, \sigma_R$ - mean-square deviations of these characteristics over the long-term period. The drought conditions with this index are defined as weak ($1.0 < S_i < 2.0$), medium ($2.0 \le S_i < 3.0$), or severe ($S_i \ge 3.0$). The summer 2010 was extremely dry (S = 8), reaching the highest value of the index for this area over the period 1985-2010 (figure 2d). RIMS also allows application of near-real time remote-sensing data to the express analysis. For example, land surface temperature, vegetation indices, and fire products currently available from Moderate Resolution Imaging Spectroradiometer (MODIS) can significantly enrich this analysis. The UNH Water Balance Model (WBM) integrated in RIMS can provide additional information about impacts of such conditions on hydrology and water management. Thus, RIMS can serve as a valuable tool for analysis of impacts of recent and ongoing extreme climatic events on other natural and social components.

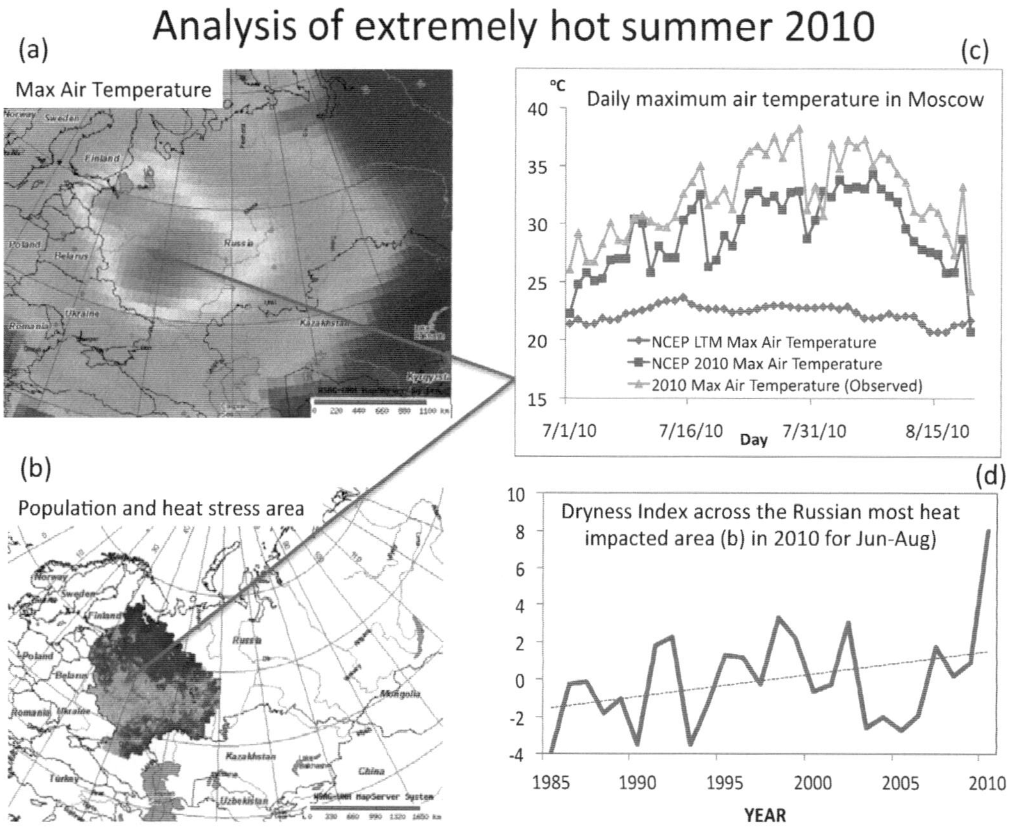


**Figure 2.** Deviation of mean maximum air temperature of July-August 2010 from long-term mean, based on NCEP reanalysis (a); highlighted heat stress area and population density (b); Intercomparison of in-situ and reanalysis air temperature data for Moscow city during 2010 heat wave; Time series of dryness index for June-August for area highlighted on (b).

*3.3. Creation and analysis of new data sets through integration of social and geophysical data*

RIMS capacity to produce combined geospatial fields from bio-geophysical assets and socioeconomic data has been widely used to evaluate and quantify human vulnerability through different water indicators [30]. RIMS data and tools were the core instrument in a global assessment that jointly considered human water security and biodiversity threats [31]. Thus, RIMS can be used as a research tool for combined analysis of environmental data from remote sensing, ground observations, various models, and socioeconomic data from regional censuses and models. RIMS aggregation procedures allow use of data with different spatial and temporal scales.

The analysis of climate change on some components of water security for Central Asia has been made using RIMS data collection and tools. Water is a key agent in Central Asia, ultimately determining human well-being, food security, and economic development. The University of New Hampshire - Water Balance Model - (WBM) was applied to understand the consequences of changes in climate, water use (figure 3a), demography, and economy on various variables and indices characterizing regional water security. The WBM accounts for sub-pixel land cover types, glacier and snow-pack accumulation/melt across sub-pixel elevation bands, anthropogenic water use (e.g., domestic and industrial water consumption, irrigation for most of existing crop types), and hydro-infrastructure for large inter-basin water transfer (e.g. Karakum Canal in the Central Asia) and reservoir/dam regulations [32]. The map (figure 3b) shows Central Asian countries and sub-country administrative units. The census data about land use, crops and irrigated area from 1980 to 2012 for the administrative units have been combined with gridded MIRCA2000 rainfed and irrigated crop data [33] to provide land use inputs for WBM simulations. The dynamics of water use for domestic, industrial and livestock needs have been simulated in WBM using country-based statistical socio-economic information along with spatially distributed population density. All WBM inputs and outputs parameters were integrated into RIMS to facilitate integrated data analysis.

The map (figure 3a) demonstrates change in water use (irrigation, domestic, industrial, livestock) across Central Asia between 2000-2014 and 1980-1985 based on WBM simulations. Inserted plots show annual changes in irrigation water demand over 1980-2014 for two sub-regions (pixels). Since 1980, irrigation water use significantly increased in Uzbekistan – where the highest population growth rate has been observed – and declined in East Kazakhstan (figure 3a). The barographs in figure 4b present official data about changes in areas of irrigated lands for administrative units. There are different patterns of these changes over 1980-2012, which reflect various socio-economic, demographic, and political issues. For example, significant drops in irrigated areas around 1990 in Osh region – Kyrgyzstan and Fergana-Uzbekistan – is probably related to the so-called Osh riots (conflict with numerous victims). Water Scarcity Index (WSI) [34] was estimated in the RIMS calculator based on the WBM simulations using only locally generated water resources (figure 3e,f) and total available water resources (including inflow) (figure 3c,d). The index combines information about water abstractions and water availability. It is defined by the intensity of water resource use, i.e. the gross freshwater abstractions as percentage of the total renewable water resources or as percentage of internal water resources. This indicator is defined by the ratio WSI=W/Q, where W are the annual freshwater abstractions and Q is the annual available water. The severity of water stress is classified in several categories (from high water stress – red to abundance of water – blue figure 3c-f). This indicator neglects temporal and spatial variations as well as water quality data. We also analyzed a number of other indexes and variables characterizing the water availability and water use in the region. There is general tendency towards significant decline in water availability by 2014 compared to 1990 (figure 3c-f). The water security situation in 2014 across the region is more stressful even despite significant political and socio-economic transformations in the region in 1990s, which led to decrease in water use. This is primarily related to change in climatic conditions and population growth.

All estimates of water stress and water availability in the Central Asia as well as computations of various water security indicators have been made within the RIMS environment using the incorporated data and tools.

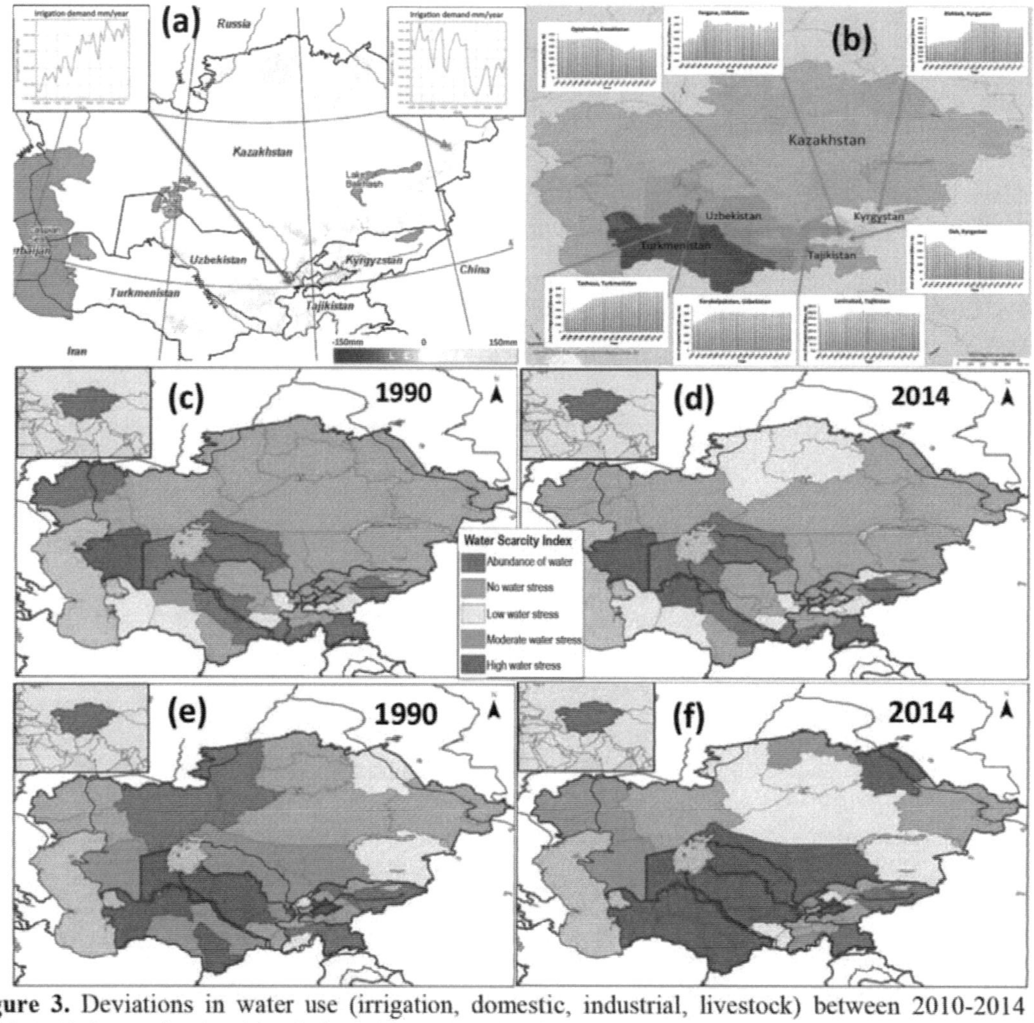


**Figure 3.** Deviations in water use (irrigation, domestic, industrial, livestock) between 2010-2014 and 1980-1985 for 6 minute grid cells based on WBM simulations; the embedded plots show opposite changes for two subregions (pixels) in annual water demand for irrigation (a); The map of administrative units of Central Asia and plots showing the dynamic of irrigated lands for several regions based on census data (b). Maps of Water Scarcity Index (WSI) based on only locally generated (c, d), and total available water resources including inflow from other sub-regions (e ,f) for 1990 and 2014.

## 4. Conclusions

The Rapid Integrated Mapping and analysis System (RIMS) originally developed at UNH has an ability to acquire near-real time data and produce express analysis of changes in the state of the terrestrial water cycle. It has been significantly enriched with data and tools to allow solutions of more sophisticated research tasks. Three different examples of specific scientific applications with RIMS have been demonstrated and discussed.

Analysis of historical changes in major components of the Eurasian pan-Arctic water budget has shown the observed increase in river flux to the Arctic Ocean. This can be partly explained by changes in precipitation and evapotranspiration primarily in the northern regions adjoining to the Arctic Ocean. The significant decline in the sea ice area during last 30 years could contribute to the wetter and colder conditions across the Arctic seacoast leading to increase in streamflow. Most noticeable changes in all major components of the terrestrial water budget in the Eurasian pan-Arctic observed since 1980 correspond to significant sea ice retreat.

Express analysis of extremely hot and dry summer of 2010 across European Russia has been made using a combination of near-real time and historical data compiled in RIMS. We found that about 60% of Russian population was impacted by this extreme event. The evaluated dryness index for summer 2010 was the highest since 1985 and its value constitutes severe drought. Thus, the climatic condition was a major contributor to extensive wildfires and decreases in vegetation and crop production (e.g. wheat yield) during 2010.

Performed integrative analysis of hydrological, water management, and population data over the last 30 years for the Central Asia was made under the RIMS. It allowed evaluating the regional water security and its dynamics due to changes in climate, water use, and demography. The water security situation from 1990 to 2014 across the study region has significantly deteriorated despite considerable decline in water use after USSR breakup. This is probably caused by fast population growth and climate change in the region.

The presented examples of the science applications demonstrate the capabilities of RIMS as a powerful instrument for various express research investigations. At the same time as it was mentioned in the Introduction the system is currently being upgraded to many modern Web server and client side features (e.g. OGC interface services, slippery maps, JQuery, etc.) that are important to be used by broader research and community. In order to facilitate the compatibility upgrade process aimed toward an increase of RIMS potential users in the science and educational fields is being carried out and discussed in another paper of this journal issue [35]. The primary goal of this project is implementation of the latest Web and GIS technologies in compliance with the Open Geospatial Consortium (OGC) standard sin RIMS codes. This effort is build on many modern solutions such as object-oriented programming, modular composition, and client-side JavaScript libraries based on the GeoExt package (http://www.geoext.org), ExtJS Framework (http://www.sencha.com/products/extjs), and OpenLayers software (http://openlayers.org). In particular, new approaches are used to facilitate and accelerate searching of required information on storage systems and provide users with distributed processing of large data sets [36]. The first results obtained in [35] are quite encouraging and show that after ongoing RIMS upgrade it will provides the opportunity to scientists, professionals and decision-makers to use various geographically distributed and georeferenced resources and processing services via web browser by integrating all listed above components in the single universal Geoinformatics system.

**Acknowledgments**

This work is funded by RF Ministry of Education and Science Project 14.613.21.0037 RFMEFI61315X0037. The online system and applications were developed under support of the National Science Foundation (NSF) under the grant ARC1204070 and the National Aeronautic and Space Agency (NASA) under the grant 14UA85.

**References**

[1] IPCC 2014 *Climate Change 2014: Impacts, Adaptation, and Vulnerability. Part B: Regional*

*Aspects. Contribution of Working Group II to the Fifth Assessment Report of the Intergovernmental Panel on Climate Change* ed V R Barros *et al* (Cambridge, United Kingdom and New York, NY, USA: Cambridge University Press)

[2] Eiken H, Walsh J E and Schlosser P 2009 *Arctic Observing Network (AON) Design and Implementation (ADI)* Project Plan 2009-09-25

[3] GCOS 2010 *Implementation Plan for the Global Observing System for Climate in Support of the UNFCCC*

[4] Rienecker M M *et al* 2008 *The GEOS-5 Data Assimilation System - Documentation of Versions 5.0.1 and 5.1.0* (*NASA GSFC Technical Report Series on Global Modeling and Data Assimilation*)

[5] Rodell M *et al* The Global Land Data Assimilation System *Bulletin of the American Meteorological Society* **85** (3) 381–394

[6] Dubois G, Cornford D, Pebesma E, Hristopulos D and Pilz J 2011 Introduction to this special issue on geoinformatics for environmental surveillance *Computers & Geosciences* **37**(3) 277–279

[7] Nativi S and Domenico B 2009 Enabling interoperability for Digital Earth: Earth Science coverage access services *International Journal of Digital Earth* **2** 79–104.

[8] Ledley T S, Dahlman L, McAuliffe C, Haddad N, Taber M R, Domenico B, Lynds S and Grogan M 2011 Making Earth Science Data Accessible and Usable in Education *Science* **333**(6051) 1838–39.

[9] Brodaric B, Fox P and McGuinness DL 2009 Geoscience knowledge representation in cyberinfrastructure *Computers & Geosciences* **35**(4) 697–99

[10] Bernholdt D, *et al* 2005 The Earth System Grid: Supporting the Next Generation of Climate Modeling Research *Proc. IEEE* **93**(3) DOI 10.1109/JPROC.2004.842745

[11] Kumar S V *et al*, 2006 Land information system: An interoperable framework for high resolution land surface modeling *Environmental Modelling & Software* **21**(10) 1402–15

[12] Allen G, Bogden P, Creager G, Decate C, Jesch C, Kaiser H, MacLaren J, Perrie W, Stone G and Zhang X 2008 Towards an integrated GIS-based coastal forecast workflow *Concurrency Computat.: Pract. Exper.* **20** 1637–51

[13] Best B D, Halpin P N, Fujioka E, Read A J, Qian S S, Hazen L J and Schick R S 2007 Geospatial web services within a scientific workflow: Predicting marine mammal habitats in a dynamic environment *Ecological Informatics* **2**(3) 210–23

[14] Pebesma E, Cornford D, Dubois G, Heuvelink G B M, Hristopulos D, Pilz J, Stöhlker U, Morin G and Skøien J O 2011 INTAMAP: The design and implementation of an interoperable automated interpolation web service *Computers & Geosciences* **37**(3) 343–52

[15] OGC Reference Model (ORM) (Open Geospatial Consortium Inc.) Document:OGC 08-062r7, v2.1, 2011-12-19, http://www.opengis.net/doc/orm/2.1

[16] Shiklomanov A I 2012 Regional Integrated Monitoring System for the Pan-Arctic (RIMS) *Report series in Aerosol Science* ISBN 978-952-5822-54-0, 128–130

[17] Prusevich A A, Shiklomanov A I and Lammers R B 2013 RIMS: An Integrated Mapping and Analysis System with Application to Siberia *Regional Environmental Changes in Siberia and Their Global Consequences* ed G Gutman and P Ya Groisman Springer) chapter 2 pp 46–50

[18] Groisman P Ya *et al* 2009 The Northern Eurasia Earth Science Partnership an Example of Science Applied to Societal Needs *Bulletin of the American Meteorological Society* **90**(5) 671–88

[19] Rawlins M A, Ye H, Yang D, Shiklomanov A and McDonald K C 2009 Divergence in seasonal hydrology across northern Eurasia: Emerging trends and water cycle linkages *Journal of Geophysical Research-Atmospheres* 114

[20] Shiklomanov A I, Lammers R B, Lettenmaier D, Polischuk Yu, Savichev O, Smith L C,

Chernokulsky A V 2013 Hydrological changes: historical analysis, contemporary status and future projections *Regional Environmental Changes in Siberia and Their Global Consequences* ed G Gutman and P Ya Groisman (Springer) chapter 4 pp 111-155

[21] Rahmstorf S 2002 Ocean circulation and climate during the past 120,000 years *Nature* **419** 207–14

[22] Shiklomanov I A, Shiklomanov A I, Lammers R B, Peterson B J and Vörösmarty C J 2000 The Dynamics of River Water Inflow to the Arctic Ocean *The Freshwater Budget of the Arctic Ocean* ed E L Lewis *et al* (Dordrecht: Kluwer Academic Press) pp 281–96

[23] Peterson B J, Holmes R M, McClelland J W, Vorosmarty C J, Lammers R B, Shiklomanov A I, Shiklomanov I A and Rahmstorf S 2002 Increasing river discharge to the Arctic Ocean *Science* **298**, 2171-73

[24] Shiklomanov A I and Lammers R B 2009 Record Russian river discharge in 2007 and the limits of analysis. *Environ. Res. Lett.* **4** 045015 doi: 10.1088/1748-9326/4/4/045015

[25] Akperov M G, and Mokhov I I 2013 Estimates of the sensitivity of cyclonic activity in the troposphere of extratropical latitudes to changes in the temperature regime *Izvestiya Atmos Oceanic Phys* **49** (2) 113–120

[26] Jaiser R, Dethloff K, Handorf D, Rinke A and Cohen J 2012 Impact of sea ice cover changes on the Northern Hemisphere atmospheric winter circulation *Tellus* A **62** 1–9

[27] Kistler R, *et al* 2001 The NCEP-NCAR 50-year reanalysis: Monthly means CD-ROM and documentation *Bulletin of the American Meteorological Society* **82**(2) 247–267

[28] CIESIN 2005 Gridded Population of the World Version 3 (GPWv3): Population Count Grid. Palisades (NY: NASA Socioeconomic Data and Applications Center (SEDAC))

[29] Bezuglova N N *et al* 2014 Analysis of relationship between the dryness index S (Ped' coefficient) and water content in upper soil layer in arid regions of Western Siberia *Problems of Regional Ecology* 6 (Nov-Dec) 68–72 ISSN 1728-323X

[30] Sullivan C *et al* 2006 Mapping the Links between Water, Poverty and Food Security *Report on the Water Indicators workshop* held at the Centre for Ecology and Hydrology (Wallingford UK 16 to 19 May 2005) GWSP Issues in GWS Research No.1 GWSP IPO Bonn Available at: www.gwsp.org

[31] Vorosmarty C J, *et al* 2010 Global threats to human water security and river biodiversity *Nature* **467**(7315) 555-61

[32] Grogan D S *et al* 2015 Quantifying the link between crop production and mined groundwater irrigation in China *Science of The Total Environment* **5**11(0) 161-75

[33] Portmann F T *et al* 2010 MIRCA2000-Global monthly irrigated and rainfed crop areas around the year 2000: A new high-resolution data set for agricultural and hydrological modeling *Global Biogeochemical Cycles* **24**

[34] Brown A and Matlock M D 2011 A review of water scarcity indices and methodologies *White paper* **106** 19

[35] Okladnikov I G, Gordov E P and Titov A G 2016 Development of climate data storage and processing model *IOP Conference Series: Earth and Environmental Science* In press

[36] Gordov E, Shiklomanov A, Okladnikov I, Prusevich A and Titov A 2016 Development of Distributed Research Center for analysis of regional climatic and environmental changes *IOP Conference Series: Earth and Environmental Science* In press

# The impact of atmospheric blocking on spatial distributions of summertime precipitation over Eurasia

**O Yu Antokhina[1], P N Antokhin[1], Yu V Martynova[2,3] and V I Mordvinov[4]**

[1] V.E. Zuev Institute of Atmospheric Optics of the Siberian Branch of the Russian Academy of Sciences, 1, pl. akad. Zueva, Tomsk, 634065, Russia
[2] Institute of Monitoring of Climatic and Ecological Systems of the Siberian Branch of the Russian Academy of Sciences, 10/3, Pr. Akademicheskii, Tomsk, 634055, Russia
[3] Siberian Regional Hydrometeorological Research Institute, 30, ul. Sovetskaya, Novosibirsk, 630099, Russia
[4] Institute of Solar-Terrestrial Physics of the Siberian Branch of the Russian Academy of Sciences, 126a ul. Lermontova, Irkutsk, 664033, Russia

E-mail: olgayumarchenko@gmail.com

**Abstract**. The correlation of precipitation anomalies over Eurasia with atmospheric blocking events was examined with ERA Interim reanalysis data. We found that, regardless of the frequency of the atmospheric blocking events, they significantly affect the distribution of rainfall over all Eurasian regions in summer, due to both the change in the westerly transport and the dominant dipole blocking structure. It is important that, depending on the blocking positions in Asia, there are heavy rainfalls in an arid zone which includes Kazakhstan, Mongolia, Northern China, and the Trans-Baikal Territory.

## 1. Introduction

Atmospheric blocking is a major large-scale phenomenon in the circulation of mid- and high-latitude atmosphere [1,2]. The mid-latitude westerly jet and the eastward progression of synoptic systems are often interrupted in long periods of atmospheric blocking. Thus, atmospheric blocking can significantly impact weather processes. The lifetime of blocking varies from a few days to a few weeks, and, therefore, they may be responsible for various extreme weather events. Blocking events (BEs) are normally associated with extreme summer droughts and snowless winter periods. In addition, they can cause heavy atmospheric precipitation, especially in summer.

The blocking formation processes and the associated weather anomalies over some areas of the Northern hemisphere have been studied in detail since the 1940s - 1950s. It has been found that strengthening of large-scale Rossby waves leads to the formation of a monopole Ω-blocking (a hot ridge in the center, cold troughs near the base) or to a dipole Rex-blocking (a closed cyclone from the equatorial side and a closed anticyclone from the polar side of the mean westerlies) [2]. Jet streams and the associated baroclinic disturbances are forced to bypass the blockings. Both factors, i.e. the change in the direction of motion of the synoptic disturbances and the descending and ascending flows in the blockings themselves, lead to the formation of both negative and positive atmospheric precipitation anomalies in the vicinity of the blockings. These anomalies are clearly visible on monthly average precipitation maps [3]. For instance, in the summer of 2003, a long-lasting blocking over Western Europe resulted in heavy rains in Eastern Europe. Also, in July 2010 a blocking over

Eastern Europe caused precipitation in Western Siberia. Besides, the significant meridional sizes of the blockings enables them to impact regions not only in the mid-latitudes but also those farther to the south [4,5].

The rainfall anomalies generated by blockings have been studied for some regions in Europe [6] and Asia [4]. In the present paper we analyze the impact of blocking processes on the atmospheric precipitation anomalies over the entire territory of Eurasia.

## 2. Method and data

In order to detect the blocking events, we used the Tibaldi and Molteni (TM) criterion [7], which is based on the estimation of the meridional gradient of the geopotential height at 500 hPa. This criterion allows us to detect blockings for a given longitude (equations (1) and (2)):

$$GHGS = \frac{Z(\varphi_0) - Z(\varphi_s)}{\varphi_0 - \varphi_s} \quad (1) \qquad GHGN = \frac{Z(\varphi_n) - Z(\varphi_0)}{\varphi_n - \varphi_0} \quad (2)$$

where Z is the geopotential height at 500 hPa, $\varphi_n$=80°N±Δ, $\varphi_0$=60°N±Δ, $\varphi_s$=40° N±Δ, Δ=-4°,0° or 4°.

The given longitude is then defined as "blocked" at some time if the following conditions are satisfied for at least one value of Δ: GHGS>0, GHGN < -10 m/deg lat.

To study the relationship between BEs and precipitation distributions, some local blocking estimates were used [2,8]. A blocking event was registered if the TM blocking condition was met at least for one day at some longitude. Some calculations of the TM criterion were performed using the ERA Interim data [9] from 1979 to 2015.

Figure 1 shows a time-longitude cross-section distribution of the blocking frequency (BF) over Eurasia in July obtained using the TM criterion. The July intensity of summertime blocking over Asia is most frequently at its maximum.

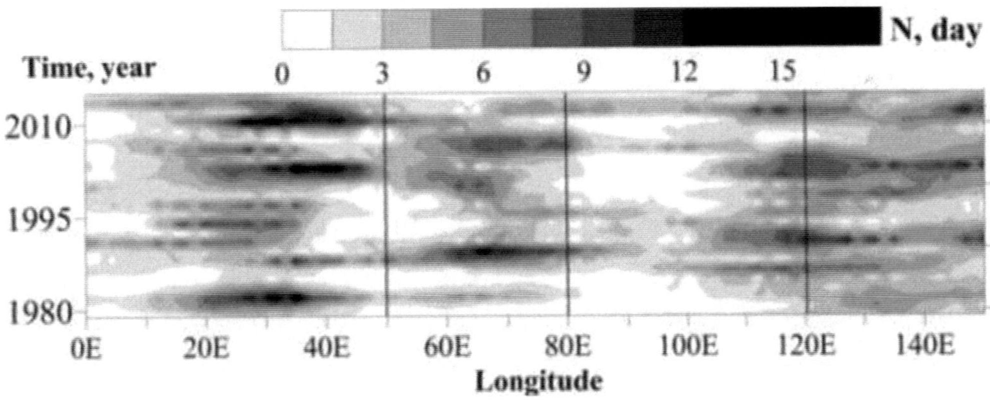


**Figure 1.** Time-longitude cross-section of BF (N) in July.

Normally, three blocking regions are distinguished over Eurasia [4]: 0-50E (E: Europe), 50-80E (US: Ural-Siberia) and 120-160E (OS: Okhotsk Sea). For these regions, the authors of [4] have investigated some characteristics of summertime blockings. For the OS and E regions, the correlations of blockings with atmospheric precipitation have also been studied [4,6]. The region from 80E to 120E, which has low BF, has been studied in much less detail. However, one can see that periods of high BF are observed even in this region (figure 1). That is why we investigated blockings in the region of 80E-120E (ES: East Siberia) too.

First we calculated the integral BF for each of the four regions, and then, the correlation coefficients between these four values and the total precipitation at a regular grid of the entire Northern hemisphere for each summer month.

## 3. Results

Figure 2 shows distributions of calculated correlation coefficients for various Eurasian regions. Positive and negative anomalies of precipitation (AP) caused by the structural features of blockings are more or less typical for all the regions. Dipole AP configurations are described in [6] for Europe and in [4] for the Okhotsk Sea region. In Europe, during the periods of blocking much precipitation falls in the south (figures 2a,c,e). During blockings in the Okhotsk Sea region, the atmospheric fronts in Japan and Korea are strengthened (figures 2h,j,l). Besides, during blocking periods in the Far East, a lot of precipitation falls over the Trans-Baikal Territory (figures 2h,j).

For the Siberian region, the correlation between blockings and precipitation anomalies has not been studied earlier, but a close examination of figure 2 shows that the impact of blockings on APs is as profound (figures 2g,I,k) as it is in the Ural-Siberian region (figures 2b,d,f). Due to the blocking process, the large-scale quadrupole anomaly structure in Eastern Siberia is even more evident than that in Europe; although the BF in Europe is much higher than that in Eastern Siberia [4]. In the course of blocking, much precipitation falls in the arid Mongolia which, basically, is anomalous. Thus, despite their low frequency, the blockings in Eastern Siberia play an important role in the region. Overall, it should be noted that the blockings in each region of Asia are accompanied by precipitation in the arid belt of Asia (Kazakhstan, Mongolia, Transbaikalia, and North China).

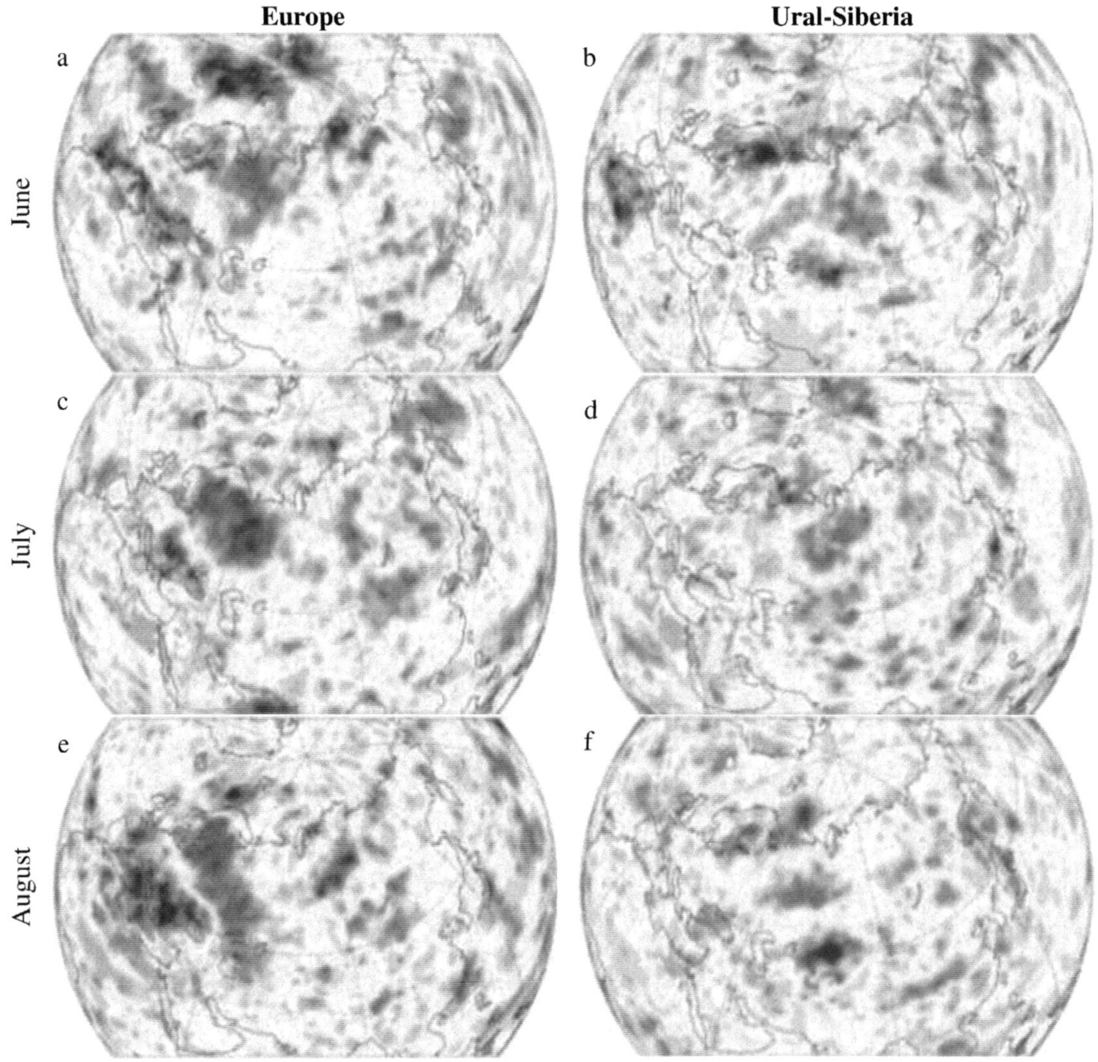

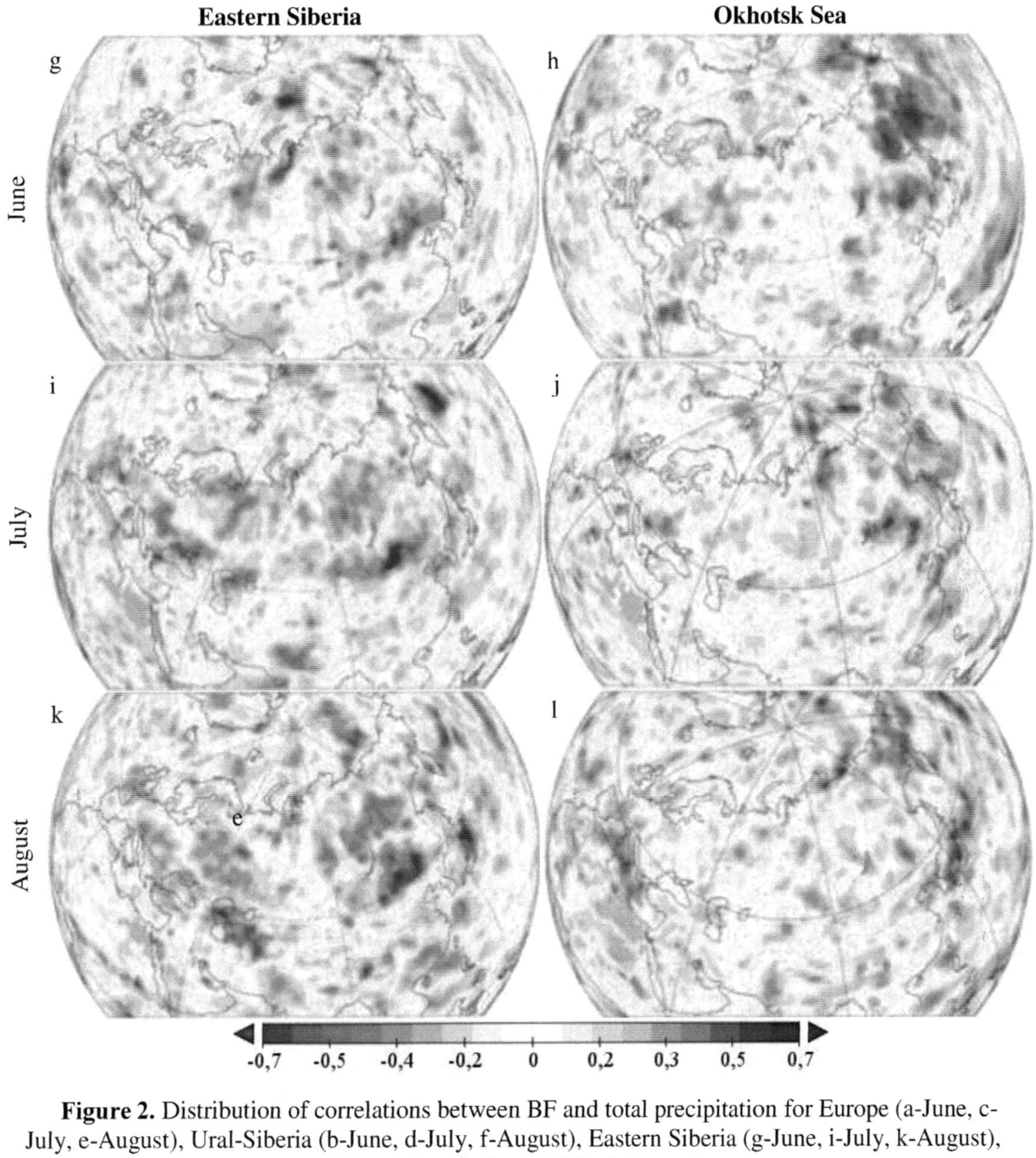


**Figure 2.** Distribution of correlations between BF and total precipitation for Europe (a-June, c-July, e-August), Ural-Siberia (b-June, d-July, f-August), Eastern Siberia (g-June, i-July, k-August), Okhotsk Sea (h-June, j-July, l-August).

All blockings in Eurasia are accompanied by positive precipitation anomalies at their north-western boundaries. These anomalies result from changes in the trajectories of baroclinic disturbances. The effect of this factor is evident in the case of East-Siberian blockings.

It is clearly seen that the northward shift of the cyclone series reduces precipitation in the southern regions of Europe. During blockings over Europe, much precipitation falls in Siberia due to the increased intensity of ultrapolar cold air intrusions. In the Far East region, special attention should be paid to the formation of positive precipitation anomalies northwards of the blocking anticyclone.

Despite the similarity of the correlation fields for specific regions in different summer months, there are some differences. These are not so evident over the entire Eurasia, but for some countries and regions they may play an essential role. For example, in August the negative anomalies resulting from a blocking anticyclone over Europe (figure 2e) are especially well-pronounced in the Volga region. When a blocking occurs in Eastern Siberia, the Central Asian republics (Kazakhstan, Uzbekistan and Turkmenistan) are areas of negative or positive anomalies in some months. For example, when a blocking takes place in July (figure 2i) a precipitation shortage is clearly seen, and, on the contrary, when a blocking takes place in August (figure 2k), positive precipitation anomalies prevail. Our main attention is focused on the arid and semiarid regions of Asia, because it was noted [10] that in the course of climatic changes of the recent decades the droughts of the arid and semiarid regions of Asia are among the most dramatic ones globally. Some changes in the atmospheric blocking may be one of the reasons of this phenomenon.

## 4. Conclusions

Atmospheric blockings can be responsible for both negative and positive atmospheric precipitation anomalies. It has been found that, regardless of frequency, atmospheric blockings have a pronounced effect on the distribution of atmospheric precipitation throughout all regions of Eurasia. The formation of positive and negative precipitation anomalies is associated both with changes in the trajectories of jet streams and with the dipole structure of blockings in summer. It is of special importance that the blockings in all regions of Asia are accompanied by large precipitation in an arid belt zone that includes Kazakhstan, Mongolia, North China, and the Trans-Baikal Territory.

## 5. Acknowledgments

This research was supported by the RAS Presidium (program no. 15), the Department of Earth Sciences, the Russian Foundation for Basic Research (grants no. 14-05-00526 and no. 14-05-00590), and the Presidential Program “Leading Scientific Schools of the Russian Federation” (grant NSh-6894.2016.5).

## References

[1] Mokhov I I, Timazhev A V and Lupo A R 2014 Changes in atmospheric blocking characteristics within Euro-Atlantic region and Northern Hemisphere as a whole in the 21st century from model simulations using RCP anthropogenic scenarios *Global Planet. Change* **122** pp 265–270

[2] Pelly J L and Hoskins B J 2003 A new perspective on blocking *J. Atmos. Sci.* **60** pp 743–755

[3] https://kunden.dwd.de/GPCC/Visualizer

[4] Park Y-J and Ahn J-B 2014 Characteristics of atmospheric circulation over East Asia associated with summer blocking *J. Geophys. Res. Atmos.* **119** pp 726–738

[5] Hong Ch-Ch, Hsu H-H, Lin N-H and Chiu H 2011 Roles of European blocking and tropical-extratropical interaction in the 2010 Pakistan flooding *Geoph. Res. Let.* **38** pp 553–597

[6] Rimbu N, Stefan S, Busuoic A and Georgescu F 2016 Links between blocking circulation and precipitation extremes over Romania in summer *Int. J. Climatol.* **36** pp 369–376

[7] Tibaldi S and Molteni F 1990 On the operational predictability of blocking *Tellus* **42A** pp 343–365

[8] Masato G, Hoskins B J and Woollings T J 2012 Wave-breaking characteristics of midlatitude blocking *Q. J. R. Meteorol. Soc.* **138** pp 1285–1296

[9] Dee D P et. al. 2011 The ERA-Interim reanalysis: configuration and performance of the data assimilation system *Q. J. R. Met. Soc.* **137** pp 553–597

[10] Berezhnykh T V, Marchenko O Yu, Mordvinov V I and Abasov N V 2012 Changes in the summertime atmospheric circulation over East Asia and formation of long-lasting low-water periods within the Selenga river basin *Geogr. Nat. Resour.* **33** pp 223–229